THE TOTAL SYNTHESIS
OF NATURAL PRODUCTS

The Total Synthesis of Natural Products

VOLUME 6

Edited by

John ApSimon

Ottawa—Carleton Institute for Research and Graduate Studies in Chemistry

and

Department of Chemistry
Carleton University, Ottawa

A WILEY-INTERSCIENCE PUBLICATION

JOHN WILEY & SONS

NEW YORK • CHICHESTER • BRISBANE • TORONTO • SINGAPORE

Library of Congress Cataloging in Publication Data:
(Revised for volume 6)

ApSimon, John.
  The total synthesis of natural products.

  Includes bibliographical references.
  1. Chemistry, Organic—Synthesis.   2. Natural products.
I. Title.
QD262.A68   547.7   72-4075

ISBN 0-471-03251-4 (v. 1)
ISBN 0-471-09900-7 (v. 6)

Printed in the United States of America

10   9   8   7   6   5   4   3   2   1

Contributors
to Volume 6

John W. ApSimon, Department of Chemistry, Carleton University, Ottawa

Kim E. Fyfe, Department of Chemistry, Carleton University, Ottawa

Austin M. Greaves, Department of Chemistry, Carleton University, Ottawa

G. Grynkiewicz, Institute of Organic Chemistry, Polish Academy of Sciences, Warsaw

A.H. Jackson, Department of Chemistry, University College, Cardiff, Wales

Saran A. Narang, Division of Biological Sciences, National Research Council of Canada, Ottawa

K.M. Smith, Department of Chemistry, University of California, Davis

Wing L. Sung, Division of Biological Sciences, National Research Council of Canada, Ottawa

David Taub, Merck Sharp and Dohme Research Laboratories, Rahway

Robert H. Wightman, Department of Chemistry, Carleton University, Ottawa

A. Zamojski, Institute of Organic Chemistry, Polish Academy of Sciences, Warsaw

Preface

The first five volumes in this series have been concerned with describing in a definitive manner the total synthetic approach to various classes of natural products.

This volume continues the series with chapters describing the reports and progress in the total synthesis of aromatic steroids, carbohydrates, genes, pyrrole pigments, and triterpenoids since the appearance of Volumes 1 and 2 some ten years ago.

There have been some delays in producing this volume at the Editor's end caused by the requirement of retyping and the structure drafting; however, this series of chapters brings the reader up to date with progress in the diverse classes of compounds examined herein. My particular thanks are due to Karl Diedrich of Carleton University for his efforts in the production of the various manuscripts.

The seventh volume in this series is in preparation and is planned for publication in about one year, covering the synthesis of diterpenes, diterpene alkaloids, macrocycles, and anthracyclinones.

JOHN APSIMON

Ottawa, Canada
January 1984

Contents

THE TOTAL SYNTHESIS
OF NATURAL PRODUCTS

The Total Synthesis of Aromatic Steroids 1972–1981

DAVID TAUB

Merck Sharp & Dohme Research Laboratories,
Rahway, New Jersey

1. INTRODUCTION

This review covers the literature published during 1972–1981 and updates the chapter on the total synthesis of naturally occurring aromatic steroids that appeared in Volume 2 (pp. 641-725) of this series.

Extensive synthetic effort has continued to be directed toward the aromatic steroids exemplified by estrone—and toward the related 19-norsteroids—not only because of their practical medical and commercial importance but because they serve admirably as templates for the display of new organic synthetic methodology.

The major innovations include:

1. Development of asymmetric syntheses involving chirality transfer to prochiral substrates, in particular the use of amino acids as catalysts in chirally directed aldol cyclizations.
2. Development of synthetic routes based on generation and intramolecular cycloaddition of orthoquinodimethanes.

2. EQUILENIN

A. Posner Synthesis

Posner and co-workers have developed an efficient (52% overall yield) conversion of 2-methyl-2-cyclopentenone into (±)-11-oxoequilenin methyl ether, **4** (Scheme 1).[1] The latter had been hydrogenolyzed earlier to equilenin methyl ether by Birch (see Volume 2, p. 660).

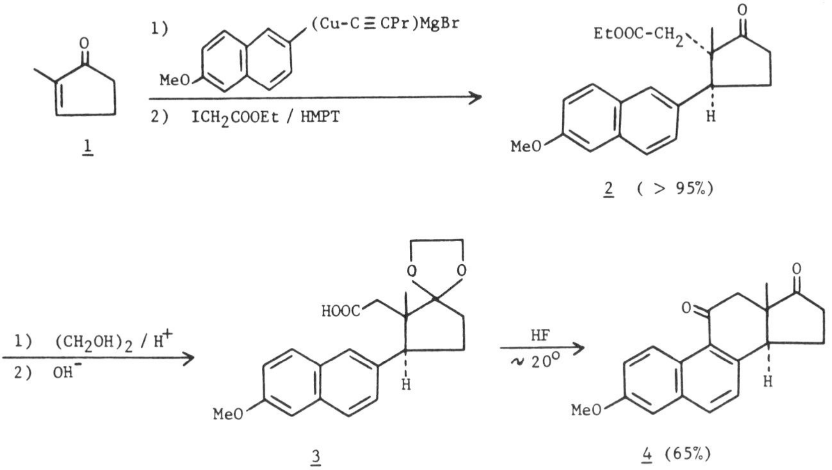

Scheme 1

2-Methyl-2-cyclopentenone **1** was treated sequentially with (6-methoxy-2-naphthyl)(1-pentynyl)coppermagnesium bromide and ethyl iodoacetate to give the stereochemically pure *trans* keto ester **2** in >95% yield (Scheme 1). The yield was considerably lower when the corresponding aryl(alkynyl)lithium cuprate and methyl bromoacetate were utilized. Analogous reactions with the smaller vinyl group instead of 6-methoxy-2-naphthyl were not as clean stereochemically, producing appreciable amounts of *cis* isomers (see below, orthoquinodimethane approach[36,40,53]). Conversion of **2** to the corresponding ethylene ketal, saponification, and Friedel–Crafts cyclization in liquid hydrogen fluoride led to (±)-11-oxoequilenin 3-methyl ether, **4**. Curiously, the cyclization yield was appreciably higher with the ethylene ketal acid **3** than with the corresponding 17-ketone.

Extension to the natural (+) series was accomplished by transfer of chirality from sulfur to carbon via (+)-2-tolylsulfinyl-2-cyclopentenone, **7**[2a] (Scheme 2). The latter was prepared in optically pure form from the ethylene ketal of 2-bromo-2-cyclopentenone **5** by lithiation and treatment with (−)-menthyl p-toluenesulfinate to yield (+) **6**, followed by deketalization. Conjugate addition of 6-methoxy-2-naphthylmagnesium bromide to (+) **7** followed by *in situ* methylation gave stereochemically pure **8a** in 42% yield along with 40-50% of unmethylated analog **8b**. More vigorous methylation conditions led to elimination of p-toluenesulfinic acid. Generation of enolate **9** with dimethylcopperlithium followed by alkylation with methyl bromoacetate then led to (+)-methyl ester **10**. The overall yield of (+) **4** is 25% based on the Friedel–Crafts procedure developed for the racemic series.

Sulfoxide (+) **7** has also been utilized in effective chiral syntheses of (3S)-2-methyl-3-vinylcyclopentanone **12**[46] and the corresponding trimethylsilyl enol

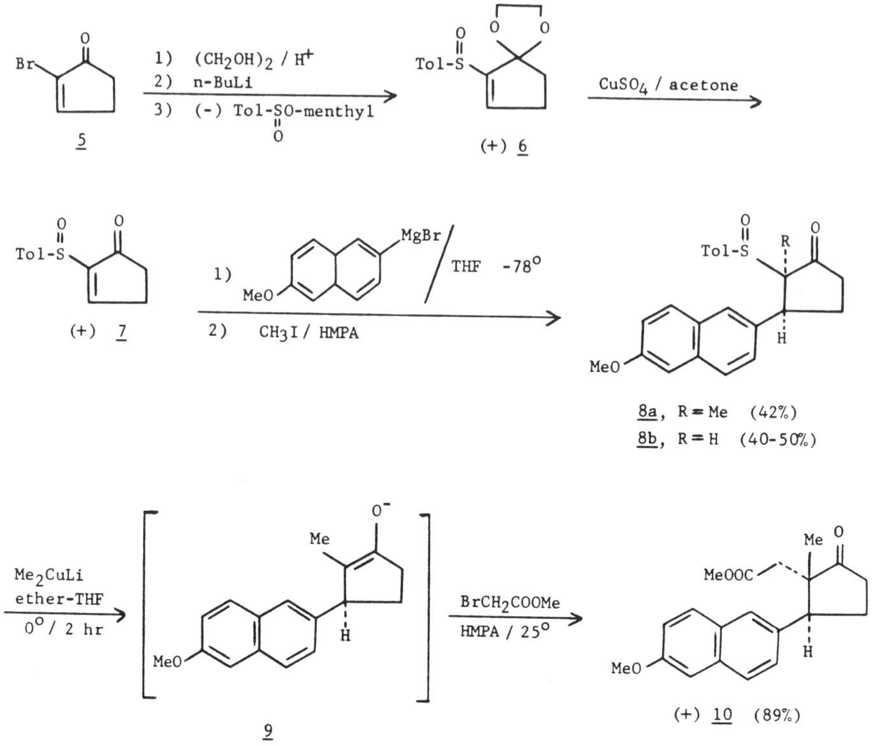

Scheme 2

ether [cf. (±) 95[35,42,2b]], ring D synthons in orthoquinodimethane approaches discussed below. Conjugate addition of vinylmagnesium bromide to (+) 7 occurred with 100% asymmetric induction when zinc bromide was added first to preform the zinc chelate.

3.   ESTRONE AND RELATED 19-NORSTEROIDS

A.   Amino Acid Mediated Asymmetric Cyclizations

(a)   Introduction

A major advance, of significance not only for steroid synthesis but also for organic synthesis in general,[3] is the finding that aldol type cyclizations of prochiral substrates can be catalyzed by chiral α-amino acids to yield chiral products of high optical purity. The discovery was made independently by groups at Schering A.G. (Berlin)[4] and Hoffman-LaRoche (Nutley, N.J.)[5] in the development of routes to chiral hydrindenones as CD part structures in CD → ABCD approaches to 19-norsteroids. Some representative examples are shown here:

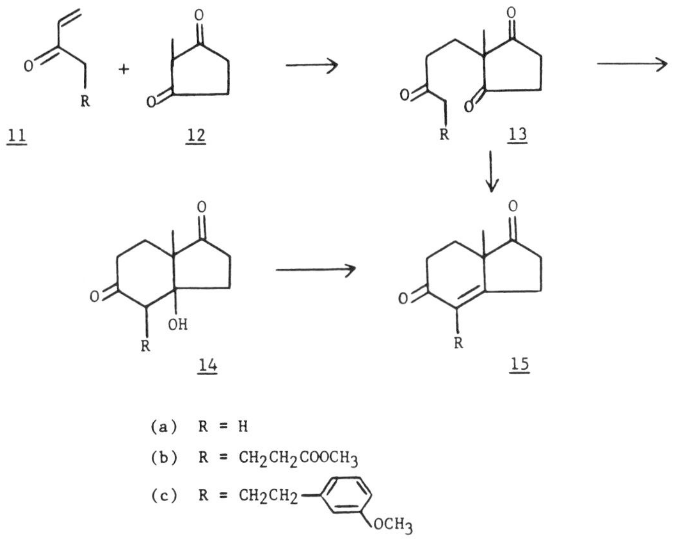

(a)   R = H

(b)   R = CH$_2$CH$_2$COOCH$_3$

(c)   R = CH$_2$CH$_2$—⟨phenyl⟩—OCH$_3$

Yields in the Michael condensation for preparation of the triones **13** were improved by operating in aqueous media and omitting basic catalysis.[4,6] For example, trione **13a** was obtained in 88% yield by stirring a mixture of methyl vinyl ketone **11a** and 2-methylcyclopentane-1,3-dione **12** in water at 20° for 5 days.[6] This Michael condensation is considered to be self-catalyzed by the acidic 1,3-dione.[6]

In the Schering route to **15a**,[4] trione **13a**, L-proline, and 1N perchloric acid (molar ratios 1:0.5:0.27) in refluxing acetonitrile for 22 hours afforded an 87% yield of (+) **15a** of 84% optical purity. Alternatively,[5,8a] **13a** in dimethyl formamide containing 0.05% water and 1% by weight of L-proline at 20° for 22.5 hours yielded ketol **14a**, which on treatment with p-toluenesulfonic acid in benzene gave (+) **15a** in 94% yield of 87% optical purity.

In the above examples, L-α-amino acids induce the natural 13β-chirality. Amines or amino acid derivatives (esters, amides) are much less effective, and a tertiary amino acid, hygrinic acid, was ineffective. For trione **13a**, secondary amino acids (e.g., proline) are best; for **13b** and **13c** [R > H, see also the Danishevsky (Scheme 8)[19] and Tsuji (Scheme 10)[22] syntheses below], a primary amino acid (e.g., L-phenylalanine) is preferred. The mechanism of the reaction has not yet been clarified.[5,7]

Applications of the chiral hydrindenone syntheses to industrially feasible routes to 19-norsteroids and their (of necessity) totally synthetic 18-ethyl counterparts (e.g., norgestrel) were then developed.

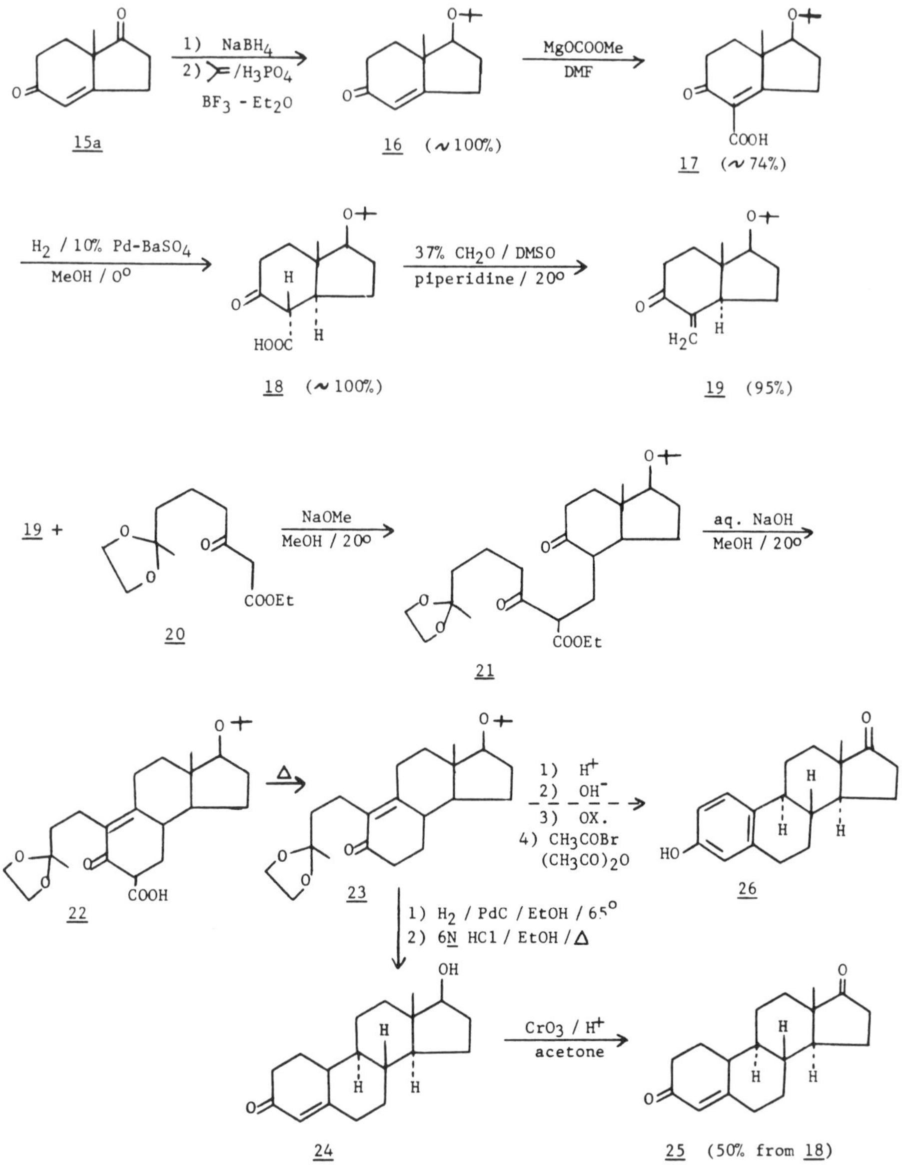

Scheme 3

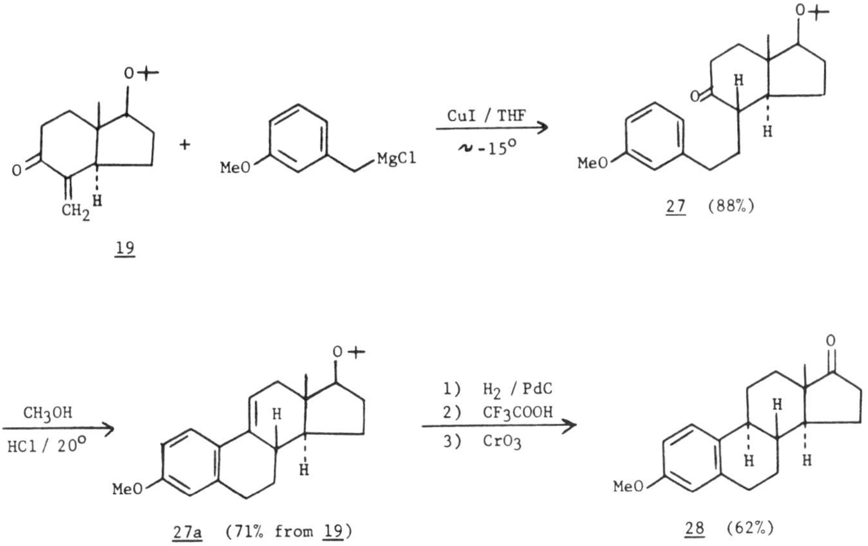

27 (88%)

27a (71% from 19)   28 (62%)

Scheme 4

(b) Hoffman-LaRoche Syntheses

At the outset of the Hoffman-LaRoche route[8] (Scheme 3), hydrindendione **15a** was converted into the corresponding 17β (steroid numbering) *t*-butyl ether **16**. Carbonation with magnesium methyl carbonate then yielded unsaturated acid **17**. The latter was the chosen intermediate because it was known that the direction of hydrogenation of the double bond in hydrindenones **15** is strongly dependent on the nature of the substituent R[9,10] and that carboxyl, ester, and methyl aryl sulfone[11a] functionality strongly favor reduction from the α-face to give the desired C/D *trans* stereochemistry. Furthermore, carboxyl was required as a removable activating group in the Mannich reaction step that followed reduction.

Hydrogenation was carried out at 0° to minimize decarboxylation of the saturated β-keto acid product **18**. Mannich reaction proceeded with *in situ* decarboxylation to afford α-methylene ketone **19**, which on Michael reaction with ketal β-keto ester **20**[8,11] yielded adduct **21**. Saponification, B ring closure, and decarboxylation then led to ketalenone **23** in high yield, which was converted into (+)-19-nortestosterone **24** and thence to (+)-19-norandrostenedione **25** in 50% yield from **18** or 27% overall yield from **12**. However, ketal hydrolysis, A ring closure, oxidation at C-17, and isomerization by the Roussel procedure (acetyl bromide–acetic anhydride in methylene chloride at 20°)[12] should yield (+)-estrone **26** efficiently.

Alternatively, Cohen et al.[13] (Scheme 4) reacted **19** with *m*-methoxybenzyl magnesium chloride in the presence of cuprous iodide to produce the 1,4-adduct

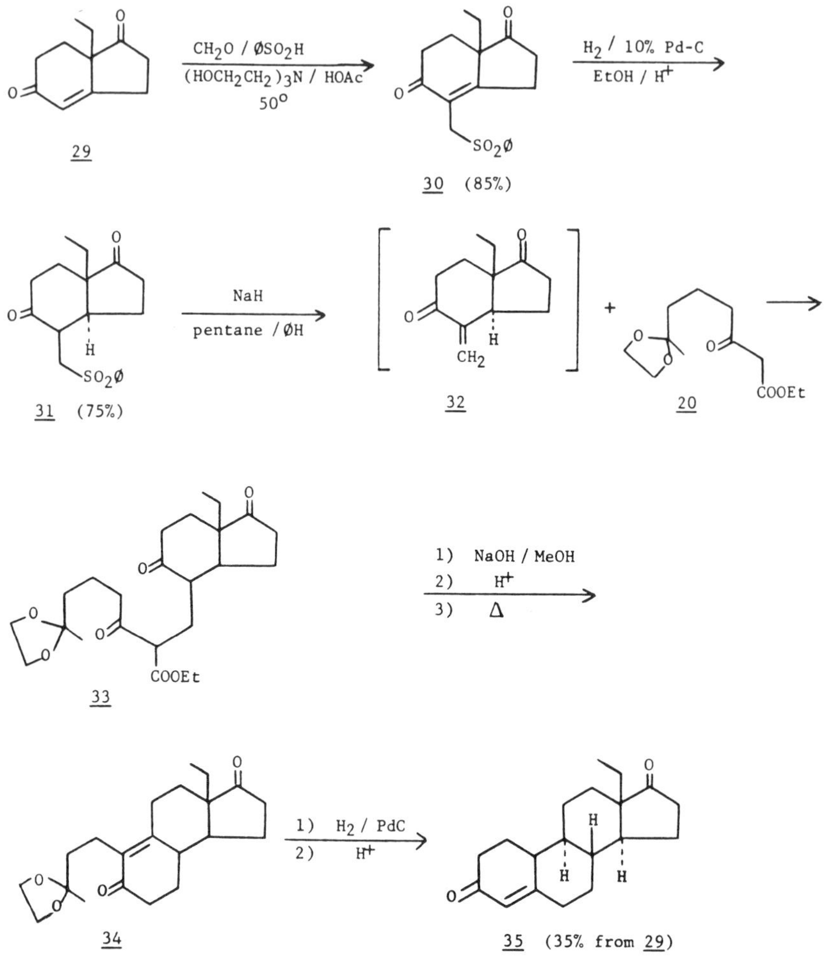

Scheme 5

27 in good yield. Acid catalyzed B ring closure, hydrogenation, and conversion to the 17-ketone then yielded (+)-estrone methyl ether 28.

(c)   Schering A.G. Syntheses

The Schering group described a route to (+)-13-β-ethylgon-4-ene-3,17-dione 35 (for conversion to norgestrel), which is clearly adaptable to estrone synthesis. The key step (Scheme 5)[11a] is a variant of the Mannich reaction involving sulfonylmethylation of 29 with formaldehyde and benzenesulfinic acid in 3:1 triethanolamine:acetic acid at 50° to yield unsaturated sulfone 30. Hydrogenation of the latter in ethanol:1% 1N hydrochloric acid gave crystalline saturated sulfone

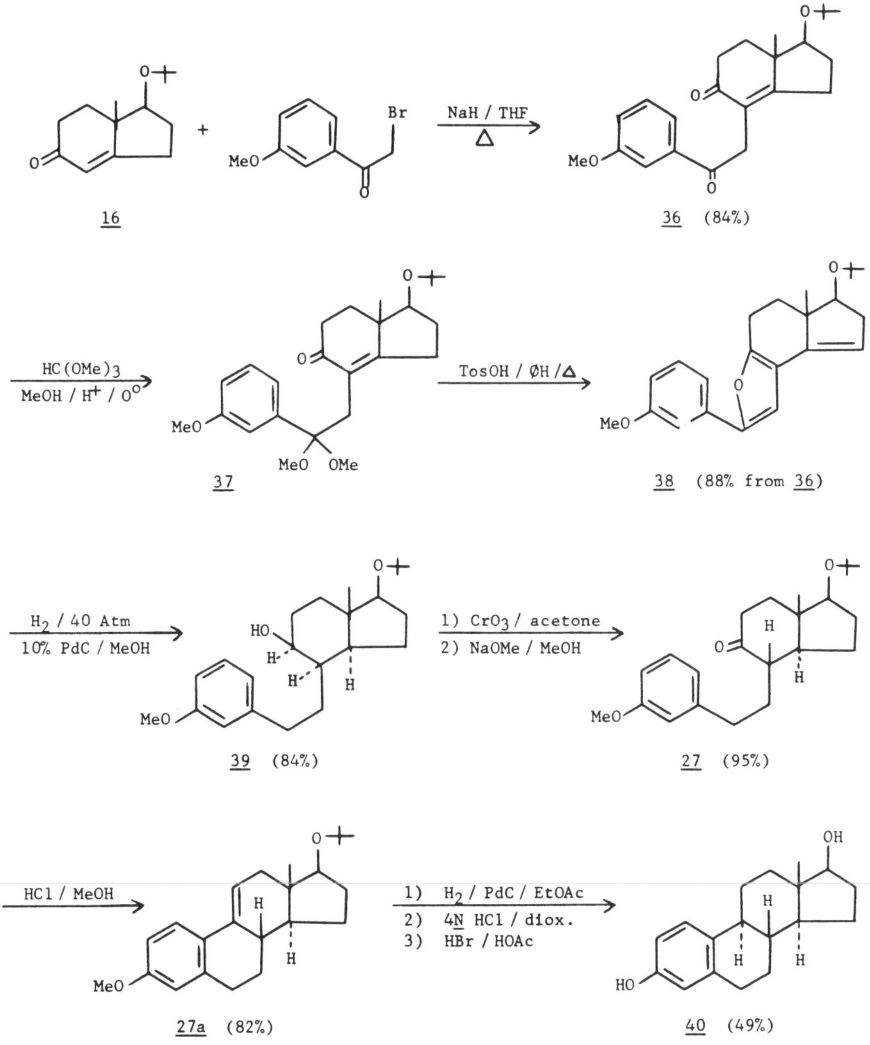

Scheme 6

31 in 75% yield, with hydrogenolysis of the allylic carbonyl group a minor side reaction. Condensation of **31** (via enedione **32**) with ketal β-keto ester **20** followed by saponification, β ring closure, and decarboxylation then gave tricyclic ketalenedione **34** (analogous to Hoffman-LaRoche intermediate **23**) and thence **35**.

The Schering chemists also reported a synthesis of (+)-estradiol **40** based on direct alkylation of the anion of (+)-**16** with m-methoxyphenacyl bromide (Scheme 6).[14] The 84% yield obtained was considerably higher than that achieved previously utilizing the less active *m*-methoxyphenethyl halide or tosylate as alkylating agent (Volume 2, pp. 712, 713). Direct acid-catalyzed formation of the indenofuran **38** was slow and occurred with some loss of the *t*-butyl group. However, prior conversion into the dimethyl acetal **37** permitted smooth furan formation under mild conditions. Hydrogenation of **38** under moderate pressure led in high yield to the 8a,14α-9β-hydroxy compound **39**, which on oxidation and side chain isomerization to the equatorial position afforded the 8β,14α-9-ketone **27**,[13] converted by standard procedures to **27a** and thence to (+)-estradiol **40**.

(d) Danishevsky Syntheses

Danishevsky and co-workers devised an ingenious synthesis of estrone (and other 19-nor-steroids) utilizing 6-substituted α-picolines as ring A synthons in a variant of the Robinson annulation process.[15] The synthesis was initially applied to (±)-D-homoestrone and an improved version was developed for (+)-estrone and related (+)-19-norandrostenones.

Model studies showed that Birch reduction of 6-substituted α-picolines and hydrolysis of the intermediate bisenamines yield 1,5-diketones, which can cyclize to enones A and/or B.[16] In fact, literature precedent[17] and experience with 1,4-diketones favored cyclization mode B (e.g., jasmone). However, the model studies showed substantial and in some cases predominant cyclization to A.

In the synthesis of (±)-D-homoestrone (Scheme 7),[18a,b] Michael addition of the monoketal **42** of the Wieland–Miescher enedione to 6-vinyl-2-methylpyridine **41** led to tricyclic adduct **43** in good yield. Reduction to the 17a-β-alcohol, double bond hydrogenation, and ketalization produced **44** with the requisite 8β,14α-stereochemistry in 58% yield. However, modification of the hydrogenation conditions from ethyl acetate-triethylamine to ethanol-perchloric acid raised the yield to 82%.[18c] Birch reduction, hydrolysis, cyclization, and ketal reversal

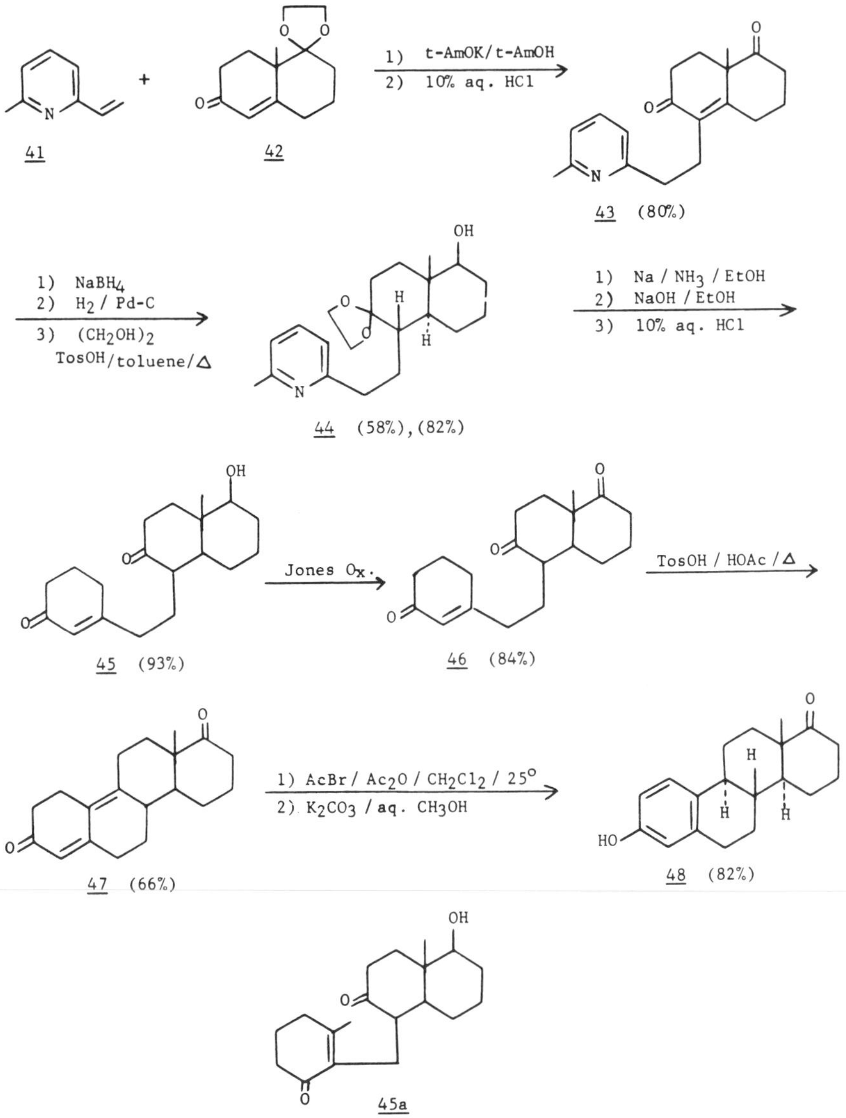

41 +   42 $\xrightarrow{\begin{array}{l}1)\ \text{t-AmOK/t-AmOH}\\2)\ 10\%\ \text{aq. HCl}\end{array}}$ 43 (80%)

$\xrightarrow{\begin{array}{l}1)\ \text{NaBH}_4\\2)\ \text{H}_2\ /\ \text{Pd-C}\\3)\ (\text{CH}_2\text{OH})_2\\ \ \ \ \text{TosOH/toluene/}\triangle\end{array}}$ 44 (58%),(82%) $\xrightarrow{\begin{array}{l}1)\ \text{Na / NH}_3\text{ / EtOH}\\2)\ \text{NaOH / EtOH}\\3)\ 10\%\ \text{aq. HCl}\end{array}}$

45 (93%) $\xrightarrow{\text{Jones O}_x.}$ 46 (84%) $\xrightarrow{\text{TosOH / HOAc / }\triangle}$

47 (66%) $\xrightarrow{\begin{array}{l}1)\ \text{AcBr/ Ac}_2\text{O / CH}_2\text{Cl}_2\text{ / 25}^\circ\\2)\ \text{K}_2\text{CO}_3\ /\ \textbf{aq.}\ \text{CH}_3\text{OH}\end{array}}$ 48 (82%)

45a

Scheme 7

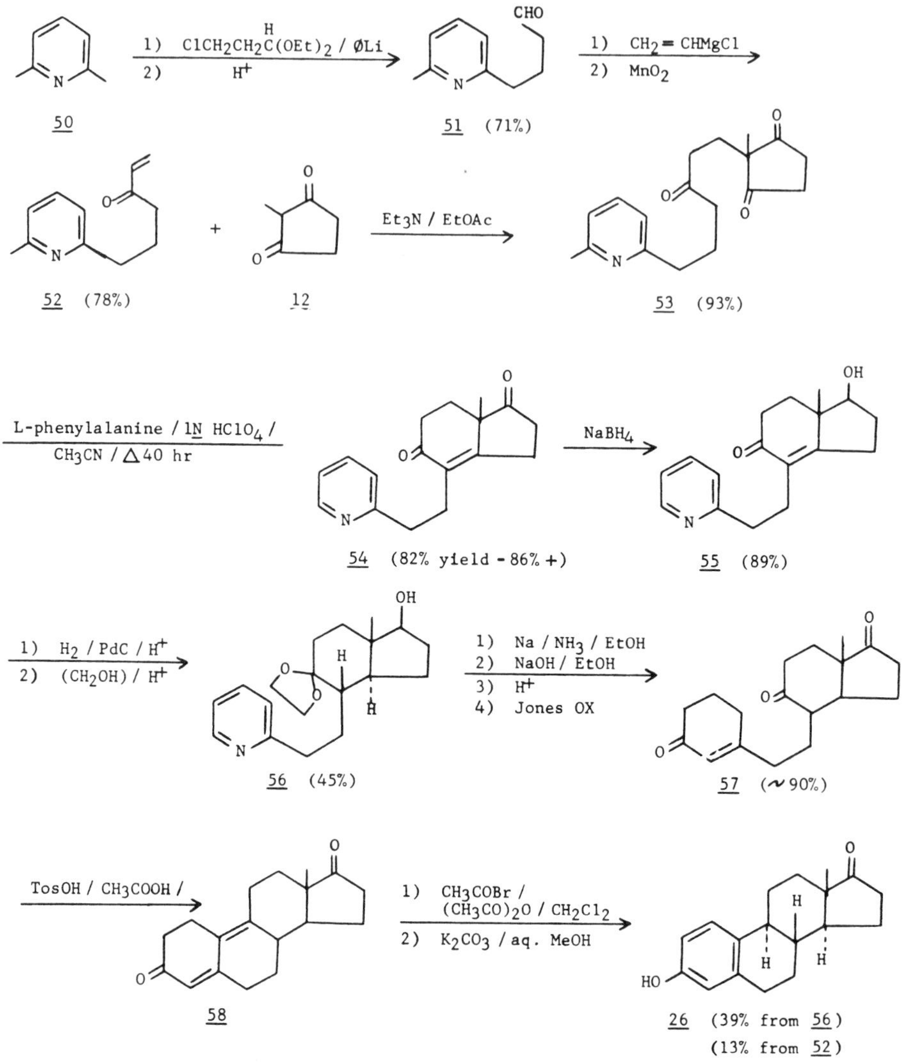

Scheme 8

then led to a single enediol **45** in 93% yield, converted to crystalline enetrione **46** and thence to (±)-D-homoestrone **48** in 21% overall yield from **42**. The absence of the alternative cyclization product **45a** anticipated from model studies may be rationalized on steric grounds.

In initial studies toward (+)-estrone, vinyl picoline **41** was condensed with hydrindenone **16** in analogy with the reaction of **41** with **42**. However, since the yield in the present case was low, an alternative route was devised (Scheme 8).[19] 2,6,-Lutidine **50** was converted in 57% yield to the enone **52** (cf. **167**, Volume 2, p. 693), which readily added 2-methylcyclopentane-1,3-dione **12** to give the prochiral bicyclic trione **53** in high yield. Asymmetric cyclization of **53** in the presence of L-phenylalanine-1N perchloric acid (molar ratio 1:1.2:0.5) in refluxing acetonitrile by the Eder–Hajos technique[4,5] led to (+) **54** of 86% optical purity in 82% chemical yield. Selective borohydride reduction to **55**, followed by catalytic hydrogenation under acidic conditions, ketalization, and chromatography gave **56** in only 45% yield along with 17% of the C/D *cis* isomer and 21% of hydrogenolysis product (**55** C$_9$–H$_2$). The present hydrogenation difficulties are in sharp contrast with the clean high yield result in the analogous step in the D-homo series. Elaboration of the ring A enone occurred unidirectionally to give **57** in 90% yield. Closure of ring B and isomerization by the Roussel procedure[12] led to (+)-estrone **26** in 39% yield from **56** or 13% yield from **52**. The low yield in hydrogenation of the 8(14) double bond unfortunately negates the high yield of the asymmetric cyclization.

(e)   Tsuji Syntheses

Tsuji and colleagues have synthesized 1,7-octadiene-3-one **62** from the readily available butadiene telomer **59**[20] and have explored the utility of **62** in natural product synthesis, including two routes to 19-norsteroids.[21,22]

Dimerization of butadiene catalyzed by palladium acetate-triphenylphosphine yielded a separable mixture of octadiene acetates **59** and **60** in high yield. Acetate

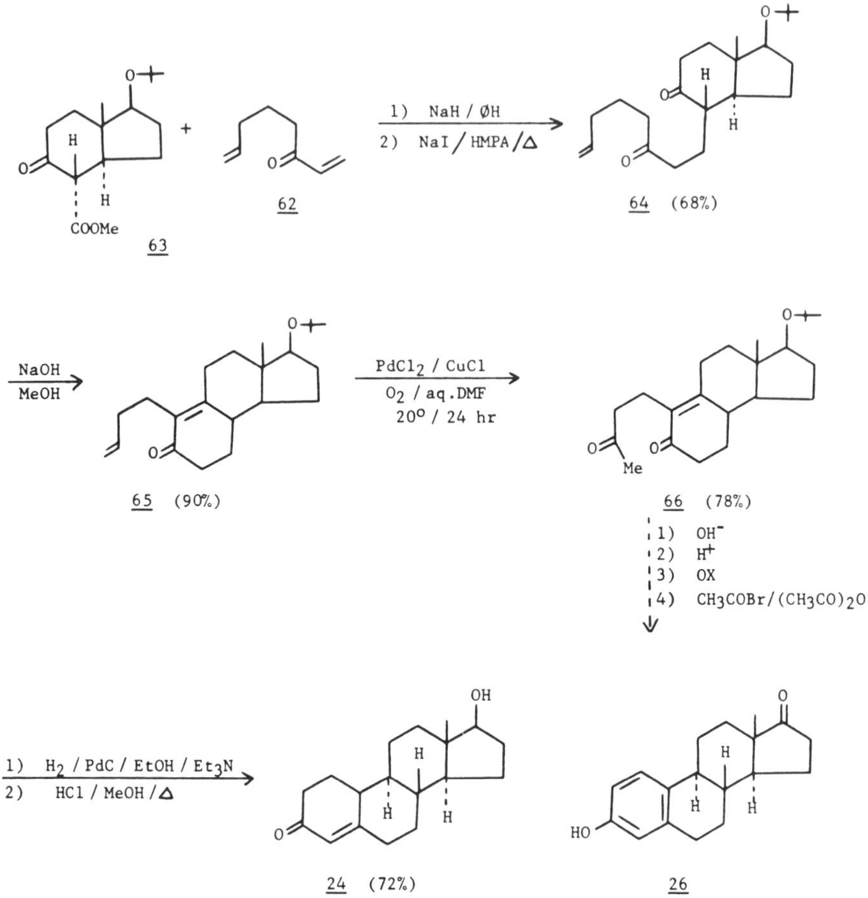

Scheme 9

60 could be rearranged to **59** by the palladium catalyst. Conversion of **59** to alcohol **61** and dehydrogenation over copper-zinc alloy in a packed column at 280-360° provided dienone **62** in good yield.[20,21,23]

The 1-ene-3-one part-structure in **62** can be utilized in Michael addition reactions and subsequently the remaining double bond can be oxidized to methyl ketone,[24] making **62** a bisannulation reagent synthetically equivalent to 7-octene-2,6-dione, as is Danishevsky's 6-vinyl-2-picoline (**41**, Scheme 7).

As shown in Scheme 9,[21] (+)-β-keto ester **63** [available by diazomethane treatment of **18** (Scheme 3)][8b] on Michael reaction with **62**, followed by decarbomethoxylation, yielded enedione **64**. Aldol cyclization to **65** and palladium chloride–cuprous chloride catalyzed oxidation of the terminal double bond then gave tricyclic enedione **66**. The latter was converted into (+)-19-nortestosterone

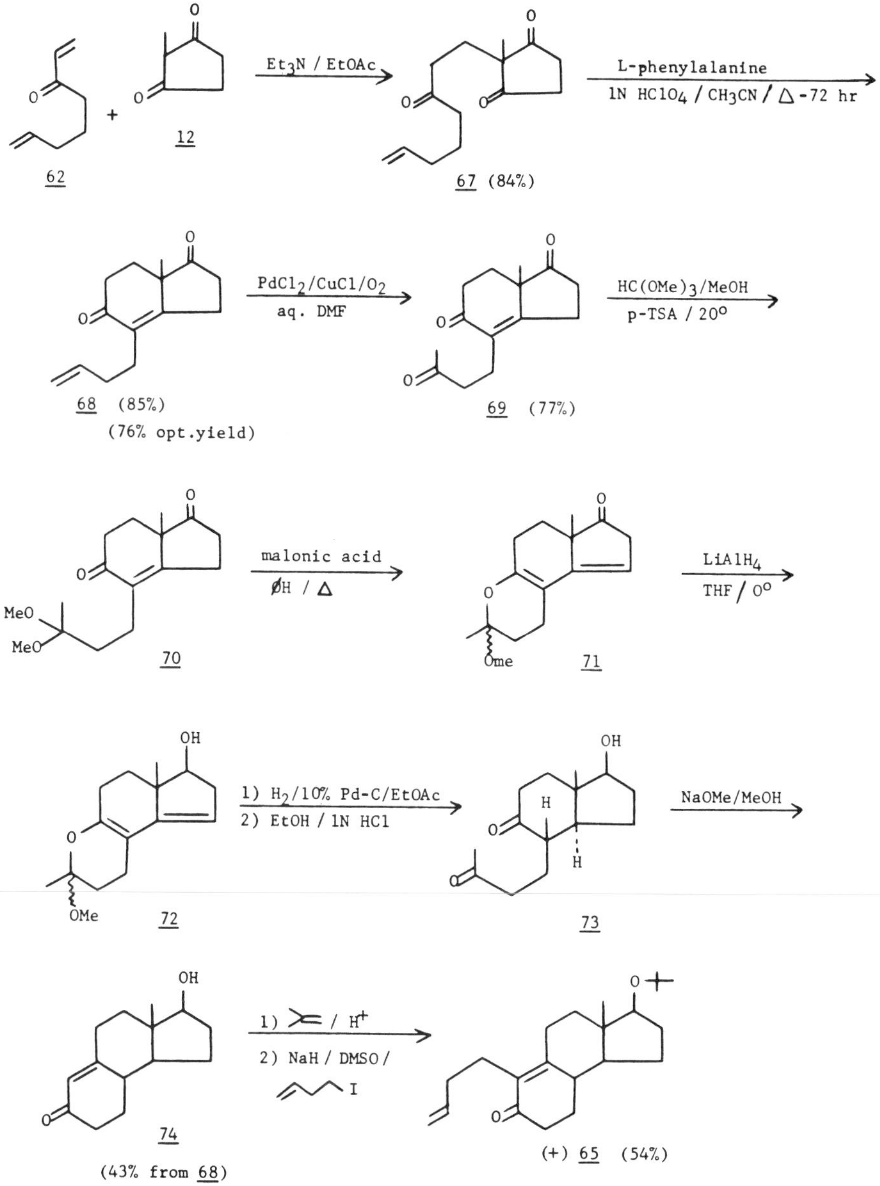

Scheme 10

24 but should also be convertible into (+)-estrone **26** as indicated earlier (cf. Scheme 3, **23** → **26**; Scheme 8, **58** → **26**).

In the route just described (CD → ABCD), the bisannulation reagent **62** is the source of rings A and B. In an alternative, somewhat more involved, approach[22] (D → BCD → ABCD; Scheme 10), reagent **62** is the source of rings B and C.

Michael reaction of **62** and 2-methylcyclopentane-1,3-dione led to adduct **67** in good yield. Asymmetric aldol cyclization with concomitant dehydration was accomplished using L-phenylalanine-1N-perchloric acid (molar ratio 1:1:0.4) in refluxing acetonitrile to produce (+)-**68** of 76% optical purity in 85% yield.[4,5,19] Palladium chloride catalyzed terminal olefin oxidation[24] then yielded enetrione **69**. The latter compound has been prepared by Eder et al.,[25] who developed an effective method for its α-face hydrogenation via the 17β-hydroxydiene system **72** (cf. Scheme 6) which in turn was obtained in good yield via intermediates **70** and **71**. Hydrogenation of the Δ^{14} double bond of **72** proceeded completely in the desired α sense, and the product was hydrolyzed to hydroxydione **73**. Base catalyzed aldol cyclization then provided the tricyclic enone **74**. Conversion of the 17β-hydroxy group to the t-butyl ether and base catalyzed addition of 3-butenyl iodide produced the butenylated ketone (+) **65** in 54% conversion yield. The latter had been produced by the earlier route (Scheme 9) and converted into (+)-19-nortestosterone **24**.

Tsuji has also prepared the trisannulation reagent **75** from dienone **62** as indicated and utilized it in a synthesis of (±)-D-homoandrost-4-ene-3,17a-dione.[26]

B. Orthoquinodimethane Approach

(a) Introduction

Thermolysis of benzocyclobutenes and trapping of the intermediate orthoquinodimethanes with external dienophiles was described some time ago by Cava[27] and by Jensen and Coleman.[28] The intramolecular version of this reaction was initially studied independently by Oppolzer and Kametani, who have reviewed its development and application to the stereospecific synthesis of polycyclic systems, including steroids.[29] The main area of interest in the steroid field has been the synthesis of estrone and related systems, although recently emphasis has shifted to saturated steroids.[30]

(b) Kametani Syntheses

In 1976 Kametani and co-workers described the initial application of o-quino-dimethane thermolysis methodology to steroid synthesis in a route to ($\pm$)-D-homoestrone methyl ether **83a**, remarkable for the high yield in the cycloaddition step, but which suffers from a low yield in the condensation reaction leading to key intermediate **81** (Scheme 11).[31]

The required 2-(4-methoxybenzocyclobutenyl)-ethyl iodide **79** was prepared by a multistep process from 2-bromo-5-methoxybenzaldehyde via 1-cyano-4-methoxybenzocyclobutene **77** as illustrated. 2-Methylcyclohexenone was then converted into 2-methyl-3-vinyl-6-n-butylthiomethylenecyclohexanone **80**. Blocking C-6 was considered necessary to insure regiospecific condensation of **79** at C-2. However, the condensation reaction afforded adduct **81** in only 16% yield. In subsequent work (with 2-methylcyclopentenone) 1:4 addition of a vinyl Grignard or lithium reagent and coupled alkylation of the intermediate enolate occurred at C-2 without the need to block C-6 (Scheme 1,[1] Scheme 14[36,40]).

Thermal cycloaddition of **81** proceeded with involvement of the olefinic bond α to the carbonyl group to give **82**.[32] A similar result was obtained with the benzylidene analog of **81**. It was therefore necessary to remove the C-6 blocking group prior to cycloaddition. Deblocked adduct **83** on refluxing in o-dichloro-benzene underwent intramolecular cycloaddition smoothly via the indicated ster-ically favorable exo transition state to produce ($\pm$)-D-homoestrone methyl ether **83a** in 95% yield. Compound **83a** had been previously converted to ($\pm$)-estrone by Johnson et al. (see Volume 2, pp. 681-682).

Although condensation leading to the five-membered ring D counterpart of **81** occurred in 26% yield, it was not possible to deblock the latter successfully to the five-membered ring D counterpart of **83**. However, a direct route utilizing five-membered ring D intermediates was achieved by Kametani in his next synthesis.

The poor yield in the above alkylation step (**79** + **80** → **81**) was improved considerably by placing the iodomethyl group on the ring D precursor and condensing the latter with 1-cyano-4-methoxybenzocyclobutene (e.g., **77** + **88** → **89**, Scheme 12). This modification and utilization of the chiral indanone **84**[8a] as starting material led to an asymmetric synthesis of (+)-estradiol **40**.[32a] Chiral indanone **84** was converted into (+)-(1S,2S,3S)-1-t-butoxy-2-(2-iodoe-thyl)-2-methyl-3-vinylcyclopentane **88** by the indicated multistep procedure.

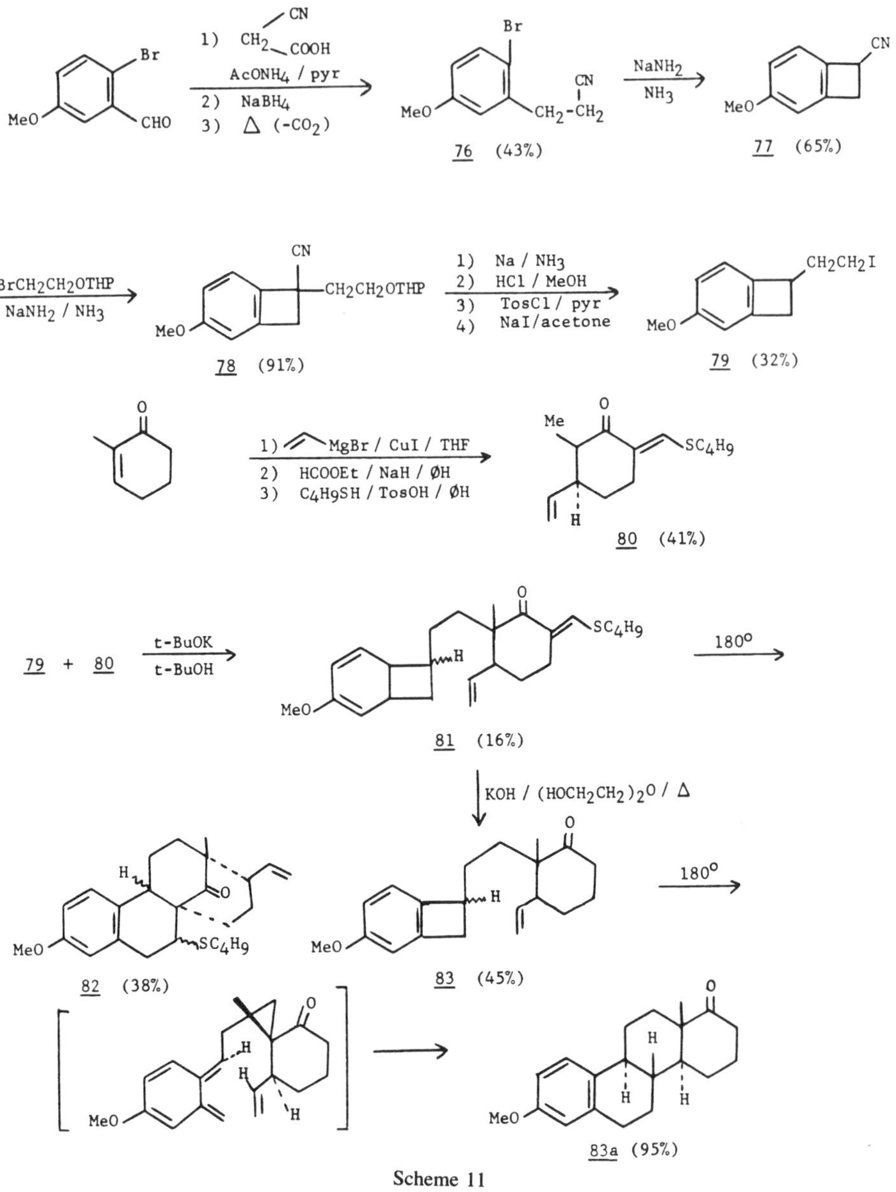

Scheme 11

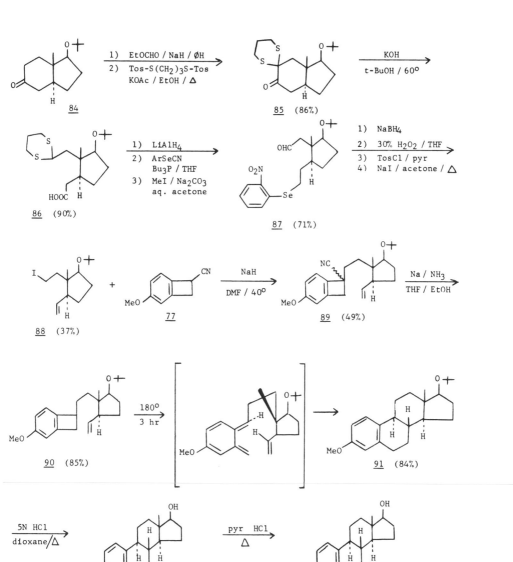

Scheme 12

Condensation of **88** with **77** and reductive removal of the cyano function gave key intermediate **90**. Thermolysis as in the previous synthesis led in 84% yield via an *exo* transition state to the chromatographically pure chiral estradiol derivative **91**, in turn deprotected to (+)-estradiol **40**.

Kametani has also described analogous routes to 14α-hydroxyestrone-3-methyl ether, (−)-3β-hydroxy-17-methoxy-D-homo-18-nor-5α-androst-13,15,17-triene (a key intermediate in Nagata's synthesis of pregnanes), as well as other ring D aromatic D-homosteroids, and their conversion into various classes of pregnanes.[29c] Generation of dienes from tetrahydrobenzocyclobutene [bicyclo(4,2,0)oct-1(6)-ene] systems has been accomplished and could lead to direct routes to saturated steroids.[33]

(c) Oppolzer Syntheses

Following earlier application of *o*-quinodimethane cycloaddition methodology to the alkaloid field,[29a] Oppolzer's initially described approach to steroids involved the reaction **93** → **94**, which apparently has not been described in detail.[34]

In contrast to the above *trans-anti-trans* cycloaddition, Oppolzer's next synthesis[35] led exclusively to the *cis-anti-trans* product **99** via an *endo* transition state (Scheme 13). Preference for either an *endo* or *exo* transition state is delicately influenced therefore by the nature of the substitution on the incipient ring C.

Since yields in joining the benzocyclobutene and ring D units via alkylation procedures[31] (Scheme 11) were low, Oppolzer examined alternative methodology for preparation of the required cycloaddition precursors. The ring D species **95** as its lithium enolate reacted efficiently with the α-bromo oxime **96** (via the corresponding nitroso olefin) to give the hemiacetal **97** in 77% yield. Thermolysis of the demethoxy analog of **97** involved the carbon–nitrogen double bond and led to the isoquinoline **101**. However, thermolyis of the oxime benzyl ether **98** gave the *cis-anti-trans* steroid **99**, in turn converted into 9β-11-ketoestrone methyl ether **100**.

Oppolzer's next approach utilized resolved acid chloride **104** as the ring D

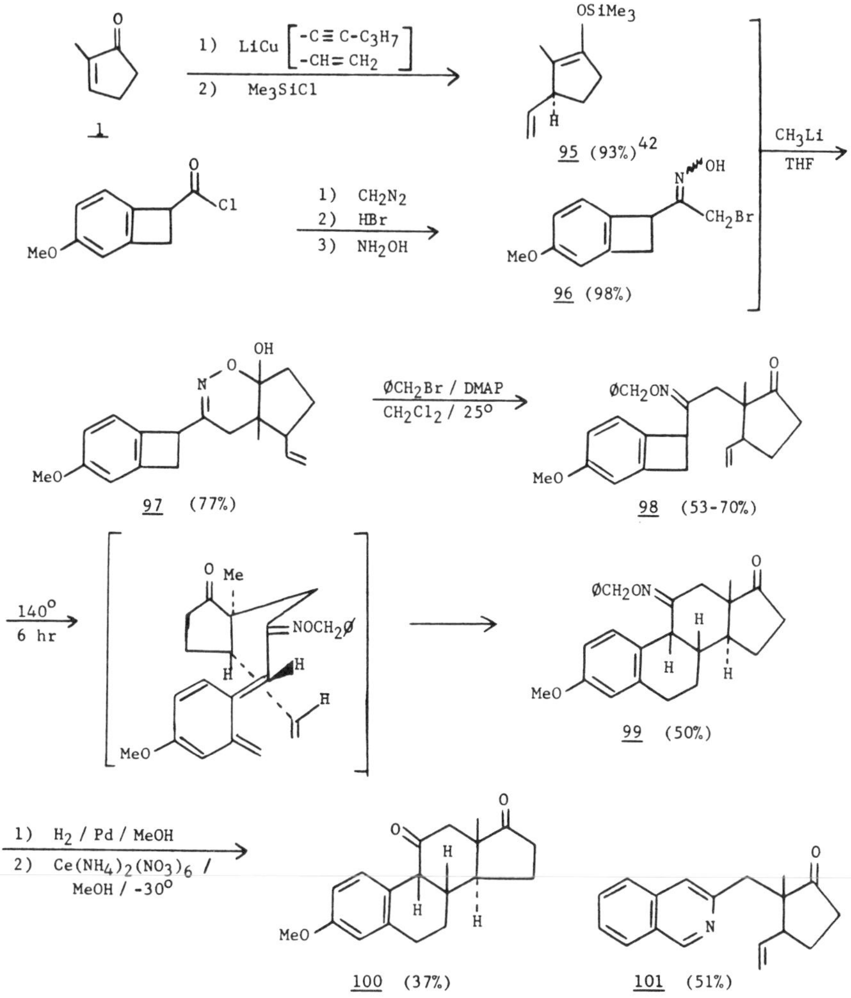

Scheme 13

component and led to (+)-11-ketoestrone methyl ether **108** (Scheme 14).[36] The cycloaddition precursor **107** differs from **98** (Scheme 13) in possessing a carbonyl group in place of o-benzyloxime functionality, and this is sufficient to favor an *exo* transition state leading to a *trans-anti-trans* product.

Vinylcuprate conjugate addition to 2-methylcyclopentenone **1** and alkylation of the intermediate enolate with methyl bromoacetate led in high yield to the vinyl ester **102** (containing ~10% of the C-2 epimer). Saponification, efficient resolution, and oxalyl chloride treatment furnished pure (+)-acid chloride **104**,

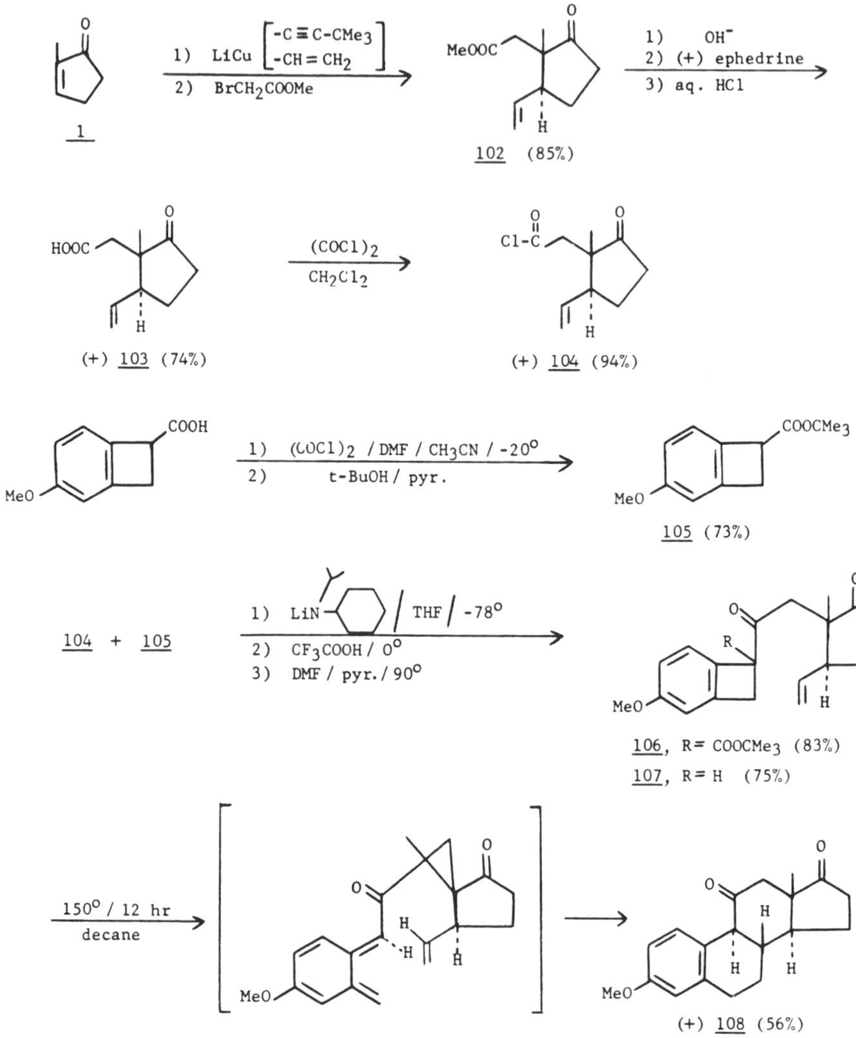

Scheme 14

which was condensed with benzocyclobutene *t*-butyl ester **105** (prepared via the acid chloride in 73% yield) to give dione ester **106**. The 83% yield in this acylation step contrasts with the considerably lower yields obtained on analogous alkylations.[31,32b] De-esterification and decarboxylation to **107** followed by thermolysis then led to the (+)-steroid **108** in 56% yield—as well as 5% of its *cis-anti-trans* isomer.

In the previously discussed routes the requisite orthoquinodimethanes were derived by thermolysis of isolated benzocyclobutene precursors. However, alternative approaches to *o*-quinodimethanes, including chelotropic elimination of sulfur dioxide, nitrogen, or carbon monoxide as well as reverse Diels–Alder processes and photoenolization of orthomethyl phenyl ketones have been described.[29a] The thermal elimination of sulfur dioxide from benzo(c)1,3-dihydrothiophene dioxide **109** and the trapping of the intermediate *o*-quinodimethane with an external dienophile was demonstrated by Cava and Deana in 1959.[37]

Oppolzer, in developing this methodology into a synthesis of (+)-estradiol, first devised a procedure for the regiospecific introduction of the ring D dienophile containing a component at C-1 in a C-5 substituted benzo(c)-1,3-dihydrothiophene dioxide. It was ascertained that sulfone **109**[37] could be monoalkylated in good yield by deprotonation with, for example, *n*-butyllithium in tetrahydrofuran at − 78° followed by addition of an alkenyl bromide.[38] Thermolysis of the alkylated sulfones **110** occurred mainly via the E dienes **111** to give the desired adducts **112** in high yields, as well as a few percent of **113** derived via 1,5 hydrogen shift in the Z isomer of **111**.

In order to promote selective deprotonation at C-1 in **109**, an electron attracting substituent is required at the para position, C-5. The nitrile group was chosen— also because it permitted ready conversion to phenolic hydroxyl at the end of the synthesis (Scheme 15).[39] Iodination of sulfone **109** followed by iodide-cyanide exchange gave the C-5 nitrile **115**. The optically active acid (+) **103** (Scheme 14), via its methyl ester (+) **102**, was converted in good yield into iodo olefin **116**.

Alkylation of sulfone **115** with **116** was carried out with two molar equivalents each of **115** and sodium hydride to provide adduct **117** as a 1:1 mixture of diestereomers at C-1 in 82% yield based on **116**. Interestingly, the C-5 nitro

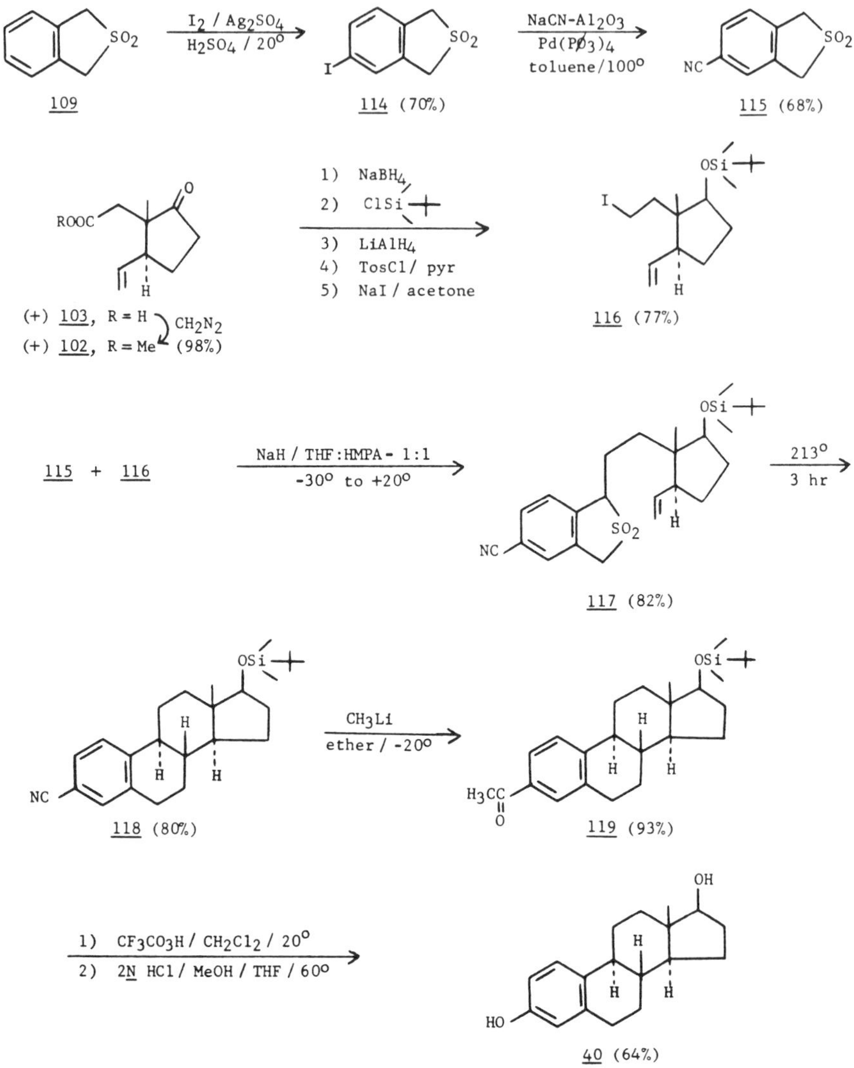

Scheme 15

analog of **115** (readily available by nitration of **109**) yielded tars on treatment with base.

Thermolysis of adduct **117** in refluxing 1,2,4-trichlorobenzene proceeded cleanly to give pure *trans-anti-trans* 3-cyanoestratriene **118** in 80% yield. Generation of 3-hydroxyl functionality was accomplished by reaction with methyl lithium to produce the 3-acetyl compound **119** followed by Baeyer–Villiger oxidation to the corresponding 3-acetate and concluding acid hydrolysis to $(+)$-estradiol **40**.

(d) Nicolau Synthesis

Nicolau and co-workers have independently reported parallel studies on the generation of o-quinodimethanes via thermal elimination of sulfur dioxide from 1-substituted benzo(c)1,3-dihydrothiophene dioxides which led to an efficient synthesis of $(\pm)$-estra-1,3,5(10)-triene-17-one.[40]

(85%)

(e) Vollhardt Synthesis

A novel procedure for generation of orthoquinodimethanes, developed and exploited by Vollhardt, is based on the co-oligomerization of α-ω-diacetylenes and monoacetylenes in the presence of cyclopentadienylcobalt dicarbonyl catalyst.[41] The reaction is carried out by addition of the diacetylene and catalyst to excess refluxing monoacetylene. When $n = 2$, the benzocyclobutene co-oligomerization product formed *in situ* is in thermal equilibrium with the corresponding o-quinodimethane, which can undergo Diels–Alder cyclization with an appropriately positioned dienophile.

For estrone synthesis the vinyldiacetylene **121** was required (Scheme 16).[42] It was obtained, as indicated from 2-methylcyclopentenone **1** and 1,5-hexadiyne, as an epimeric mixture at C-4, but with the key cyclopentanone substituents *trans*. In the alkylation step a 3:1 molar ratio of **120** to **95** was utilized. Co-oligomerization of **121** with refluxing bistrimethylsilylacetylene in the presence of five mol percent of $CpCo(CO)_2$ catalyst led to the *trans-anti-trans* bistrimethylsilylestratriene **122** in 71% yield. It is noteworthy that five carbon–carbon bonds

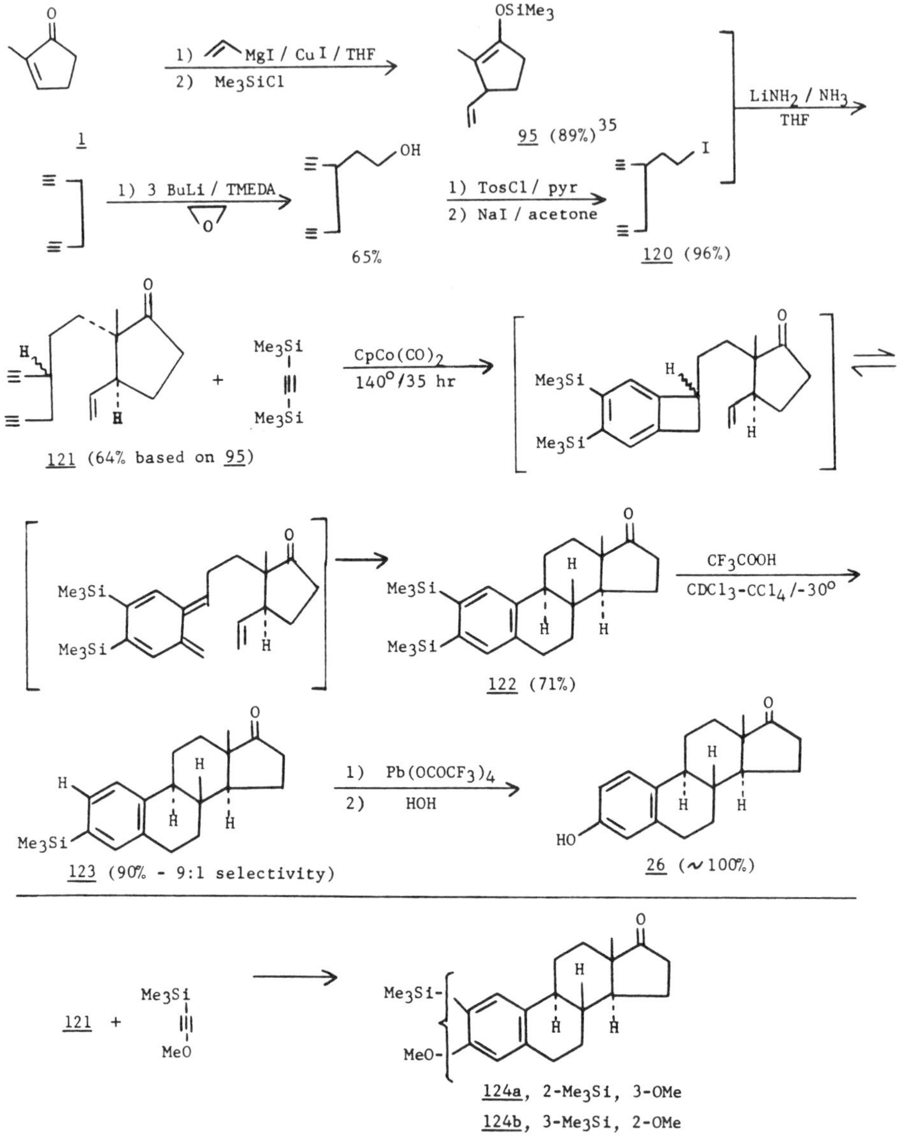

Scheme 16

are formed in this step. Increased reactivity to electrophilic reagents at C-2 relative to C-3 permitted selective protonolysis to **123**.[43] Essentially quantitative oxidative aryl–silicon cleavage then led to (±)-estrone **26**.[43]

A more direct route to 3-methoxyestratrienes via reaction of methoxytrimethylsilylacetylene with **121** was not practical as the co-oligomerization proceeded in 31% yield to a 2:1 mixture of **124a** and **124b** (Scheme 16).[43]

(f) Quinkert Synthesis

Photolysis of orthotoluyl ketones has been utilized to generate the corresponding transient dienols (hydroxy o-quinodimethanes), which have been trapped by reaction with external or internal dienophiles.[29a,44]

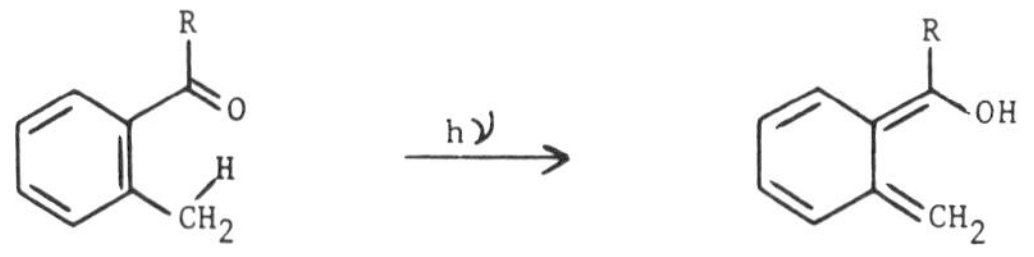

Based on this photoenolization process, Quinkert has devised syntheses of (±)[45] and (+)[46] estrone from simple starting materials, notable also for a novel route to the ring D moiety **128** (Scheme 17).

The aromatic component **125** was prepared efficiently in four steps from m-cresol methyl ether. Synthesis of the D ring component proceeded via the vinylcyclopropane diester **126** obtained by condensation of dimethyl malonate with 1,4-dibromo-2-butene. Reaction of **126**, via its homoenolate anion, with methyl dimethyl malonate and ensuing Dieckmann cyclization and decarbomethoxylation led to **127** and thence to the required methylvinylcyclopentanone **128**. Michael addition of **128** to vinylsilylketone **125** took place regiospecifically and trans to the vinyl group to give the photolysis precursor **129**. Exposure of **129** to long wavelength uv light at 98° in methylcyclohexane containing pyridine and 2,4,6-trimethylphenol yielded **130** and a minor amount of the 9β-hydroxy isomer. Dehydration then led to **131** containing 5% of the isomeric Δ-8 olefin—both previously converted into (±)-estrone methyl ether **28** and thence (±)-estrone by known methodology.

Extension of the synthesis to (+)-estrone required preparation of the chiral ring D component (+) **128**. This was accomplished in three ways, as follows[46]:

1. Resolution of vinylcyclopropane diacid **126a** with brucine to produce (+) **126a**. Chirality is retained in the ring expansion step. It is assumed that the

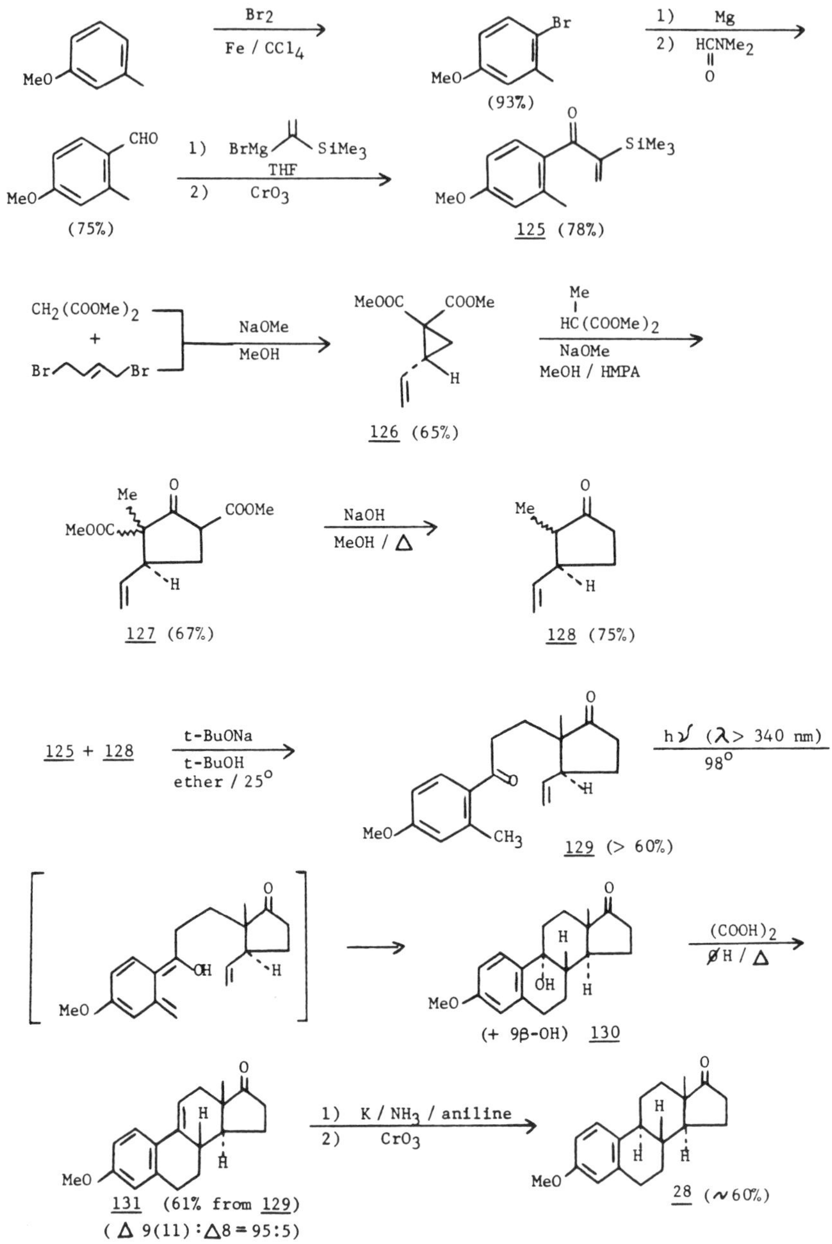

Br₂ / Fe / CCl₄ →

Br_2
Fe / CCl_4

MeO ... (93%)

1) Mg
2) HCNMe₂ ‖ O

CHO
MeO (75%)

1) BrMg / SiMe₃ THF
2) CrO₃

O
SiMe₃
MeO
125 (78%)

CH₂(COOMe)₂ + Br⌒⌒⌒Br

NaOMe
MeOH

MeOOC COOMe

126 (65%)

Me
HC(COOMe)₂
NaOMe
MeOH / HMPA

Me
MeOOC
COOMe
O
127 (67%)

NaOH
MeOH / Δ

Me
O
H
128 (75%)

125 + 128

t-BuONa
t-BuOH
ether / 25°

O
MeO CH₃
129 (> 60%)

hν (λ > 340 nm)
98°

MeO OH
(bracketed intermediate)

MeO OH H
(+ 9β-OH) 130

(COOH)₂
ØH / Δ

MeO
131 (61% from 129)
(Δ 9(11) : Δ8 = 95:5)

1) K / NH₃ / aniline
2) CrO₃

MeO
28 (∼60%)

Scheme 17

28

configuration at C-2 is inverted on addition of methyl dimethyl malonate to the corresponding diester (+) **126** at the outset of this reaction.

(+) <u>126a</u>

2. An alternative synthesis of (+) **128** from the bicyclo-(2,2,1)heptanecarboxylic acid (−) **132** obtained by resolution of (±) **132** with (+)-phenethylamine [*cf.* Grieco synthesis (Scheme 18)[48]].

3. Use of chiral malonic esters in the reaction with 1,4-dibromo-2-butene to provide (+) **126**. About 80% optical enrichment was achieved.

(g) Grieco Synthesis

The use of bicyclo(2,2,1)-heptane derivatives in natural product synthesis has been explored extensively by Grieco and co-workers.[47] They have devised a synthesis of (±)-estrone based on orthoquinodimethane methodology in which the ring D component is derived stereospecifically from the bicyclo-(2,2,1)heptane species **134** (Scheme 18).[48] The latter and related bicycloheptanes are readily available from norbornadiene by a sequence which features the Prins reaction, devised independently by groups led by Corey[49a] and Sutherland[49b] in conjunction with their studies on prostaglandin synthesis.

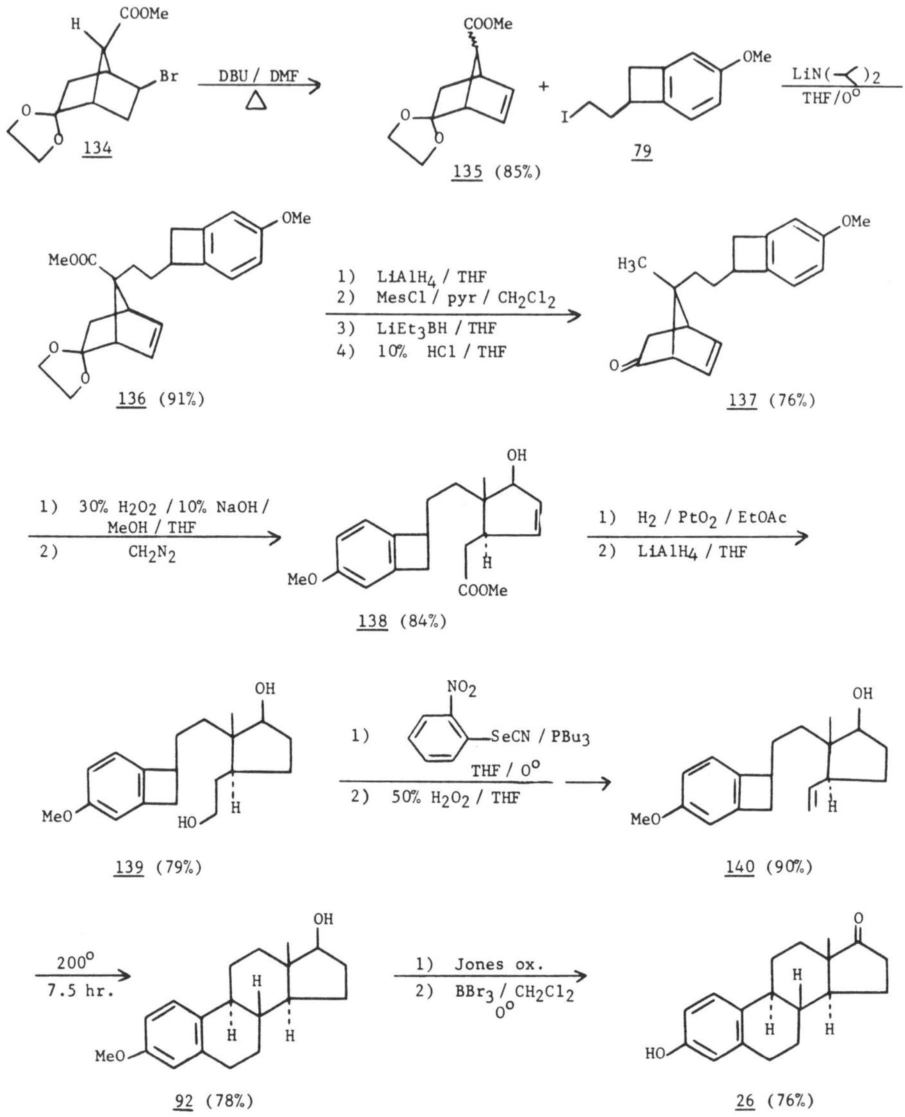

134

DBU / DMF △

135 (85%)

+

79

LiN(—◁)$_2$
THF / 0°

136 (91%)

1) LiAlH$_4$ / THF
2) MesCl / pyr / CH$_2$Cl$_2$
3) LiEt$_3$BH / THF
4) 10% HCl / THF

137 (76%)

1) 30% H$_2$O$_2$ / 10% NaOH / MeOH / THF
2) CH$_2$N$_2$

138 (84%)

1) H$_2$ / PtO$_2$ / EtOAc
2) LiAlH$_4$ / THF

139 (79%)

1) ⬡—SeCN / PBu$_3$ (NO$_2$)
   THF / 0°
2) 50% H$_2$O$_2$ / THF

140 (90%)

200°
7.5 hr.

92 (78%)

1) Jones ox.
2) BBr$_3$ / CH$_2$Cl$_2$
   0°

26 (76%)

Scheme 18

30

Bromoketoacid **133** on diazomethane esterification and ketalization was converted into bromoketalester **134**. Elimination of hydrogen bromide yielded **135**, which was alkylated stereospecifically with 1-(2-iodoethyl)-4-methoxybenzocyclobutene **79**, prepared earlier by Kametani (Scheme 11),[31] to give **136** in surprisingly high yield. Reduction of ester functionality to methyl and deketalization led to **137**. The latter on Baeyer–Villiger oxidation with 30% hydrogen peroxide and 10% sodium hydroxide in aqueous methanol-tetrahydrofuran (conditions developed for similar bicyclo-(2,2,1)-heptenones) followed by esterification, yielded cyclopentenol ester **138** with the ring D substituents possessing the required stereochemistry. Interestingly, the saturated counterpart of **137** failed to undergo Baeyer-Villiger oxidation to the corresponding lactone or hydroxy acid.

Hydrogenation of **138** and conversion of the acetic ester side chain into vinyl led to **140**, which on thermolysis gave *trans-anti-trans* steroid **92** ((±)-estradiol-3-methyl ether) as the sole isolable product. Oxidation and ether hydrolysis then yielded (±)-estrone **26**. Extension to (+)-estrone should be straightforward, since the optically active forms of **134** are readily available.

(h)   Tsuji Synthesis

Tsuji and his group have developed a palladium catalyzed cyclization reaction that has led to a simple synthesis of 2-carbomethoxy-3-vinylcyclopentanone **142**,[50] which they have utilized in syntheses of cyclopentanoid natural products.[51] In the steroid field, **142** served as a ring D component in an orthoquinodimethane-based route to (±)-18-hydroxyestrone. The (+)-diastereomer is a component of pregnancy urine. (±)-Estradiol-3-methyl ether was also obtained (Scheme 19).[52]

3-Keto-8-phenoxy-6-octenoate **141** was prepared by condensation of 1-chloro-4-phenoxy-2-butene with the dianion of methyl acetoacetate. Cyclization of **141** was carried out in refluxing acetonitrile (1 hour) in the presence of 5-10 mol% of palladium acetate and 20-35 mol% of triphenyl phosphine to give **142** in 59% isolated yield (along with ~13% of 2-carbomethoxycyclohept-4-enone).[50] Al-

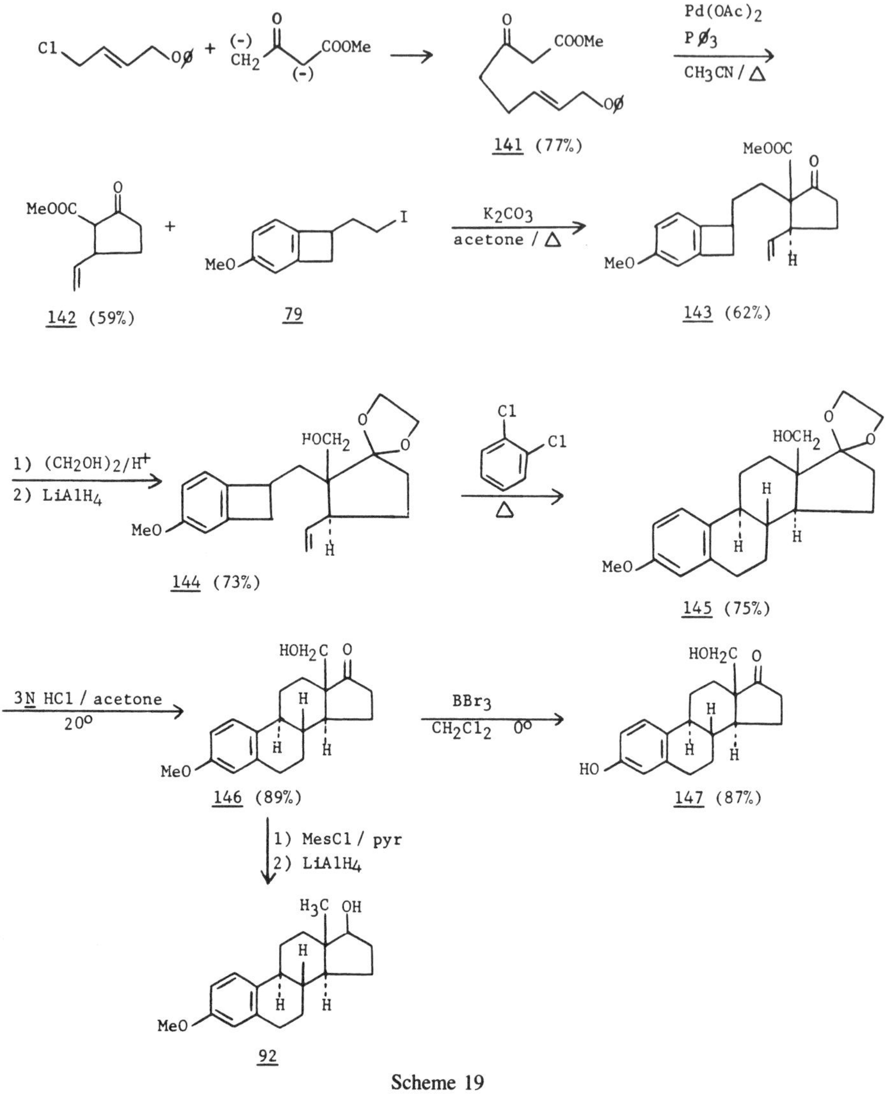

Scheme 19

kylation of **142** with 1-(2-iodoethyl)-4-methoxybenzocyclobutene **79** (Scheme 11)[31] gave adduct **143** in 62% yield. A minor amount of *cis* isomer was easily separated from **143** chromatographically. Ketalization and reduction of the ester group produced primary alcohol **144**, which was thermolyzed in *o*-dichloroben-zene in 75% yield to a single product **145**. Deprotection then afforded (±)-18-hydroxyestrone-3-methyl ether **146** and (±)-18-hydroxyestrone **147**. Lithium aluminum hydride reduction of the mesylate of **146** produced (±)-estradiol-3-methyl ether, confirming the *trans-anti-trans* stereochemistry of **146** and **147**.

(i) Saegusa Synthesis

Saegusa and co-workers have shown that orthoquinodimethanes can be generated by fluoride ion desilylation of ortho-(α-trimethylsilylalkyl)benzyltrimethyl ammonium salts at temperatures in the 20° to 80° range,[53] strikingly lower than the *ca*. 180-220° range required to generate orthoquinodimethanes from benzocyclobutenes or benzodihydrothiophene dioxides.

Based on this methodology, syntheses of (±)-estrone methyl ether and (±)-6β-methylestra-1,3,5(10)-triene-17-one were then developed.[54] In the estrone synthesis (Scheme 20), the aromatic component **149** was obtained in high yield from 3-methyl-4-chlorophenol. The trimethylsilylmethylene moiety was introduced in >90% yield by nickel catalyzed cross-coupling of trimethylsilylmethyl magnesium chloride and aryl chloride **148**.

Following past precedent,[36,40] preparation of the ring D component **152** originated from 2-methylcyclopentenone **1** via **150** prepared by 1:4 addition of vinyl magnesium bromide followed by addition of *t*-butylbromoacetate. Use of the *t*-butyl ester may have contributed to greater stereospecificity in the *trans* addition of the acetic ester side chain at C-2 relative to the C-3 vinyl group, since these substituents were >96% *trans* in **150**. Lower stereospecificity was reported for addition of methylbromoacetate (~ 88%)[36] and ethyl bromoacetate (~78%).[40] Conversion of **150** to the methyl ester, ketalization with 2,2-dimethylpropane-1,3-diol, hydride reduction, tosylation, and displacement by bromide then led to **152** in 47% yield from **1**.

Generation of the silicon stabilized benzyl carbanion **153** from **149** required the presence of hexamethylphosphotriamide. In its absence lithiation occurred at least in part in the aromatic ring. Alkylation of **153** with **152** produced **154** in 94% yield as a 2:1 isomer mixture at the benzylic site. Methiodide formation, addition of caesium fluoride, and refluxing in acetonitrile for 1.5 hours then afforded (±)-estrone methyl ether **28** (containing 7-8% of the C-9 β-H isomer) in 86% yield. Pure **28** was obtained on recrystallization.

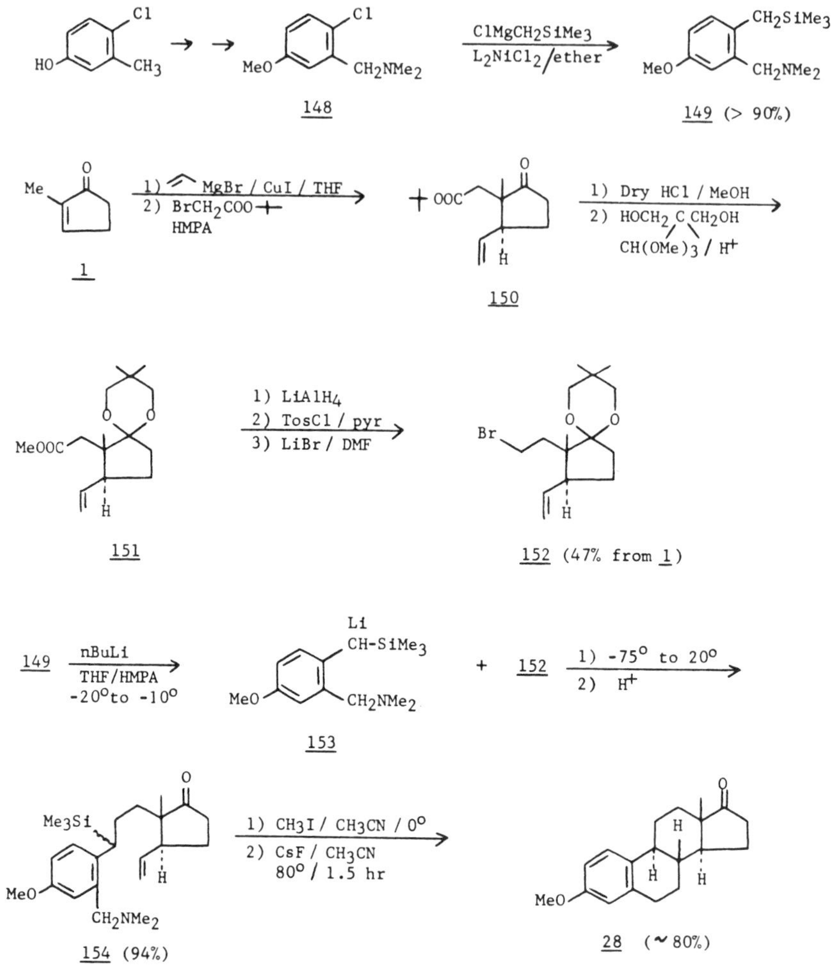

Scheme 20

(j) Magnus Synthesis

Magnus et al. have devised a route to ($\pm$)-11α-hydroxyestrone 3-methyl ether which, like the Saegusa synthesis (Scheme 20),[54] makes use of fluoride mediated elimination of the trimethylsilyl group, but which is coupled with alternative methodology for *in situ* orthoquinodimethane generation (Scheme 21).[55]

The aromatic component **159** was prepared from 4-methoxybenzoic acid via oxazolines **155**, **156** and **157** and trimethylsilylmethylbenzaldehyde **158**. The

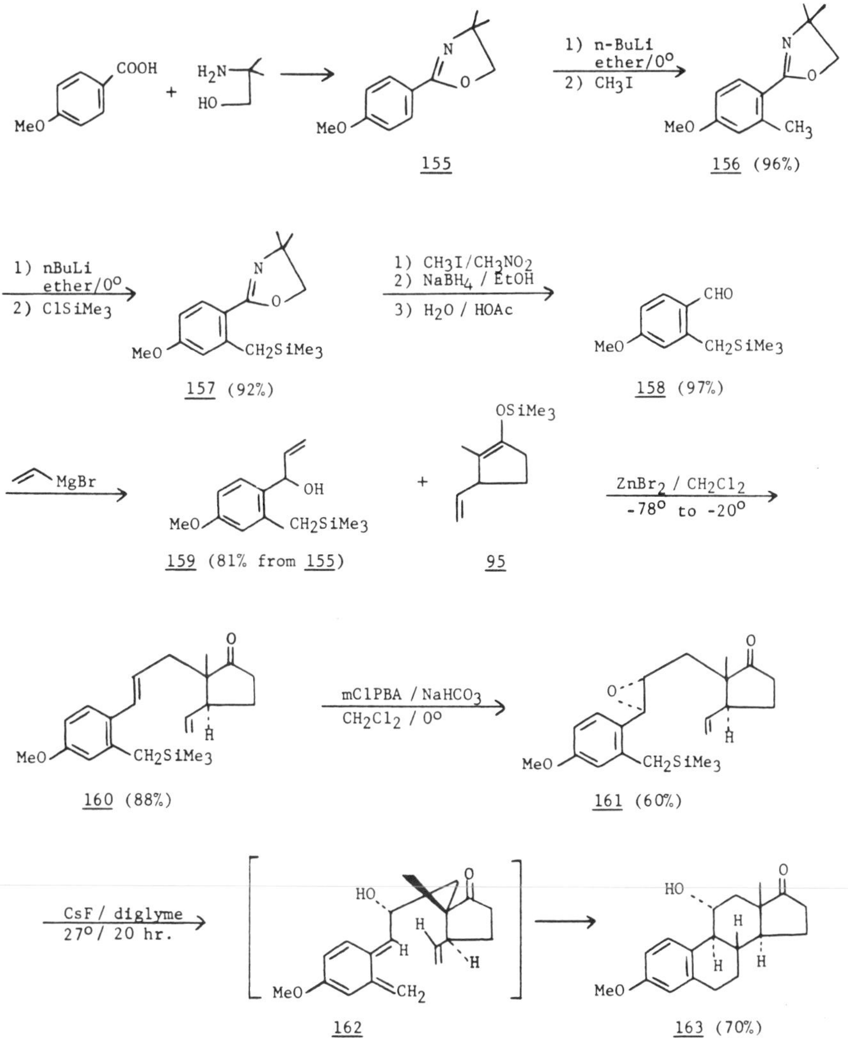

Scheme 21

latter gave vinylcarbinol **159** in 81% overall yield from **155** on reaction with vinyl magnesium bromide. In a variant of the Torgov reaction (Volume 2, p. 698) coupling of vinyl carbinol **159** with the known ring D component, tri-methylsilyl enol ether **95** (Scheme 13[35] and Scheme 16[42]), catalyzed by zinc bromide, produced adduct **160** regiospecifically in 88% yield as a 4:1 *trans:cis* mixture with respect to the substituents on C-13 and C-14 (steroid numbering).

It was now necessary to attach an appropriate leaving group at C-9. This was done by preparing the 9α,11α-oxide **161** in 60% yield by selective reaction with *m*-chloroperbenzoic acid. Treatment of **161** in diglyme with caesium fluoride at 27° for 20 hours then led, via orthoquinodimethane **162**, to (±)-11α-hydroxyestrone methyl ether **163** in 70% yield. The exceptionally mild conditions for this cycloaddition and the use of the Torgov reaction for coupling the aromatic and ring D components are noteworthy. An alternative *a priori* cycloaddition possibility is that of **160** with pickup of an electrophile at C-11.

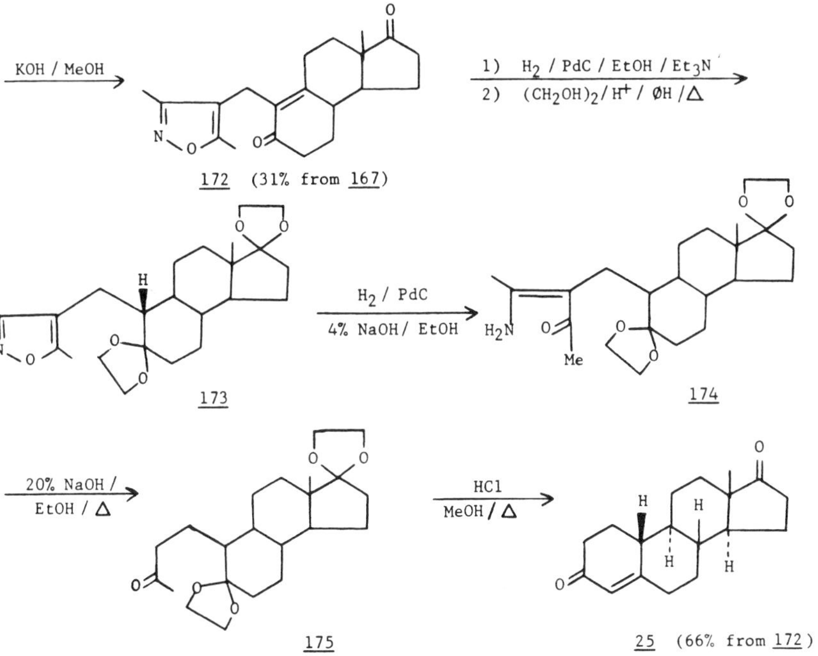

KOH / MeOH →

1) H$_2$ / PdC / EtOH / Et$_3$N
2) (CH$_2$OH)$_2$/ H$^+$ / ØH /△ →

<u>172</u> (31% from <u>167</u>)

H$_2$ / PdC
4% NaOH / EtOH →

H$_2$N

Me

<u>173</u>

<u>174</u>

20% NaOH /
EtOH / △ →

HCl
MeOH / △ →

H   H

H   H

<u>175</u>

<u>25</u> (66% from <u>172</u>)

C. Miscellaneous

(a) Saucy Synthesis

Scott, Borer, and Saucy have described a route to (+)-19-norandrost-4-ene-3,17-dione **25** based on condensation of the chiral trisannulating agent **167** (which utilizes the 3,5-dimethylisoxazole moiety[56] as the ring A precursor) with 2-methylcyclopentane-1,3-dione **12** (Scheme 22).[57] This route evolved from analogous approaches described earlier (see Volume 2, p. 717) and was applied initially in the (±) series.[58]

   3,5-Dimethyl-4-chloromethylisoxazole and triphenylphosphine in refluxing toluene gave the phosphonium salt **164**, which as the corresponding ylid, was condensed with acrolein dimer to afford the Wittig product **165** in good yield. Hydration, oxidation, and hydrogenation then led to lactone **166**. Low-temperature Grignard reaction of the latter with vinylmagnesium chloride yielded the corresponding vinyl ketone to which was added (−)-α-phenethylamine. The diastereomeric Mannich bases were separated by seeding and fractional crystallization from isopropyl ether and the required 2S, 6R diastereomer **167** was obtained in 70% yield (of the theoretical 50%).

Mannich base **167**, on heating with 2-methylcyclopentane-1,3-dione **12** in toluene-pyridine-acetic acid, yielded dienol ether **169** that was ~72% 13β-methyl and 28% 13α-methyl. Improved asymmetric induction was attained by first converting the Mannich base to the corresponding bulkier methiodide and condensing the latter with **12** in refluxing aqueous t-butanol to yield ketol **168**. This intermediate in benzene containing p-toluenesulfonic acid at 20° was converted into dienol ether **169** that was 89% 13β-methyl:11% 13α-methyl in 46% chemical yield from Mannich base **167**. Lithium aluminum hydride reduction and catalytic hydrogenation then afforded **170**, which on hydration, oxidation, and base catalyzed cyclization gave pure (+)-enedione **172** in 31% yield from Mannich base **167**.

In the concluding phase of the synthesis, hydrogenation and bisketal formation produced **173**, which on base catalyzed hydrogenolysis of the isoxazole ring and basic hydrolysis yielded **175**. Acid catalyzed deketalization and cyclization then gave (+)-19-norandrost-4-ene-3,17-dione **25** in 66% yield from enedione **172**.

Conducting the hydrogenolytic opening of the isoxazole ring on the bisketal **173** resulted in higher yields than when carried out on the corresponding bisketone. In the latter case the carbinolamine intermediate **176** is subject to dehydration to synthetically useless dihydropyridine type by-products.

176

(b) Johnson Synthesis

As part of his extensive program on steroid synthesis based on biomimetic olefin cyclization processes, W. S. Johnson has developed a stereospecific synthesis of (±)-estrone (Scheme 23).[59]

m-Alkoxyphenylpropionaldehyde **177** (R = CH$_3$, CH$_3$OCH$_2$) and the previously prepared phosphonium iodide **180**[60] were condensed by the Schlosser modification of the Wittig reaction to give olefin (>98% E) **181** (R = CH$_3$, CH$_3$OCH$_2$). Ketal hydrolysis, cyclodehydration, and reduction then led to key intermediate **183** (R = CH$_3$, H). Stannic chloride catalyzed cyclization in methylene chloride at −100° gave the isomeric tetracyclic compounds **184** (R = CH$_3$)

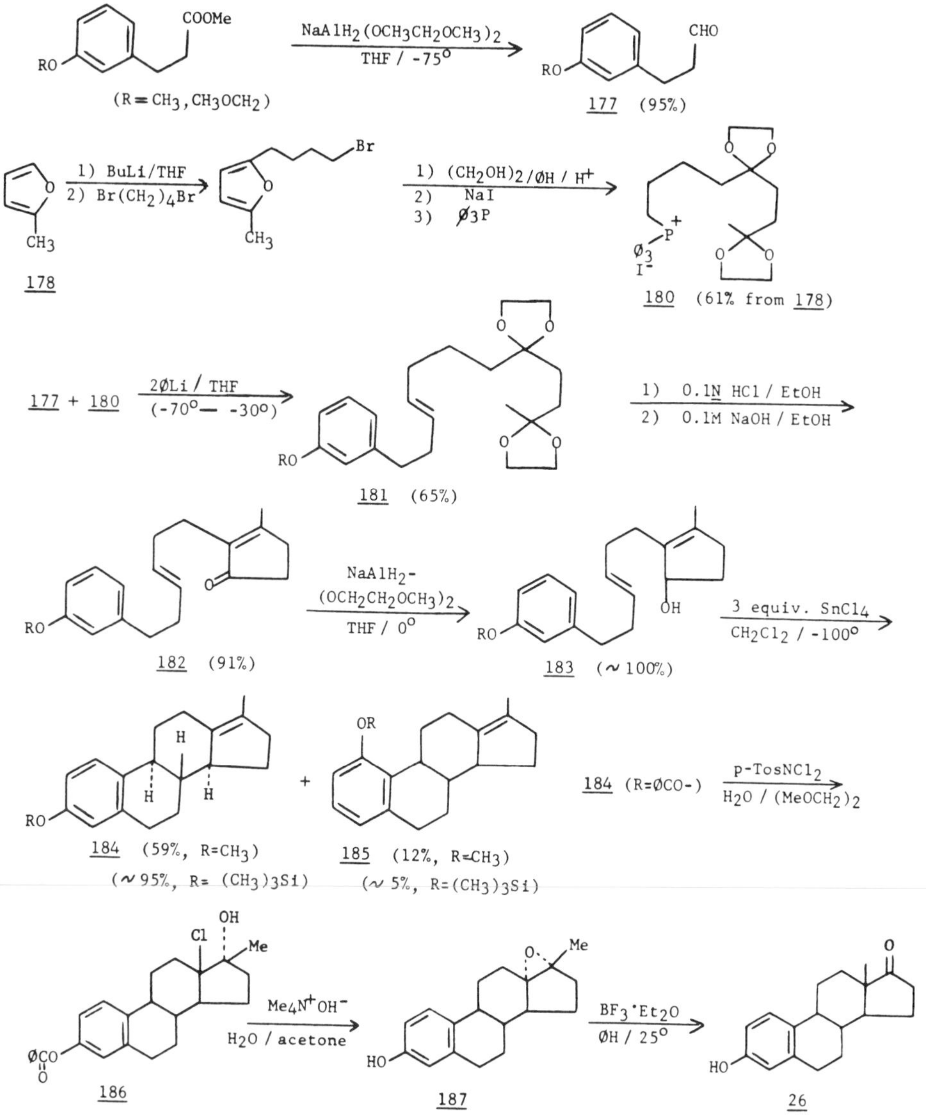

$$RO-\text{(aryl)}-CH_2CH_2-COOMe \quad (R = CH_3, CH_3OCH_2)$$

$$\xrightarrow[\text{THF / }-75^\circ]{\text{NaAlH}_2(\text{OCH}_3\text{CH}_2\text{OCH}_3)_2}$$

177 (95%)

178

$$\xrightarrow[\text{2) Br(CH}_2)_4\text{Br}]{\text{1) BuLi/THF}}$$

$$\xrightarrow[\text{2) \quad NaI}]{\text{1) (CH}_2\text{OH)}_2/\varnothing\text{H / H}^+}$$
$$\text{3) \quad }\varnothing_3\text{P}$$

180 (61% from **178**)

$$\textbf{177} + \textbf{180} \xrightarrow[(-70^\circ\text{---}-30^\circ)]{2\varnothing\text{Li / THF}}$$

181 (65%)

$$\xrightarrow[\text{2) \quad 0.1M NaOH / EtOH}]{\text{1) \quad 0.1N HCl / EtOH}}$$

182 (91%)

$$\xrightarrow[\text{THF / }0^\circ]{\text{NaAlH}_2-(\text{OCH}_2\text{CH}_2\text{OCH}_3)_2}$$

183 ($\sim$100%)

$$\xrightarrow[\text{CH}_2\text{Cl}_2 / -100^\circ]{\text{3 equiv. SnCl}_4}$$

184 (59%, R=CH$_3$)
($\sim$95%, R= (CH$_3$)$_3$Si)

+

185 (12%, R=CH$_3$)
($\sim$5%, R=(CH$_3$)$_3$Si)

184 (R=$\varnothing$CO-) $\xrightarrow[\text{H}_2\text{O / (MeOCH}_2)_2]{\text{p-TosNCl}_2}$

186 $\xrightarrow[\text{H}_2\text{O / acetone}]{\text{Me}_4\text{N}^+\text{OH}^-}$ **187** $\xrightarrow[\varnothing\text{H / 25}^\circ]{\text{BF}_3\cdot\text{Et}_2\text{O}}$ **26**

Scheme 23

39

and 185 (R = CH$_3$) in the ratio 4.3:1, isolated respectively in yields of 59% and 12%. The para:ortho isomer ratio and the rate of cyclization were found to be dependent on the nature of the C-3 substituent and the C-14 leaving group.[61]

The best results were achieved with the 3-trimethylsilyl ether of 183, which gave 184 (R = Me$_3$Si) and 185 (R = Me$_3$Si) in the ratio 20:1 in essentially quantitative yield. Intermediate 184 was conveniently isolated as the crystalline 3-benzoate.

To complete the synthesis, the required 13α,17α-epoxide 187 was best obtained via the chlorohydrin 186 (and/or the isomeric 13α-hydroxy-17β-chloro structure) rather than by direct epoxidation, which gave mainly the β-epoxide. Boron trifluoride catalyzed rearrangement of 187 then gave (±)-estrone 26 in 22% overall yield from 182 (R = H).

(c) Daniewski Synthesis

A novel and effective synthesis of (−)-3-methoxyestra-1,3,5(10),8,14-pentaene-17-one 193, a key intermediate in the Smith and Torgov estrone syntheses (Volume 2, pp. 697, 701, 705), conceived by Daniewski (Scheme 24),[62] utilizes an organoborane-diazoketone condensation[63] as its key step. (−)-Chloroacid 189, derived via Michael reaction of methyl 2-chloroacrylate with 2-methylcy-clopentane-1,3-dione 12, ester hydrolysis and optical resolution, was converted into chlorodiazoketone 190. Reaction of 190 with tris-(m-methoxyphen-ethyl)borane followed by dechlorination yielded 192. Brief treatment of the latter with acetic acid containing a little perchloric acid led to (−)-estrapentaene 193 as the optically pure 13β-methyl enantiomer. The borane-diazoketone reaction took place with remarkably high asymmetric induction.[7a,62b]

(d) Bryson Synthesis

In another application of borane chemistry to steroid synthesis, Bryson developed a synthesis of (±)-estrone methyl ether in which ring C is constructed by hy-droboration-carbonylation of an appropriate 1:4-diene (Scheme 25).[64] The basic methodology was demonstrated earlier in a route to 17-desoxyestrone methyl ether.[65]

Condensation of the sodium enolate of acetylacetone 194 with carbethoxy-cyclopropyltriphenyl tetrafluoborate 195[66] led in high yield to acetylcyclopen-tenyl ester 196 and thence to ketal aldehyde 197. The latter with m-methoxy-phenethyl magnesium chloride 198 yielded allylic alcohol 199, which on Claisen rearrangement with dimethylacetamide dimethylacetal afforded amide 200. Conversion to amine, amine oxide, and Cope elimination then led to the desired 1,4-diene 201. Hydroboration with thexylborane followed by carbonylation by the Pelter procedure[67] produced hydrindanone 203 with the required 8β,14α-

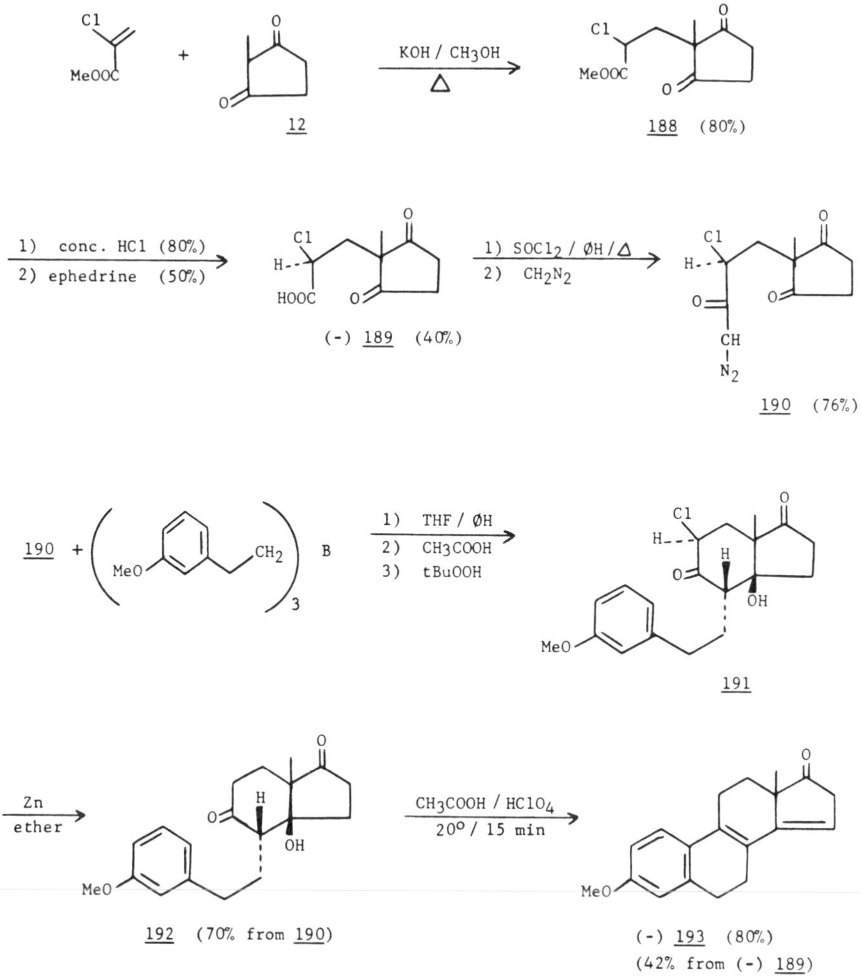

Scheme 24

stereochemistry. Acid catalyzed ring closure and ketal hydrolysis followed by hydrogenation led to **204** and thence to (±)-estrone methyl ether **28**.

(e) Posner Synthesis

Posner has devised a novel one-pot (but low yield) synthesis of (±)-9(11)-dehydro-8-epiestrone methyl ether **210** by a coupled sequence of two Michael reactions and a Wittig reaction (tandem Michael–Michael ring closure) (Scheme 26).[68] 6-Methoxy-1-tetralone **205** was converted, via trimethylsilyl enol ether

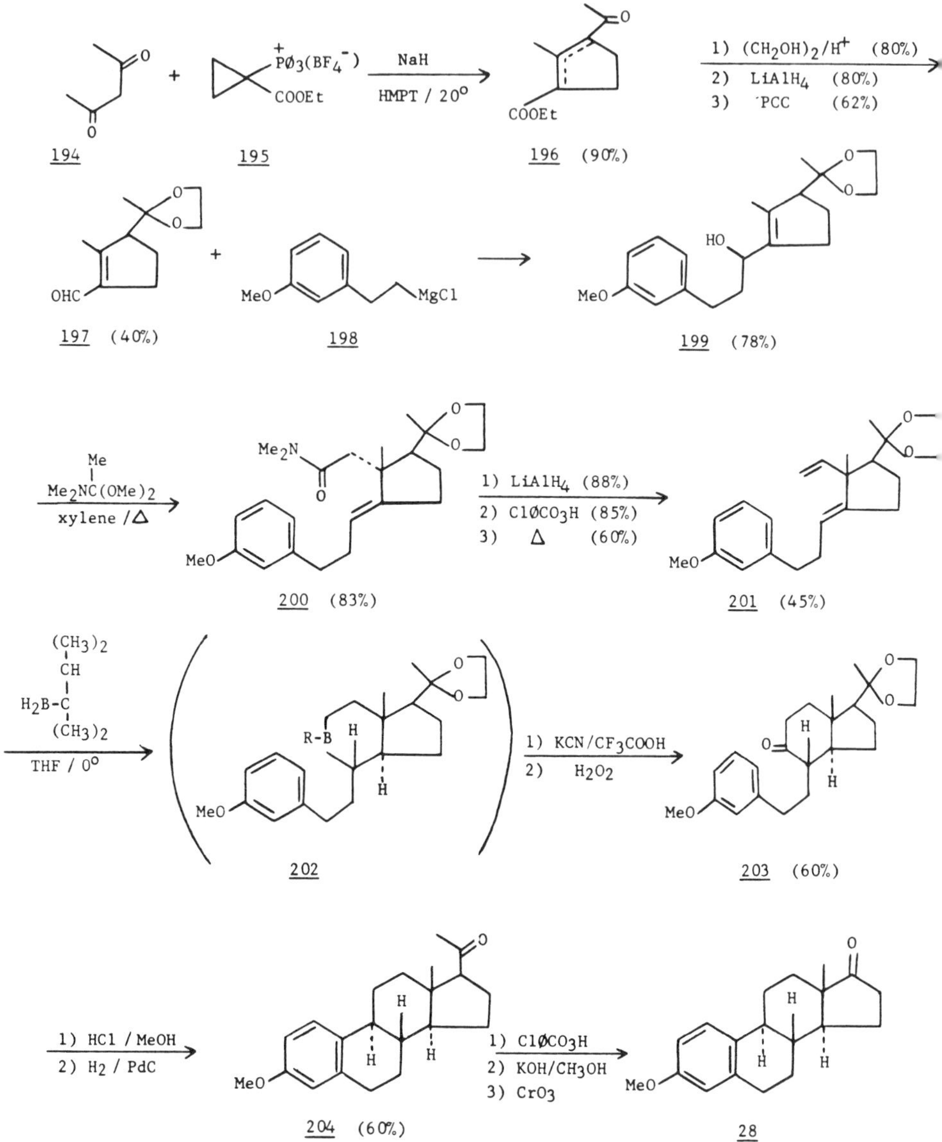

Scheme 25

42

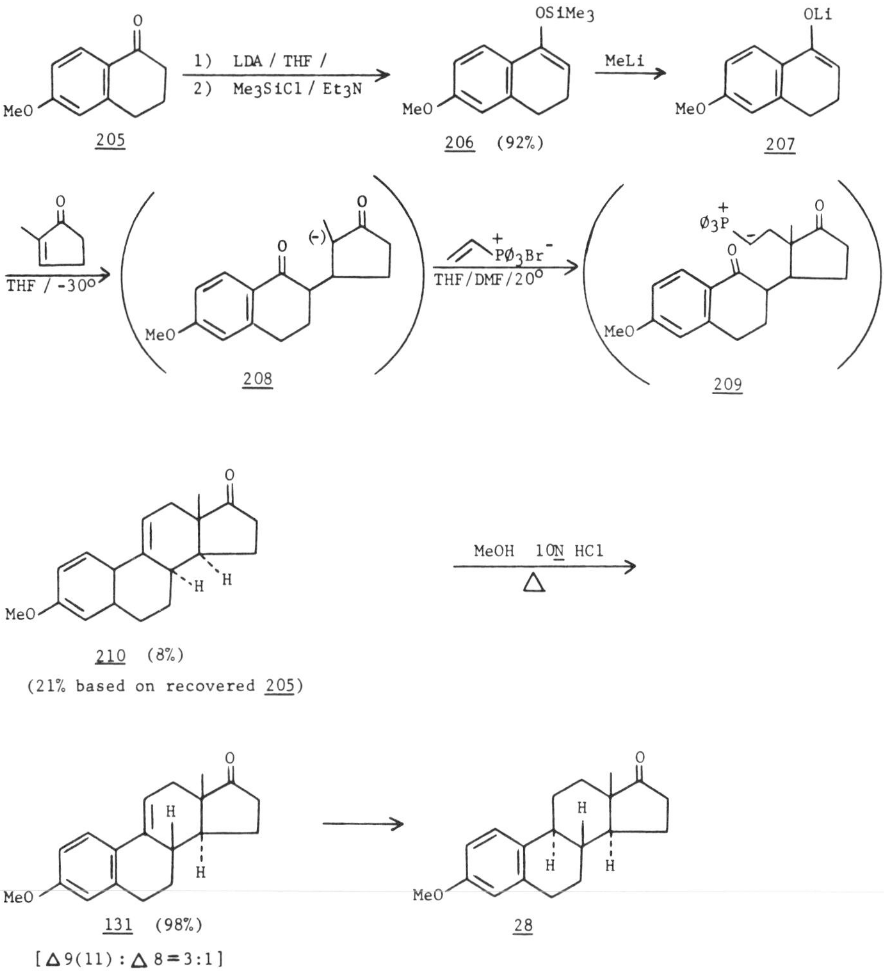

Scheme 26

206, into its lithium enolate **207**, which was condensed sequentially with 2-methyl-2-cyclopentenone and vinyltriphenylphosphonium bromide to give, via enolate **208** and ylid **209**, (±)-9-(11)-dehydro-8-epiestrone methyl ether **210** in 8% yield (21% corrected for recovered **205**). The low yield of **210** is believed to result from competing reaction of ylid **209** with vinyltriphenylphosphonium bromide leading to polymerization. Equilibration of **210** in refluxing methanolic hydrochloric acid led in high yield to a 3:1 mixture of (±)-9(11)-dehydroestrone methyl ether **131** and its 8-dehydro counterpart, which had been converted previously into (±)-estrone methyl ether **28** (see Quinkert synthesis **131** → **28**).

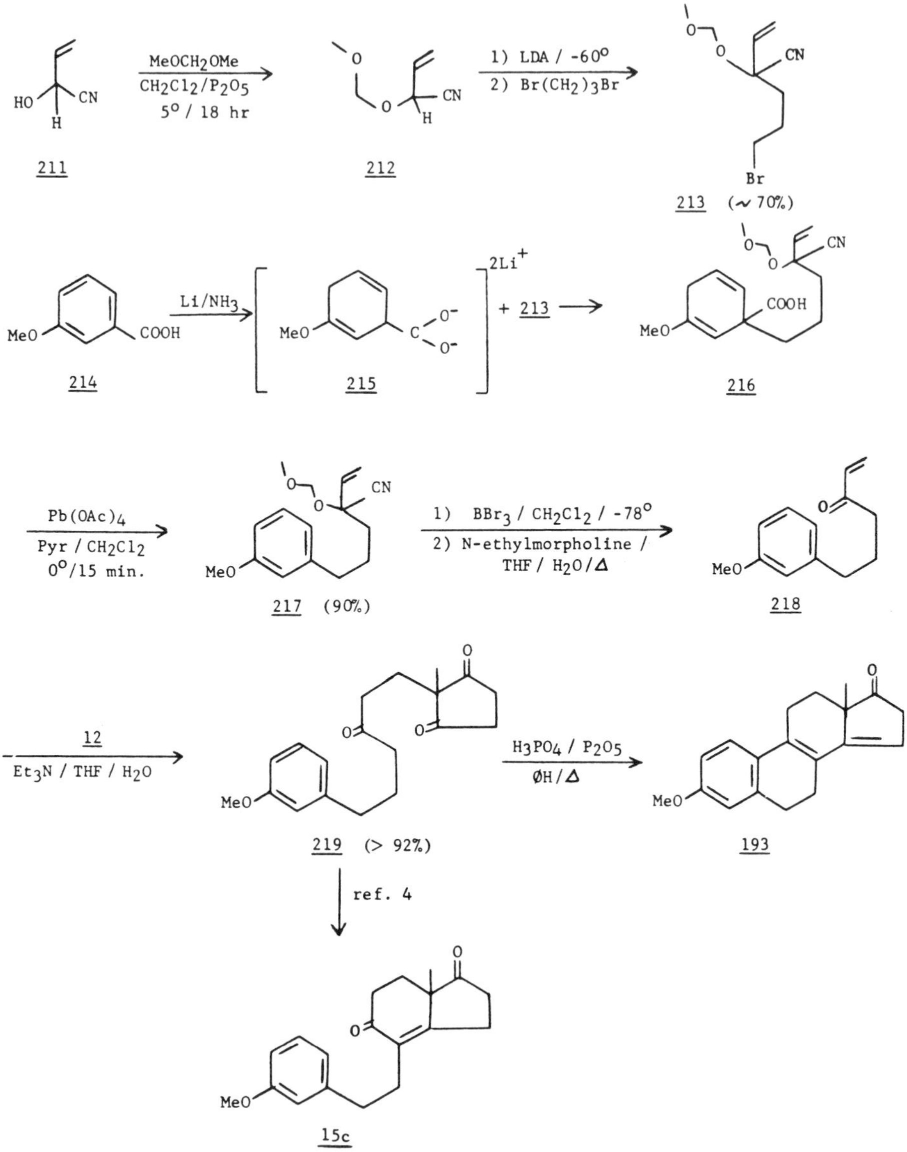

Scheme 27

(f) Mander Synthesis

Mander has devised a highly convergent approach to estrone and analogs based on hypothetical structural units *a, b,* and *c* and the indicated bond connections.

a   b   c

The strategy was reduced to practice as shown in Scheme 27 with syntheses of prochiral trione **219** and its 1,3-dimethoxy analog.[69]

Acrolein cyanhydrin **211** was converted into its methoxymethyl ether **212**, in turn alkylated with 1,3-dibromopropane to give the key 6-carbon synthon **213**. *m*-Methoxybenzoic acid was utilized as the ring A component. Reductive alkylation via dianion **215** led to dihydroaromatic intermediate **216** which was decarboxylated with ensuing aromatization using lead tetraacetate (or, alternatively, anodic oxidation) to afford **217** in excellent yield. Unmasking of the carbonyl function generated known enone **218**[70,71] (see Volume 2, pp. 693, 697), which was condensed with 2-methyl-1,3-cyclopentanedione **12** to produce prochiral trione **219**[70] in high yield. The latter had been cyclized previously to (+) **15c**[4] and thence converted into estrone.[70] Alternatively, trione **219** was cyclodehydrated to the well-known estrone precursor estrapentaene **193** (see Volume 2, pp. 697, 701, 705).

(g) Ziegler Synthesis

Ziegler and Wang have achieved a synthesis of (±)-estrone from 6-methoxytetralone based on Cope rearrangement and polyene cyclization methodology (Scheme 28).[72]

Unsaturated aldehyde **221** was prepared efficiently from 6-methoxytetralone **205** via unsaturated nitrile **220**. Conversion of **221** into the corresponding trimethylsilyl cyanohydrin **222** and alkylation with 1-chloro-2,6-dimethyl-2(E), 6-heptadiene led to aryltriene **223**. The latter underwent Cope rearrangement at 160° to **224** without interference from the alternative higher energy rearrangement pathway involving the prohomo D ring.

Treatment of **224** with potassium fluoride in refluxing methanol led to the methyl esters **225** and **226** (9αH:9βH = 1:4)—presumably via ketonitrile in-

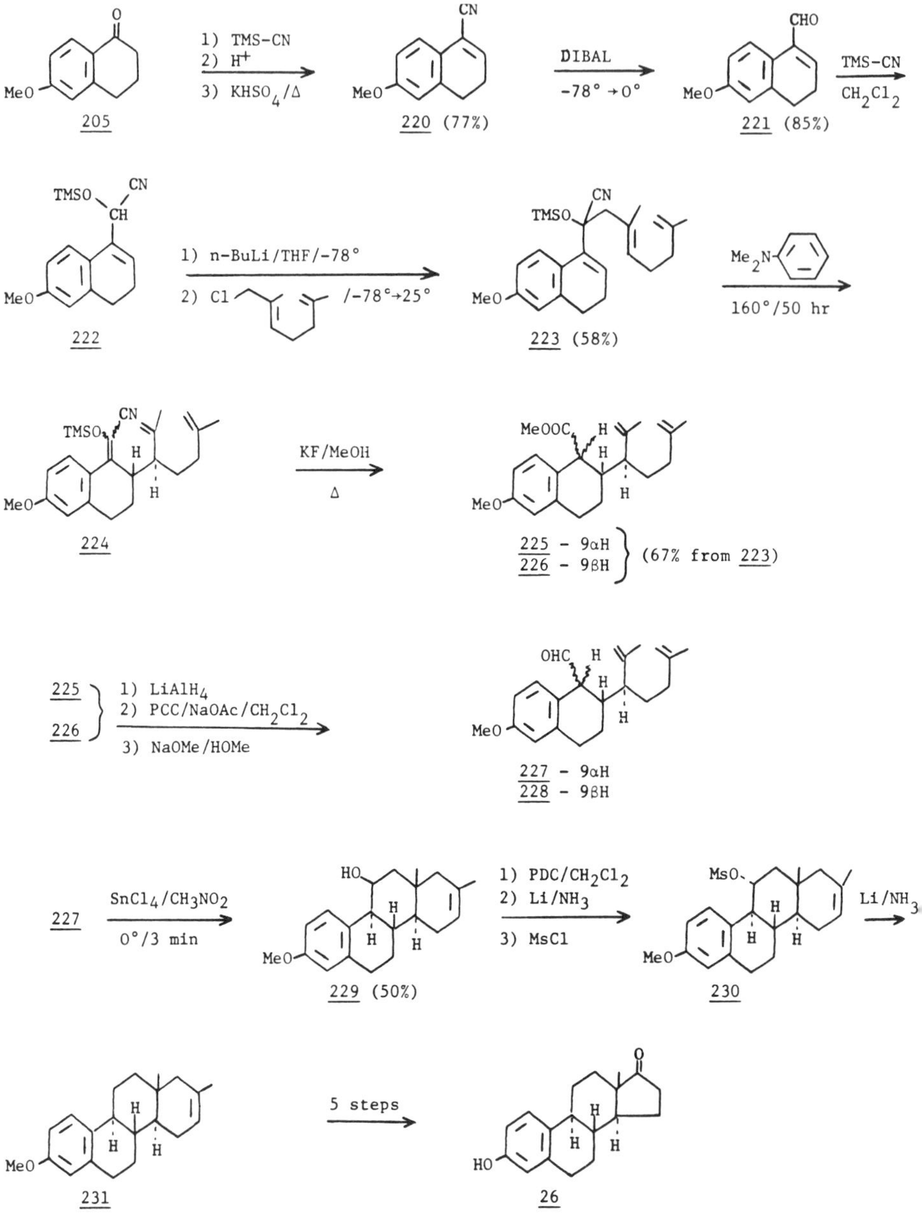

Scheme 28

termediates. Conversion to the corresponding aldehydes and equilibration yielded **227** and **228** (9αH:9βH = 92:8). Stannic chloride catalyzed cyclization of the predominant 9α-H aldehyde **227** produced the 11β-hydroxy D-homosteroid **229**. For conversion to estrone, ring C functionality was best removed via the 11-ketone and 11α-mesylate **230**. Lithium-ammonia reduction of the latter provided **231**, which had previously been converted into (±)-estrone methyl ether in four steps by Valenta et al. (See Volume 2, pp. 672-673) and thence into (±)-estrone.[73,74]

ACKNOWLEDGMENT

The expert typing by Rita Pollard of the text and formulations is gratefully acknowledged.

REFERENCES

1. C. M. Lentz and G. H. Posner, *Tetrahedron Lett.*, 3769 (1978); G. H. Posner, M. J. Chapdelaine, and C. M. Lentz, *J. Org. Chem.*, **44**, 3661 (1978).
2. (a) G. H. Posner, J. P. Mallamo, and K. Miura, *J. Am. Chem. Soc.*, **103**, 2886 (1981). (b) G. H. Posner, M. Hulce, J. P. Mallamo, S. A. Drexler, and J. Clardy, *J. Org. Chem.*, **46**, 5244 (1981).
3. For a recent example, see R. B. Woodward et al., *J. Am. Chem. Soc.*, **103**, 3210 (1981).
4. U. Eder, G. Sauer, and R. Wiechert, *Angew. Chem. Int. Ed.*, **10**, 496 (1971).
5. Z. G. Hajos and E. R. Parrish, German Patent 2,102,623 (1971), *Chem. Abstr.*, **75**, 129,414r (1971); *J. Org. Chem.*, **39**, 1615 (1974).
6. Z. G. Hajos and D. R. Parrish, *J. Org. Chem.*, **39**, 1612 (1974).
7. For discussion, see (a) N. Cohen, *Acc. Chem. Res.*, **9**, 412 (1976); (b) M. E. Jung. *Tetrahedron*, **32**, 3 (1976).
8. (a) R. A. Micheli, Z. G. Hajos, N. Cohen, D. R. Parrish, L. A. Portland, W. Sciamanna, M. A. Scott, and P. A. Wehrli, *J. Org. Chem.*, **40**, 675 (1975). (b) An earlier version of this approach, developed in the (±) series, was described by Z. G. Hajos and D. R. Parrish, *J. Org. Chem.*, **38**, 3244 (1973).
9. G. Nominé, G. Amiard, and V. Torelli, *Bull. Soc. Chim. France*, 3664 (1968).
10. Z. G. Hajos and D. R. Parrish, *J. Org. Chem.*, **38**, 3239 (1973).
11. G. Stork and J. d'Angelo, *J. Am. Chem. Soc.*, **96**, 7115 (1974).
11a. G. Sauer, U. Eder, G. Haffer, G. Neef, and R. Wiechert, *Angew. Chem. Int. Ed.*, **14**, 417 (1975).
12. Roussel-Uclaf, Belgian Patent 634,308 (1963) (see Vol. 2, p. 710); Danishevsky syntheses (Schemes 7[18a,b],8[19]).
13. N. Cohen, G. L. Banner, W. F. Eichel, D. R. Parrish, G. Saucy, J. M. Cassal, W. Meier, and A. Fürst, *J. Org. Chem.*, **40**, 681 (1975).
14. U. Eder, H. Gibian, G. Haffer, G. Neef, G. Sauer, and R. Wiechert, *Chem. Ber.*, **109**, 2948 (1976); U. Eder, *J. Steroid Biochem.*, **11**, 55 (1979).

15. For a review on heterocyclic compounds as annulation agents, see T. Kametani and H. Nemeto, *Heterocycles,* **10,** 349 (1978).

16. (a) S. Danishevsky and R. Cavanagh, *J. Am. Chem. Soc.,* **90,** 520 (1968). (b) S. Danishevsky and A. Zimmer, *J. Org. Chem.,* **41,** 4059 (1976).

17. G. Stork and R. Borch, *J. Am. Chem. Soc.,* **86,** 935 (1964).

18. (a) S. Danishevsky and A. Nagel, *J. Chem. Soc., Chem. Commun.,* 373 (1972). (b) S. Danishevsky, P. Cain, and A. Nagel, *J. Am. Chem. Soc.,* **97,** 380 (1975). (c) T. C. McKenzie, *J. Org. Chem.,* **39,** 629 (1974).

19. S. Danishevsky and P. Cain, *J. Org. Chem.,* **39,** 2925 (1974); *J. Am. Chem. Soc.,* **97,** 5282 (1975); *ibid.,* **98,** 4975 (1976).

20. (a) S. Takahashi, T. Shibano, and N. Hagihara, *Tetrahedron Lett.,* 2451 (1967); (b) W. E. Walker, R. M. Manyik, K. E. Adkins, and M. L. Farmer, *Tetrahedron Lett.,* 3817 (1970). (c) J. Tsuji, *Acc. Chem. Res.,* **6,** 8 (1973).

21. J. Tsuji, I. Shimizu, H. Suzuki, and Y. Naito, *J. Am. Chem. Soc.,* **101,** 5071 (1979).

22. I. Shimizu, Y. Naito, and J. Tsuji, *Tetrahedron Lett.,* 487 (1980).

23. T. Takahashi, K. Kasuga, M. Takahashi, and J. Tsuji, *J. Am. Chem. Soc.,* **101,** 5072 (1979).

24. J. Tsuji, I. Shimizu, and K. Yamamoto, *Tetrahedron Lett.,* 2975 (1976).

25. U. Eder, G. Sauer, J. Ruppert, G. Haffer, and R. Wiechert, *Chem. Ber.,* **108,** 2673 (1975).

26. J. Tsuji, Y. Kobayashi, and T. Takahashi, *Tetrahedron Lett.,* 483 (1980).

27. M. P. Cava and D. R. Napier, *J. Am. Chem. Soc.,* **79,** 1701 (1957); M. P. Cava, A. A. Deana, and K. Muth, *ibid.,* **81,** 6458 (1959).

28. F. R. Jensen and W. E. Coleman, *J. Am. Chem. Soc.,* **80,** 6149 (1958).

29. For reviews, see (a) W. Oppolzer, *Synthesis,* 793 (1978); (b) T. Kametani, *Pure Appl. Chem.,* **51,** 747 (1979); (c) T. Kametani and H. Nemoto, *Tetrahedron,* **37,** 3 (1981); (d) R. L. Funk and K. P. C. Vollhardt, *Chem. Soc. Rev.,* **9,** 41 (1980).

30. For a recent example of the latter, the total synthesis of (+) chenodeoxycholic acid, see T. Kametani, K. Suzuki, and H. Nemoto, *J. Am. Chem. Soc.,* **103,** 2890 (1981).

31. T. Kametani, H. Nemoto, H. Ishikawa, K. Shiroyama, and K. Fukumoto, *J. Am. Chem. Soc.,* **98,** 3378 (1976); T. Kametani, H. Nemoto, H. Ishikawa, K. Shiroyama, H. Matsumoto, and K. Fukumoto, *ibid.,* **99,** 3461 (1977).

32. Analogous adducts have been utilized in diterpenoid syntheses, e.g., T. Kametani, H. Matsumoto, T. Honda, and K. Fukumoto, *Tetrahedron Lett.,* **22,** 2379 (1981).

32a. T. Kametani, H. Matsumoto, H. Nemoto, and K. Fukumoto, *J. Am. Chem. Soc.,* **100,** 6218 (1978); *Tetrahedron Lett.,* 2425 (1978).

33. T. Kametani, M. Tsubuki, H. Nemoto, and K. Suzuki, *J. Am. Chem. Soc.,* **103,** 1256 (1981).

34. W. Oppolzer and M. Petrzilka, unpublished work; W. Oppolzer, *Angew. Chem. Int. Ed.,* **16,** 10 (1977).

35. W. Oppolzer, M. Petrzilka, and K. Bättig, *Helv. Chim. Acta,* **60,** 2964 (1977).

36. W. Oppolzer, K. Bättig, and M. Petrzilka, *Helv. Chim. Acta,* **61,** 1945 (1978).

37. M. P. Cava and A. A. Deana, *J. Am. Chem. Soc.,* **81,** 4266 (1959).

38. W. Oppolzer, D. A. Roberts, and T. G. C. Bird, *Helv. Chim. Acta,* **62,** 2017 (1979).

39. W. Oppolzer and D. A. Roberts, *Helv. Chim. Acta,* **63,** 1703 (1980).

40. K. C. Nicolau and W. E. Barnette, *J. Chem. Soc., Chem. Commun.,* 1119 (1979); K. C. Nicolau, W. E. Barnette, and P. Ma, *J. Org. Chem.,* **45,** 1463 (1980).

41. K. P. C. Vollhardt, *Acc. Chem. Res.,* **10,** 1 (1977).

42. R. L. Funk and K. P. C. Vollhardt, *J. Am. Chem. Soc.,* **99,** 5483 (1977).

43. R. L. Funk and K. P. C. Vollhardt, *J. Am. Chem. Soc.*, **101**, 215 (1979). For an alternative approach see E. D. Sternberg and K. P. C. Vollhardt, *J. Org. Chem.*, **47**, 3447 (1982).

44. P. G. Sammes, *Tetrahedron*, **32**, 405 (1976).

45. (a) G. Quinkert, *Chimia*, **31**, 225 (1977). (b) G. Quinkert, W. D. Weber, U. Schwartz, and G. Dürner, *Angew. Chem. Int. Ed.*, **19**, 1027 (1980). (c) G. Quinkert, W. D. Weber, U. Schwartz, H. Stark, H. Baier, and G. Dürner, *Liebigs Ann. Chem.*, 2335 (1981).

46. G. Quinkert, U. Schwartz, H. Stark, W. D. Weber, H. Baier, F. Adam, and G. Dürner, *Angew. Chem. Int. Ed.*, **19**, 1029 (1980); *Liebigs Ann. Chem.*, 1999 (1982).

47. For a recent example in the sesquiterpene field, see P. A. Grieco and Y. Ohfune, *J. Org. Chem.*, **45**, 2251 (1980).

48. P. A. Grieco, T. Takigawa, and W. A. Schillinger, *J. Org. Chem.*, **45**, 2247 (1980).

49. (a) J. S. Bindra, A. Grodski, T. K. Schaaf, and E. J. Corey, *J. Am. Chem. Soc.*, **95**, 7522 (1973). (b) R. Peel and J. K. Sutherland, *J. Chem. Soc., Chem. Commun.*, 151 (1974).

50. J. Tsuji, Y. Kobayashi, H. Kataoka, and T. Takahashi, *Tetrahedron Lett.*, **21**, 1475 (1980).

51. For a recent example, see Y. Kobayashi and J. Tsuji, *Tetrahedron Lett.*, **22**, 4295 (1981).

52. J. Tsuji, H. Okumoto, Y. Kobayashi, and T. Takahashi, *Tetrahedron Lett.*, **22**, 1357 (1981).

53. Y. Ito, M. Nakatsuka, and T. Saegusa, *J. Am. Chem. Soc.*, **102**, 863 (1980).

54. Y. Ito, M. Nakatsuka, and T. Saegusa, *J. Am. Chem. Soc.*, **103**, 476 (1981).

55. S. Djuric, T. Sarkar, and P. Magnus, *J. Am. Chem. Soc.*, **102**, 6885 (1980).

56. G. Stork, S. Danishevsky, and M. Ohashi, *J. Am. Chem. Soc.* **89**, 5459 (1967).

57. J. W. Scott, R. Borer, and G. Saucy, *J. Org. Chem.*, **37**, 1652 (1972).

58. J. W. Scott and G. Saucy, *J. Org. Chem.*, **37**, 1652 (1972).

59. P. A. Bartlett and W. J. Johnson, *J. Am. Chem. Soc.*, **95**, 7501 (1973).

60. W. S. Johnson, M. B. Gravestock, and B. E. McCarry, *J. Am. Chem. Soc.*, **93**, 4332 (1971).

61. P. A. Bartlett, J. I. Brauman, W. S. Johnson, and P. A. Volkmann, *J. Am. Chem. Soc.*, **95**, 7502 (1973).

62. (a) A. R. Daniewski and M. Kócor, *J. Org. Chem.*, **40**, 3136 (1975). (b) A. R. Daniewski, *J. Org. Chem.*, **40**, 3135 (1975).

63. J. Hooz and S. Linke, *J. Am. Chem. Soc.*, **90**, 5936 (1968).

64. T. A. Bryson and C. J. Reichel, *Tetrahedron Lett.*, **21**, 2381 (1980).

65. T. A. Bryson and W. E. Pye, *J. Org. Chem.*, **42**, 3215 (1977).

66. Method of P. L. Fuchs, *J. Am. Chem. Soc.*, **96**, 1607 (1974).

67. A. Pelter, M. G. Hutchings, and K. Smith, *J. Chem. Soc., Chem. Commun.*, 1529 (1970); *ibid.*, 1048 (1971).

68. G. H. Posner, J. F. Mallamo, and A. Y. Black, *Tetrahedron*, **37**, 3921 (1981).

69. L. N. Mander and J. V. Turner, *Tetrahedron Lett.*, **22**, 3683 (1981).

70. G. H. Douglas, J. M. H. Graves, D. Hartley, G. A. Hughes, B. J. McLoughlin, J. Siddall, and H. Smith, *J. Chem. Soc.*, 5072 (1963).

71. A. Horeau, L. Ménager, and H. Kagan, *Bull. Soc. Chim. France*, 3571 (1971).

72. F. E. Ziegler and T. F. Wang, *Tetrahedron Lett.*, **22**, 1179 (1981).

73. (a) R. A. Dickinson, R. Kubela, G. A. MacAlpine, Z. Stojanac, and Z. Valenta, *Can. J. Chem.*, **50**, 2377 (1972). (b) J. Das, R. Kubela, G. A. MacAlpine, Z. Stojanac, and Z. Valenta, *Can. J. Chem.*, **57**, 3308 (1979).

74. For a synthesis of (±) estrone methyl ether based on the tandem Cope-Claisen rearrangement see F. E. Ziegler and H. Lim, *J. Org. Chem.*, **47**, 5230 (1982).

Gene Synthesis

SARAN A. NARANG and WING L. SUNG

*Division of Biological Sciences,
National Research Council of Canada,
Ottawa, Canada*

ROBERT H. WIGHTMAN

*Department of Chemistry,
Carleton University,
Ottawa, Canada*

1. INTRODUCTION

The recent development of rapid and efficient techniques for the chemical synthesis of oligonucleotides, coupled with recombinant DNA methodology, has made the laboratory synthesis of physiologically functional genetic material a practicality. From a chemical viewpoint, oligonucleotides are very sensitive molecules possessing diverse functions, uncommon solubility, and difficult purification problems in the conventional organic sense. Even more difficult is their structural analysis by conventional physicochemical methods. In spite of these drawbacks, their synthesis can be a fascinating challenge to organic chemists because these are the "master molecules" of all living organisms. These molecules contain not only the structural information to direct the synthesis of all proteins but also the regulatory signals for controlling metabolism (Figure 1).

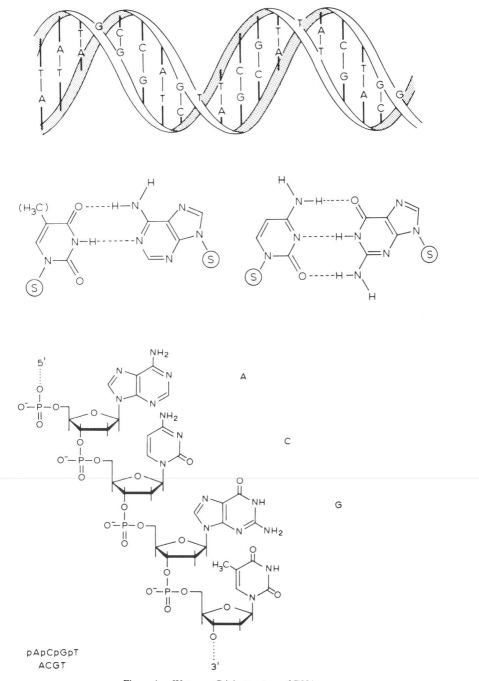

pApCpGpT
ACGT

Figure 1. Watson—Crick structure of DNA.

Our original contribution in this series[1] essentially reviewed the field up to 1970; it seems appropriate now to highlight the developments and applications in synthetic oligonucleotides up to 1982. Several recent reviews [2-7] have adequately outlined chemical developments during the past decade. Consequently, we mention only the most recent modifications. We are here concerned primarily with the so-called "triester" approach and include some selected examples of important molecular biological techniques and applications involving mainly "synthetic genes" of the deoxy type. Coverage of the literature is not exhaustive, but we have tried to include leading references into 1982 for all major researchers in this area.

2. SYNTHETIC STRATEGY

The chemical synthesis of polynucleotides beyond 40 units long is not possible with any of the present methodologies, at least for the time being. Decreasing yields dictate a practical compromise for chemical synthesis of oligonucleotides up to about 20 units. A strategy has been developed for the synthesis of longer chains. Short synthetic oligonucleotides (usually 10–20 bases in length) are then joined, with the use of DNA ligase enzyme in the presence of a complementary strand or template. The sequence of such a gene is preceded and terminated in specific endonuclease-restriction site sequence. The complete synthetic sequence is ligated to a plasmid and cloned in *Escherichia coli*. The cloned DNA is selected and analyzed for the presence of a correct sequence. The importance of this method of DNA synthesis can hardly be overstressed to synthetic chemists. Once the specific sequences have been assembled by a well-defined chemicoenzymatic method and cloned in *E. coli,* the bacterium containing chimeric DNA ensures their permanent availability—an expected but nevertheless dramatic feature of DNA structure, namely, the ability to guide its own replication.

3. CHEMICAL SYNTHESIS

The fundamental objective in oligonucleotide synthesis is the formation of an ester linkage between an activated phosphoric acid function of one nucleotide with the hydroxyl group of another nucleoside/nucleotide, thus ultimately forming the natural phosphodiester bridge between the 5'-hydroxyl of one nucleoside unit and the 3'-hydroxyl of the next. To achieve these ends, the nucleic acid chemist must devise (i) selective blocking/deblocking procedures for a primary and a secondary hydroxyl (or two in ribonucleosides), a primary amino group, and often two of the three dissociations of phosphate. In addition, he or she must be cognizant of the variable labilities of certain purines and pyrimidines and their glycosidic bonds plus the variable reactivities of mono-, di-, and trisubstituted phosphate. One must surmount some frustrating stereochemical and neighbouring group effects especially noticeable as the oligonucleotide chain grows. All protecting groups must be removed without causing chain cleavages and with minimal formation of side products. The required product must be separated from these almost identical impurities and methods devised to prove the structure unambiguously.

A. Historical Background

The pioneering research from Todd's laboratory[8] in the 1950s was followed by Khorana's phosphodiester approach, which dominated the field of oligonucleo-tide synthesis for almost 20 years beginning in the late 1950s. This group was mainly responsible for developing important techniques such as protecting groups, phosphorylating procedures, condensation reaction, and enzymatic characteri-zation, which played a major role in the further development of modern chem-istry. Although the phosphodiester approach was laborious and time consuming, Khorana's group accomplished the total synthesis of a biologically active tyrosine tRNA gene[9] as a climax. By mid-1960s, other organic chemists such as Let-singer,[10] Reese,[11] and Eckstein[12] reintroduced the phosphotriester approach, in which the third phosphate dissociation was masked during synthesis. It was not until 1973, when the phosphotriester approach was further modified by Narang[13] and Cramers[14] independently, that its practicality was established in achieving the total synthesis of a biologically active genetic element.[15] This method has now become the more practical and simpler approach, as evident from a recent international symposium series on synthesis.[16]

(a) Phosphodiester Approach

In the traditional method of oligonucleotide synthesis, the so-called phospho-diester approach developed and championed by Khorana and his co-workers,[17] the 5'-phosphate of a commercially available nucleotide, after suitable protection of other functional groups, is condensed with the 3'-hydroxyl of another protected nucleoside or nucleotide using dicyclohexylcarbodiimide or the arylsulfonyl chlo-rides, and the chain can then be elongated by further stepwise block condensation. Longish reaction times, drastically decreasing yields as the chain length grew unless large excesses of one component of the condensation were provided, and time-consuming purification procedures via anion exchange column chromatog-raphy make this approach a tedious one. Recent improvements by using solvent extraction procedures[18] or high performance liquid chromatography (hplc) techniques[19] have helped somewhat to alleviate this problem. But a computer-

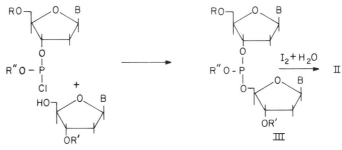

Phosphodiester Approach

Phosphotriester Approach

Phosphite Approach

Figure 2.   Representation of the major approaches to oligonucleotide synthesis.

optimized synthetic path[20] predicted 20 man-years to synthesize a gene of approximately 100–150 units by this approach.

(b) Phosphotriester Approach

A possible solution to some of the problems inherent in the phosphodiester approach (water solubility, ion exchange chromatography, and lower yields) would accrue if the third dissociation of phosphate was masked, thus creating a neutral organic molecule amenable to the more standard manipulations of organic chemistry. Actually, this approach was first introduced by Michelson and Todd[8] in 1955.

The basic principle of the triester method is to mask each internucleotidic phosphodiester function by a suitable protecting group during the course of building a defined sequence. After completing synthesis, all the protecting groups are removed at the final step to give a deoxyoligonucleotide containing each internucleotidic $3' \rightarrow 5'$ phosphodiester linkage. The main advantages of this method include the opportunity for large-scale (50–75 g) synthesis, significantly shorter time periods, especially in the purification steps, and high yields using almost stoichiometric amounts of reactants. This is probably due to the absence of any endo-P-O⁻ groups in the oligonucleotide chain, thus avoiding chain scission and pyrophosphate formation.

The triester method as originally reported[10-12] involved the phosphorylation of the 3'-hydroxyl group of a 5'-protected nucleoside with a substituted phosphate followed by subsequent condensation with a primary 5'-hydroxyl of a second nucleoside. This is essentially a one-pot procedure. However, it was observed that owing to incomplete phosphorylation in the first stage, subsequent coupling with 5'-protected nucleoside led to a complicated reaction mixture including the so-called 3'-3' and 5'-5' coupled products. Since these mixtures could not be completely resolved on conventional silica gel columns, the advantages of large-scale synthesis and high yield of product were somewhat nullified.[21]

(c) Modified Phosphotriester Approach

To overcome this difficulty, the "one-pot" synthetic approach, which includes phosphorylation and coupling steps, was modified to a "two-step" sequential procedure.[13,14] The basic feature of this approach is to start the synthesis of an oligonucleotide from a totally protected mononucleotide containing a fully masked 3'-phosphotriester group. Since the resulting intermediate oligonucleotides contain a fully masked 3'-phosphate group, the necessity for a phosphorylation step at each condensation stage is eliminated, thus simplifying the approach. (See Figure 3.)

Such a starting material was prepared by treating 5'-dimethoxytrityl N-acyl

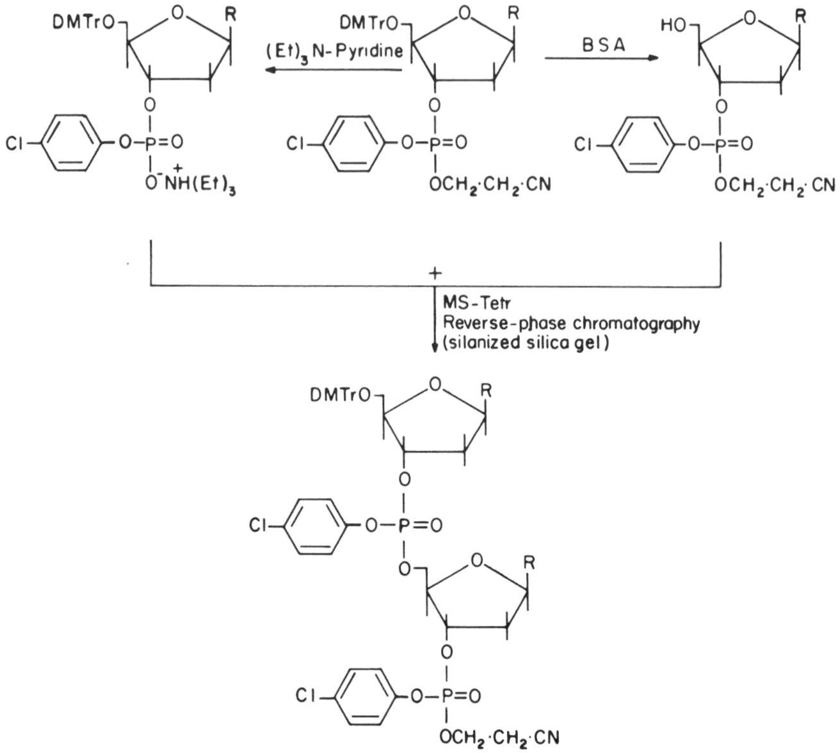

DMTr = Dimethoxytrityl
BSA = Benzenesulfonic acid
Ms-Tetr = Mesitylenesulfonyl tetrozolide

Figure 3.   Modified phosphotriester approach.

deoxyribonucleoside with *p*-chlorophenyl phosphoryl ditriazolides[22] (Figure 3) followed by cyanoethylation reaction. The fully protected monomer is easily purified and permits chain elongation from either end. At each coupling step, the β-cyanoethyl group was removed selectively on treatment with triethylamine-anhydrous pyridine[23,24] or diisopropylamine-anhydrous pyridine.[25] Under these conditions all the other base labile (*N*-acyl and *p*-chlorophenyl) protecting groups were intact. Recently the above starting material without the β-cyanoethyl group as a barium[26] or triethylammmonium[27] salt has also been used directly in the coupling reaction. In practice, longer chains are built by block additions starting from the dinucleotide unit. After each cycle the desired fully protected product can be purified on a short column with medium and high pressure techniques, reversed phase technique on RP-2,[28] and a simple but very effective chromatography on deactivated silica gel with an aqueous solvent system such as acetone-water-ethyl acetate.[29]

B.   Phosphite-Triester Approach

By taking advantage of the extreme reactivity of phosphite reagents, Letsinger[30] introduced a phosphite-triester approach (Figure 2) to decrease appreciably the phosphorylation-condensation time period in building longer chains. Thus two units can be joined together in a matter of minutes rather than hours and the internucleotidic trivalent phosphorus bond oxidized easily to the required phosphotriester by aqueous iodine. This approach requires that a phosphorus be introduced at each cycle, and this limits single base addition. The phosphite reagent could be a dichloro-[31] or ditetrazole[32] derivative.

C.   Polymer Support Synthesis

In analogy with peptide synthesis, much effort has been extended to perfect a polymer supported oligonucleotide synthesis (Figure 4). In the diester approach, two major obstacles are incompatibility of the polymers with the necessary solvents and, perhaps more important, yields in the phosphorylation/condensation stage of ~80% or less.[33] This of course means the creation of increasing amounts of impurities as the chain grows. Some of these problems have been overcome by applying the phosphotriester[34,35] or phosphite[32,36] triester methodologies on new polymers and by blocking off any unreacted chains after each cycle. Recently, the faster synthesis of oligonucleotides up to 31 units long[37] has been accomplished by block condensation on polymer support. The preparation of these blocks, however, were carried out using the solution phosphotriester method.

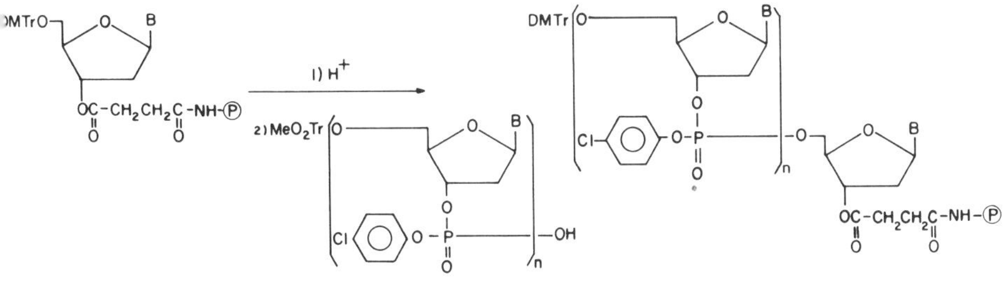

Figure 4.   Phosphotriester-polymer support synthesis.

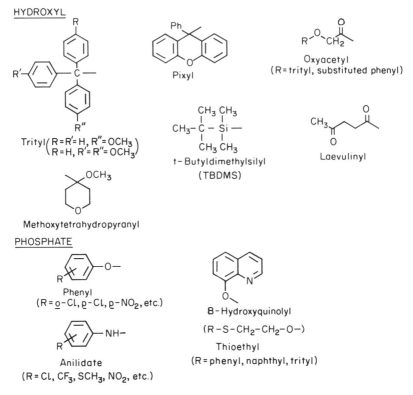

Figure 5.   Some new protecting groups for the hydroxyl and phosphate functions.

Although the yields of each coupling reaction are not quantitative, the final compound could still be isolated by HPLC or preparative gel electrophoresis.

D.   Protecting Groups

(a)   Primary Hydroxyl

The mono- and dimethoxytrityl groups continue to be the most popular acid labile protecting groups for the 5'-hydroxyl. Aromatic sulfonic acids,[38] trifluoroacetic acid,[34] and/or zinc bromide[32] have been shown to be milder and more selective deblocking agents than aqueous acetic acid. The pixyl group[39] has been suggested as a useful variant with a lability similar to the dimethoxytrityl group.

Several examples of base labile groups have been reported especially for use in oligoribonucleotide work, where an acid labile group has been used for protection of the 2'-hydroxyl. These are usually bulky ester groups such as the trityloxyacetyl or substituted phenoxyacetyl.[40] The laevulinyl group, selectively removed by hydrazine, has also been used at this position.[41]

(b) Secondary Hydroxyls

Protection of the 3'-hydroxyl in oligonucleotide synthesis must in general be complementary to the groups on the 5'-hydroxyl and phosphate. The most common base-labile groups are acetyl or benzoyl ester, and the more popular acid-sensitive groups are the tetrahydropyranyl ethers or variations thereof,[42] t-butyldimethylsilyl (TBDMS) ethers and laevulinyl esters.

By far the more complicated problem arises in ribonucleotide chemistry. Not only would it be desirable to specifically differentiate the 2'- from the 3'-hydroxyl but any protecting group on the 2'-hydroxyl must be (i) stable enough to remain intact throughout an entire chain synthesis, (ii) innocuous enough not to interfere with subsequent reactions at the 3'-hydroxyl or phosphate functionalities, and (iii) labile enough to be completely removed after synthesis of the chain (increased steric hindrance notwithstanding) under conditions that will not affect any 2'- to 3'-phosphate migration. No ideal protecting group has yet evolved, since only certain limits of mild acid or base can be tolerated before 3'- to 2'-O migration of phosphate occurs. Thus both acetyl esters[43] and tetrahydropyranyl esters[44] or their variations[45] have been shown to be acceptable in chain synthesis. Other groups requiring specific deblocking conditions have also been reported recently, such as benzyl ethers (hydrogenolysis),[46] o-NO$_2$ benzyl ethers (photolysis),[47,48] amino acylation,[49] and TBDMS ethers (fluoride displacement).[50]

The specific and direct chemical differentiation of the 2'- and 3'-secondary hydroxyls has also not been achieved. Some selective reactivity has been achieved with TBDMS chloride, silver nitrate as catalyst,[51] o-NO$_2$ phenyl diazomethane,[52] or o-NO$_2$ benzyl bromide either directly on the nucleoside[53] or via the 2',3'-0 dibutylstannylene derivatives.[47] Selective deacetylation with hydrazine has also been reported,[54] but this may be an artifact of chromatography.[55] Fractional crystallization of the mixture of 2'- and 3'-O acetates derived from hydrolysis of the 2',3'-O-methoxyethylidene derivations provides the 3'-isomer in <75% yield.[56] Thus, there are no ideal procedures for producing the so-called internal units for oligoribonucleotide synthesis; all require chromatography or crystallization with isolated yields in the 50–70% range. Terminal units at the 3'-end of the growing chain can simply be protected by two ester groups[57] or some cis-diol blocking group, for example, substituted benzylidenes or isopropylidene. Curiously, the only specific method of differentiating the 2'- and 3'-hydroxyls and at the same time providing a 3'-phosphate is not extensively used, namely enzymatic cleavage of the 2',3'-cyclic phosphate.[58]

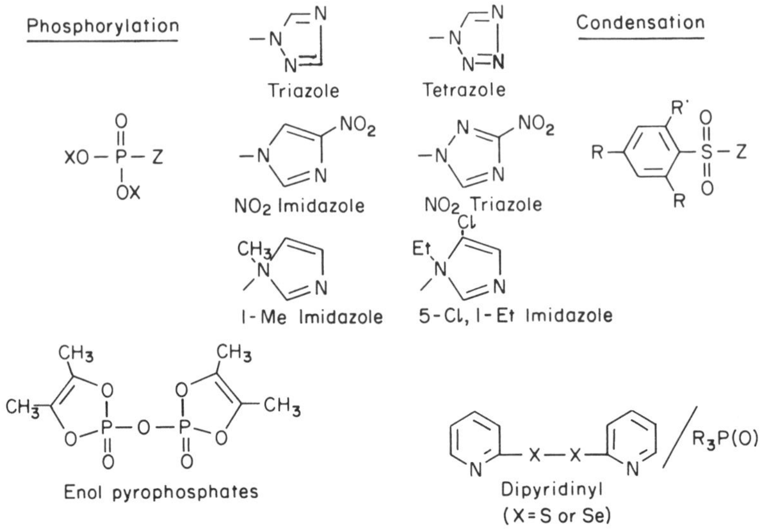

Phosphorylation   Condensation

Triazole   Tetrazole

$$XO - \overset{\overset{\displaystyle O}{\|}}{\underset{\underset{\displaystyle OX}{|}}{P}} - Z$$

NO$_2$ Imidazole   NO$_2$ Triazole

1 - Me Imidazole   5 - Cl, 1 - Et Imidazole

Enol pyrophosphates

Dipyridinyl
(X = S or Se)

Figure 6.   Some new phosphorylation/condensation reagents for oligonucleotide synthesis.

(c) Primary Amino

Although evidence has been presented that protection of the exocyclic amino group is not necessary to direct the coupling reactions for adenosine and guanosine (but it is definitely necessary for cytosine), the use of N-protected monomeric units is almost universally observed. The main reasons have to do with increased solubility, but one should be aware of complications during the phosphorylation reaction.[59,60] Amino protection is evidently not necessary with the phosphite-coupling method at lower temperature.[30]

Although a certain number of more sophisticated N-protecting groups have been reported, for example, tert. butyloxycarbonyl (tBOC), dimethylaminoacetals, by far the most common amino-protecting groups are anisoyl for cytosine, benzoyl for adenosine, and isobutyryl for guanosine. These represent the best compromises between stability and ease of removal for each of the heterocycles. Removal is usually accomplished by ammonolysis. Selective removal of N-acyl group can also be achieved with ethylenediamine-phenol.[61]

(d) Phosphate

With the increasing popularity of the triester method has come a revitalized interest in developing new blocking/deblocking procedures for one or more of the phosphate dissociations.[62] Although its lability can cause complications in extended syntheses, the β-cyanoethyl group is still extensively used, especially since the discovery that it can be removed by organic amines[23-25] rather than aqueous NaOH. Other groups removed via a β-elimination mechanism include the trichloroethyl ester with a variety of new milder conditions for its removal involving variations of the zinc catalyzed method,[43] hydrogenation,[63] and fluoride ion.[64]

Probably the most commonly employed groups requiring "direct" displacement are the substituted (o-Cl, p-Cl, o-Br, p-No$_2$, etc.) phenyl esters.[11,21,65] Deblocking with ammonium hydroxide[38,66] or fluoride[64] ion can lead to varying and unwanted amounts or internucleotide bond cleavage.[67,68] Oximate[69] or thiophenoxide[70,71] nucleophiles and transition metal ions[72] have been proposed as solutions to this problem. Groups requiring prior activation before direct displacement include substituted anilidates with isoamyl nitrite[73] and certain amidates by protonation.[74] The 8-hydroxy quinolyl group, selectively removable with copper ions, has been used as a phosphate protecting group in oligonucleotide synthesis.[75] Various alkyl esters, methyl, or benzyl[70] have also been proposed. It is the hope here to break a C-O rather than a P-O bond and thereby avoid competing internucleotide bond breakage.

E.   New Condensation and Phosphorylation Reagents

The development by Khorana and his co-workers of dicyclohexylcarbodiimide (DCC), mesitylenesulfonyl chloride (MS), and triisopropylbenzenesulfonyl chloride (TPS) as reasonably effective condensing reagents has played a significant role in the synthesis of polynucleotides by the diester method. In the case of the triester synthetic approach, these reagents were found to be unsatisfactory. TPS causes extensive sulfonation, whereas DCC does not activate the phosphodiester function; therefore a search for new condensing reagents was initiated by our group in 1974. We speculated on overcoming this problem by using another arylsulfonyl derivative with a better or less innocuous leaving group. p-Toluenesulfonylimidazole[76] was reported to be a condensing reagent but its rate of coupling was very slow. N-acetyltriazole was reported to be hydrolyzed six times faster at room temperature than N-acetyl imidazole;[77] this observation prompted us to investigate the reactivity of various arylsulfonyl triazole[78] and tetrazole[79] derivatives. The rate of the coupling reaction was higher, and the reaction mixture was cleaner with a minimum of sulfonation reaction when compared with products using arylsulfonyl chloride. Recently, nitrotriazole,[80] 1-hydroxybenzotriazole,[81] N-methylimidazole,[82] and 5-(pyridin-2yl)-tetrazole[83] have been used.

Knorre and Zarytova[84] studied the mechanism and postulated the formation of phosphodiester tetrazolide as a reactive intermediate that participates in the

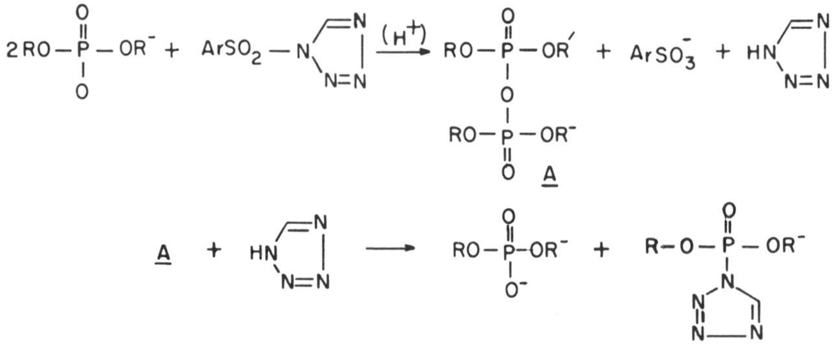

Figure 7.   Mechanism of phosphotriester linkage formation.

phosphorylation step with alcohol. It seems reasonable to suggest that tetrazole liberated in the first stage reacts with the diester component giving rise to a reactive intermediate which should be a powerful phosphorylating reagent due to the strong electronegativity of the tetrazole moiety.

F. Triazole As Phosphate Activator

Phosphorylation is one of the most crucial steps in the synthesis of polynucleotides by the triester approach. We developed bis(triazolyl)p-chlorophenylphosphate[22] as a powerful phosphorylating reagent, which is apparently a doubly activated reagent with triazole but will give only a diesterified product; thus the formation of 3'-3' and 5'-5' dimer formation is avoided.

G. General Method of Oligonucleotide Synthesis by the Modified Phosphotriester Approach

From a practical point of view, the availability of all 16 dinucleotide blocks made possible a rapid buildup of any deoxyribonucleotide sequence. The basic strategy of a synthetic plan is to extend the chain from the 3'-end towards the 5'-end by block condensation in the presence of arylsulfonyl tetrazole[37] as a coupling reagent (Figure 8). The reaction is generally over within 30 minutes; after workup, a pure, fully protected product was isolated by short-column medium-pressure chromatography on RP-2 absorbent or by a simple but very effective chromatography on deactivated silica gel with aqueous solvent systems such as acetone-water-dichloromethane.

Figure 8. General approach of deoxyoligonucleotide synthesis by the modified phosphotriester approach.

1. MS-Tetr = Mesitylenesulfonyl tetrazole
2. BSA = Benzenesulfonic acid
3. RP-2 = Silanized silica gel
4. R = p-chlorophenyl

H.   Final Deblocking Steps

The most critical step in the success of the triester approach is the final and complete removal of all protecting groups from the fully protected oligomer to yield a component containing natural 3′-5′ phosphodiester linkages, which has introduced one additional complication. Because of the complementary nature of the various protecting groups, several sequential steps must often be employed. Conditions for the removal of one group must be compatible with other functions: acid for detritylation can cause depurination; Zn catalyzed conditions for removing trichloroethyl group can cause problems with cytidine base; conditions to remove the triester phosphate blocking group can cause chain cleavages, and so on. The order of deblocking, that is, the removal of the 5′-trityl group followed by the removal of base-labile groups or the reverse of this order, is also a concern because rearrangement of the nucleotide sequence can occur. The best compromise to date appears to be very specific two-step ammonia treatment[29] or oximate ion[70] followed by acid treatment.

The final compounds are then isolated by any of the fast selective procedures available. These include thin-layer chromatography (tlc) on polyethyleneimine (PEI) cellulose plates, slab preparative gel electrophoresis, and hplc on Permaphase AAX or Partisil 10Sax column.

I.   Formation of Side Products

During the course of synthesis, the formation of various side products has been observed when a large excess of condensing reagents, arylsulfonyl tetrazole, or 3-nitro-1,2,4-triazole derivatives in the presence of large excess phosphodiester component were used for a prolonged period of time.[85] The nature of these side reactions is the incorporation of tetrazole moiety in uracil, thymidine, and guanosine bases, which is apparently reversed during the deblocking of oligonucleotides with *syn*-4-nitrobenzaldoximate ion. Similar side products were also formed when *p*-chlorophenyl phosphoryl ditetrazolide[25,86] was used as a phosphorylating reagent. In practice, it is advisable to use a slight excess of the coupling reagent for 30 minutes in order to avoid any modification of the heterocyclic bases.

J.   Sequence Analysis of Fully Protected Oligomers by Mass Spectrometry

A mass spectrometic method for determining the sequence and molecular weight of protected oligonucleotides containing phosphotriester groups has been developed using californium-252 plasma desorption techniques.[87] It has been dem-

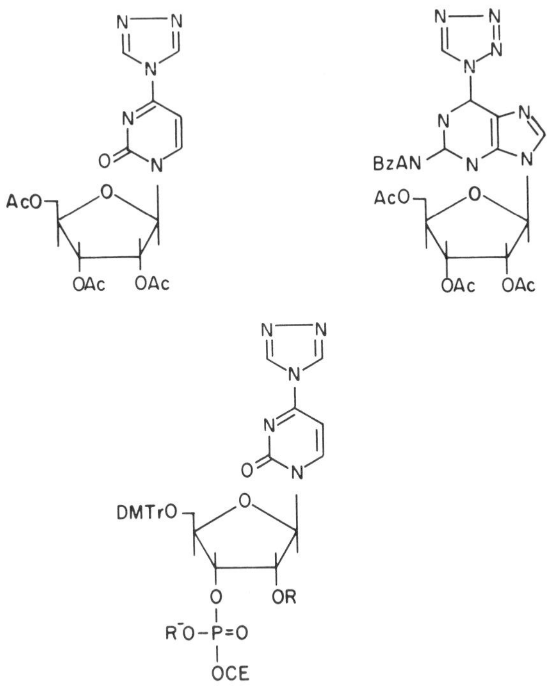

Figure 9.   Formation of side products.

onstrated that ^{252}Cf PDMS can produce positive and negative molecular ions and fragment ions of large organic molecules. The ionization process and subsequent desorption of the ions depends upon the interaction of nuclear fission fragments from californium-252 in thin solid films of these compounds. The fast chemical reaction that occurs in the solid film upon excitation by fission fragments leads to the formation of ions that are desorbed by ensuing shock waves. The negative ion mass spectra are characterized by a nested set of fragment ions extending from the 3'- or 5'-terminal nucleotides to the opposite terminal nucleotide, thereby identifying the sequence.[88]

K.   Sequence Analysis of Unprotected Oligomers Containing 3' → 5' Phosphodiester Linkages

The development of rapid and unambiguous methods for the sequence determination of DNA and RNA was one of the great breakthroughs in molecular biology of the 1970s. The techniques are so simple that DNA sequencing is now used as an indirect method of peptide sequencing. Obviously these procedures

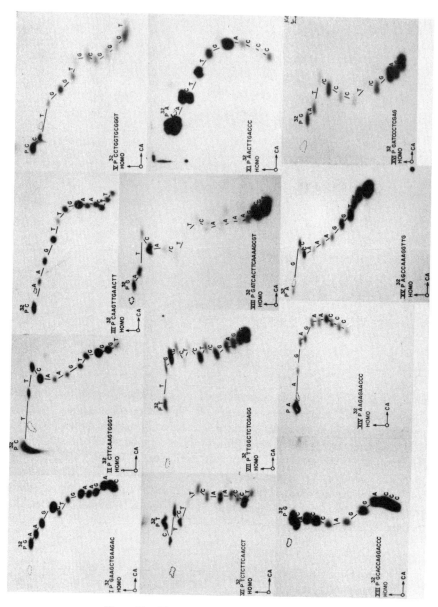

Figure 10. Fingerprints by mobility-shift method.

are of critical importance for synthetic oligomers as well as natural material. The three most important methods are outlined below.

(a) Mobility-Shift Procedure

The sequence determination of synthetic oligomers up to 20 units long can be carried out by the mobility-shift (wandering spot) method developed by Sanger[89] and Wu.[90] It involves labeling of the 5'-end of a synthetic compound with T_4-polynucleotide kinase and $[\gamma\text{-}^{32}P]$-ATP. The labeled oligomer is partially digested by snake phosphodiesterase to produce labeled sequential degradation products down to mononucleotides. These sequential partial degradation products can be fractionated two-dimensionally, first by cellulose acetate electrophoresis at pH 3.5, then by homochromatographic techniques on DEAE-cellulose. The sequence of an oligomer is determined by the characteristic mobility shifts of the labeled degradation products after exposing an audioradiogram (Figure 10).

(b) Base-Specific Chemical Cleavage

Maxam and Gilbert[91] developed a base-specific chemical cleavage method in which a terminal labeled DNA is partially cleaved at each of the four bases in four different reactions. The products were fractionated according to their size on a slab gel, and then the sequences are read from an autoradiogram by simply noting which base-specific agent cleaved at each successive nucleotide along the strand, the so-called "ladder technique." Each chemical scheme, which cleaves at one or two of the four bases, involves three consecutive steps: modification of a base, removal of the modified base from its sugar, and DNA strand scission at that sugar. The technique can sequence chains up to 500 or more units long in one run (see as an example the sequence analysis of human preproinsulin, Figure 11).

(c) "Plus-Minus" Method

The plus-minus method makes use of the ability of DNA polymerases to synthesize accurately a complementary radioactive copy of a single-stranded DNA template using DNA fragments as primers.[92] The synthesis is carried out in the presence of the four deoxyribonucleotide triphosphates, one or more of which is a $\alpha\text{-}^{32}P$-labeled, and in turn with each dideoxy or arabinose nucleotide triphosphate in separate incubations. There is in each reaction, therefore, a base-specific partial incorporation of a termination analog onto the 3'-ends of the extending transcripts througout the sequence. Partial fractionation by gel electrophoresis of the size range of terminated labeled transcripts from each reaction, each with the common 5'-end of the primer, allows a sequence to be deduced from the gel electrophoretic fingerprint.

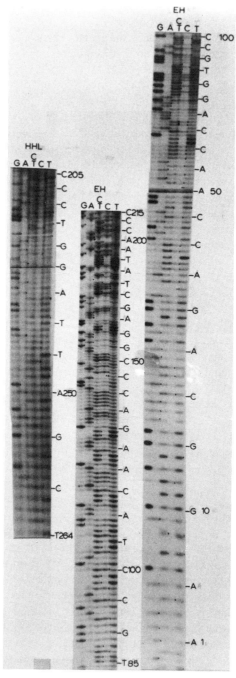

Figure 11.   DNA sequence of synthetic human proinsulin by Gilbert–Maxam method (gel-pattern).

4. ENZYMATIC SYNTHESIS

A. Synthesis of Double-Stranded DNA

Once the short segments comprising the DNA sequence of both the strands are synthesized, the next task is aligning them in proper order and linking them covalently. A strategy has been developed, mainly by Khorana and his group,[93] whereby short synthetic oligonucleotides are joined by DNA ligase enzyme,[94] in the presence of a complementary strand or template (Figure 12A). This enzyme obtained from both uninfected and T_4-phage infected *E. coli* catalyzes the covalent joining of the two molecules of deoxyribooligonucleotides or DNA between the 5'-phosphate from one molecule to the 3'-hydroxyl of another to form a phosphodiester bond in the presence of a third molecule that acts as a complementary strand near the junction of linkage. A minimum of four overlapping nucleotide pairs are required on each side of the junction to provide sufficient template interaction for the joining of deoxyribooligonucleotides with DNA ligase enzyme. This enzyme can also join fragments containing mismatched base pairs,[95] and therefore it is important to sequence the product of each enzymatic joining reaction. This method has been used to construct a double-stranded DNA molecule that includes the *E. coli* precursor tyrosine tRNA gene[9] together with its promoter, somatostatin,[96] A and B chains of human insulin,[97,98] human proinsulin,[99] leucocyte interferon[100] and partial sequence of human growth hormone.[101]

B. Enzymatic Repair of Partial Duplex

An alternative to this approach is the synthesis of a partial strand (upper and lower) by DNA ligase reaction. The resulting single strands are purified by gel electrophoresis under denaturing conditions. Since the 3'-end of the upper strand is complementary to the 3'-end of the lower strand by an overlapping stretch of nucleotide, a partial duplex is formed. With the single strand region as template, the remainder of the duplex is enzymatically synthesized with AMV reverse transcriptase and the four deoxynucleoside triphosphates to yield a complete duplex DNA[101] (Figure 12B). Using this approach, a 63-bases-long duplex for human insulin gene A was synthesized and confirmed by its DNA sequence method.[98]

C. Cloning of Synthetic Genes

During the past several years, useful techniques have been developed for the *in vitro* joining of the duplex DNA segments to vehicle DNA molecules capable of independent replication in a host cell. The cloning vehicle may be a plasmid

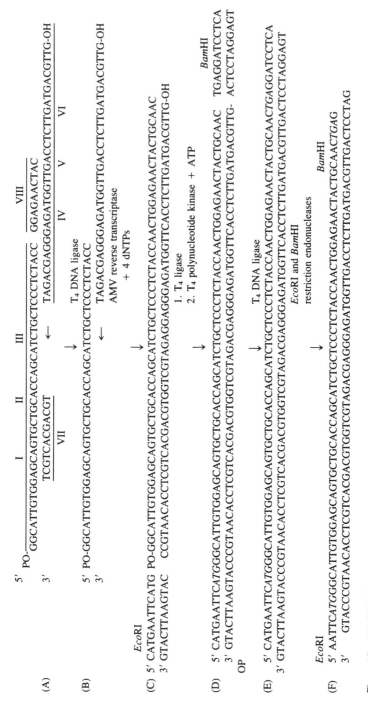

Figure 12. (A) DNA ligase reaction; (B) DNA synthesis by repair; (C) and (D) blunt-end ligation to give synthetic duplex (E) containing restriction enzyme sites; "trimming" by restriction enzymes to give synthetic duplex with exposed recognition sites (F) for insertion in plasmid.

72

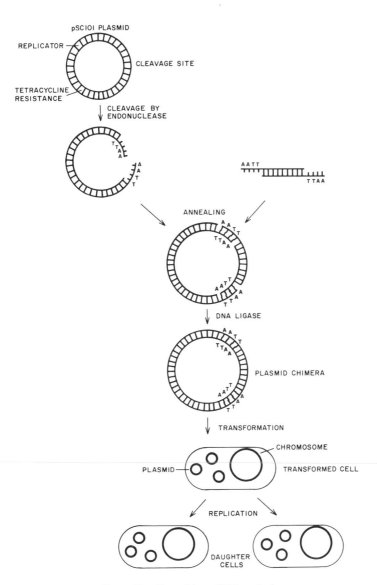

Figure 13. Recombinant DNA method.

DNA, a phase λ DNA, or an SV40 DNA, which are cleaved asymmetrically at specific sequences by a certain restriction enzyme.[102] After joining the DNA segment to the cloning vehicle, the resulting hybrid DNA (known as a chimera) can be used to transform a suitable cell.[103] The hybrid DNA can then be selected from among the transform cells known as clones, and its expression studied in terms of DNA replication, transcription, or translation.

The cloning of a DNA molecule requires three essential steps. First, the DNA sequence to be cloned must be prepared with specific unpaired or "sticky" sequences at each end recognized by a particular restriction enzyme. Next, the DNA fragment thus obtained is joined to the cloning vehicle DNA by means of DNA ligase. This "match-up" is directed by the complementary cohesive ends present at the termini of both the DNA fragment and the cloning vehicle. Finally, the hybrid DNA is introduced into a functional living cell and a suitable genetic method is used for selecting the clone containing the specific hybrid DNA.

(a) Specific Method Involving Synthetic Cohesive Ends

A DNA segment for cloning can be chemically synthesized to include a protruding single-stranded sequence that corresponds to the recognition sequence of a restriction endonuclease. For example, a *lac* operator DNA duplex has been chemically synthesized to include at each end a protruding $5'd(pA\text{-}A\text{-}T\text{-}T)$ sequence corresponding to part of the recognition sequence of *EcoRI* restriction endonuclease. A molecule of circular pMB9 plasmid DNA is cut once by an *EcoRI* restriction endonuclease to produce a linear pMB9 DNA with a protruding $5'd(pA\text{-}A\text{-}T\text{-}T)$ sequence at each end. New Watson–Crick H-bonds between the protruding sequences direct the orientation of the two molecules, which are joined covalently using T_4-DNA ligase to produce circular hybrid *lac*-pMB9 DNA. This hybrid is capable of transforming competent *E. coli* cells and expressing its biological activity *in vivo*.[104]

(b) General Method Involving Linkers

A more general method has been developed[105,106] in which a chemically synthesized linker oligonucleotide is used to create cohesive ends at the termini of blunt ended DNA molecule. Thus, any double-stranded DNA molecules can be cloned. The principle behind the method is shown in Figure 15. In the first step, a blunt-end DNA molecule is joined end-to-end to a synthetic linker using the blunt-end ligation activity of DNA ligase. The resulting molecule, with linker molecules added to each end of the original, is digested by a suitable restriction enzyme, in this case to create the *Hind*III endonuclease cohesive ends, $5'\text{-}(pA\text{-}G\text{-}C\text{-}T)$. Alternatively, a ready-made *Bam*I adaptor[107] can also be used. As no digestion with *Bam*I endonuclease is required to create the cohesive end, this

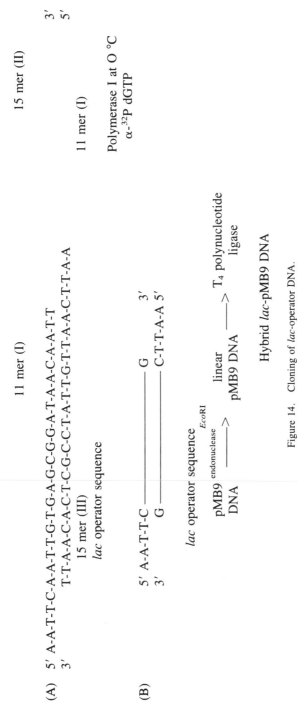

(A) 5' A-A-T-T-C-A-A-T-T-G-T-G-A-G-C-G-G-A-T-A-A-C-A-A-T-T   15 mer (II)
    3'   T-T-A-A-C-A-C-T-C-G-C-C-T-A-T-T-G-T-T-A-A-C-T-T-A-A   3'
                                                                          5'
              15 mer (III)
              lac operator sequence   11 mer (I)

                                                              Polymerase I at 0 °C
                                                              α-³²P dGTP

(B) 5' A-A-T-T-C ——————————— G   3'
    3'   G ——————————— C-T-T-A-A 5'

    lac operator sequence
                      *Eco*RI
    pMB9   endonuclease   linear
    DNA   —————————>   pMB9 DNA   —————————>   Hybrid *lac*-pMB9 DNA
                                                 T₄ polynucleotide
                                                 ligase

              Figure 14.   Cloning of *lac*-operator DNA.

75

5' pA-A-T-T-G-A-G-C-G-G-A-T-A-A-C-A-A-T-T + 5' pC-C-G-G-A-T-C-C-G-G
3' T-T-A-A-C-A-C-T-C-G-C-C-T-A-T-T-G-T-T-A-Ap 3' G-G-C-C-T-A-G-G-C-Cp

(A) *lac* operator duplex   (B) decanucleotide duplex (*Bam*I linker molecule)

T_4 ligase

(C) 5' pC-C-G-G-A-T-C-C-G-G   C-C-G-G-A-T-C-C-G-G 3'
    3" G-G-C-C-T-A-G-G-C-C   G-G-C-C-T-A-G-G-C-Cp 5'

*Bam*I endonuclease

(D) 5' pG-A-T-C-C-G-G   C-C-G   3'
    3'   G-C-C   G-G-C-C-T-A-Gp   5'
                     *Bam*I

pMB9 DNA $\xrightarrow{\text{endonuclease}}$ linear pMB9 DNA $\longrightarrow$ ↓ T_4 polynucleotide ligase
                                                                       hybrid *lac*-pMB9 DNA

For *Hin*III linker molecule:

(B) 5' pA-C-A-A-G-C-T-T-G-T
    3' T-G-T-T-C-G-A-A-C-Ap 5'

(D) 5' pA-G-C-T-T-G-T _____ A-C-A   3'
    3'   A-C-A _____ T-G-T-T-C-G-Ap   5'

Figure 15. Cloning of *lac*-operator DNA by blunt-end ligation.

Mbo II

5' TCTTCGAAGA
3' *AGAAGCTTCT* +

retrieving adaptor

1   8
5' AATTCATG   cloned   CATGAATT
3' TTAAGTAC   gene   GTACTTAA

cloned gene with cohesive ends repaired

1   8

$\dfrac{\text{Blunt-end}}{\text{ligation}} > \dfrac{\text{cleave with}}{\text{MboII}} >$ 5' TCTTCGAAGAAATTCATG
                                          3' AGAAGCTTCTTTAAGTA + $\begin{matrix}\text{cloned} & \text{C}\\ \text{C} & \text{gene}\end{matrix}$

Figure 16.   Retrieval of a gene from a cloning vehicle.

adaptor can be used for cloning DNA molecules that have internal *Bam*I sites. In the final step, this molecule is joined to a DNA vehicle (cut by the same endonuclease) to produce a hybrid DNA.

(c)   Retrieval of the Gene from the Cloning Vehicle

A general method has been developed to recover an intact gene from the cloning vehicle using synthetic retrieval adaptor[98] as shown in Figure 16.

The eight extra nucleotides at the 5'-end (left-hand side) of the insulin A-chain gene (which include the start signal) have been removed by using a *Mbo*II trimming adaptor. The specific DNA in a cloning vehicle is cut by a specific restriction enzyme and then repaired in the presence of AMV reverse transcriptase. It is next blunt-end ligated with a decanucleotide *Mbo*II adaptor. Cleavage of DNA with *Mbo*II restriction enzyme (which cleaves DNA at the arrows, eight nucleotides away from recognition sequence GAAGA) followed by *E. coli* DNA polymerase I (in the presence of [α-^{32}P] dCTP to ensure that the end of DNA is flushed), resulted in the removal of the eight extra base pairs from the A-chain gene. This strategy was used in the recovery of insulin A-chain[98] to be used in the synthesis of proinsulin gene.[99]

5.   BIOLOGICAL ROLE OF SYNTHETIC DEOXYRIBOOLIGONUCLEOTIDES

In the field of molecular biology, the availability of a well-defined sequence of oligomers offers a novel approach to understanding various biological processes at the molecular level. For example, synthesis of all the 64 possible triribonucleotides and DNA-like polymers containing repeating base sequences were successfully used in the elucidation of the genetic code.[3] In recent years, synthetic oligomers have been widely used as exemplified below.

A. Probes for the Gene

Synthetic oligomers can be used as specific probes for studies of various aspects of gene structure and function. Synthetic fragments corresponding to portions of several genes have been synthesized, for example, endolysin,[108] yeast iso-1-cytochrome C,[109] gastrin,[110] insulin,[111] interferons,[112] human β_2-microglobin,[113] and preproeukaphalin.[114] These oligomers can be used as primers for determining the primary DNA sequences by the primer extension method and also as hybridization probes for the isolation and cloning of single-copy genes in mammalian cells.

B. Protein–DNA Interaction

One of the most basic problems in molecular biology is to understand the mechanism of protein-nucleic acid interactions, that is, how do the specific amino acids of a protein molecule interact with or recognize a specific nucleotide sequence of the DNA? The sequences of a number of interesting segments of DNA—such as operator, promoter, and restriction endonuclease recognition sites—have been elucidated in recent years. For studying the point-to-point interaction of nucleic acids with proteins, large amounts of the DNA fragments are required. DNA fragments may be isolated following suitable restriction enzyme digestion, incorporation into bacterial plasmids, and amplification. Alternatively, they can be chemically synthesized.[15]

C. Tools in the Molecular Cloning of DNA

During the past several years, useful techniques have been developed for the *in vitro* joining of DNA fragments to vehicle DNA molecules capable of independent replication. The resulting hybrid DNA can be used to transform a suitable cell. Synthetic linkers[105] or adaptors 8-16 bases long containing various restriction site sequences have been synthesized and successfully used in cloning experiments or in changing a restriction in a plasmid DNA (see Section 4C).

D. Site-Specific Mutagenesis

A synthetic oligodeoxynucleotide mismatched at a single nucleotide to a specific complementary site on a wild-type circular ϕX174 DNA can be used to produce a defined point mutation, thus *in vitro* incorporation into a closed circular duplex DNA by elongation with DNA polymerase and ligation followed by transfection of *E. coli*[115-117] can generate any specific mutation at any given site. Recently a functional human suppressor tRNA has been synthesized using the M-13 system with synthetic primer techniques.[118]

E. Inhibition of Rous Sarcoma Virus Replication and Cell Transformation by a Specific Oligonucleotide

The synthetic tridecamer, which is complementary to 13 bases of the 3'- and 5'-reiterated terminal sequences of *Rous sarcoma* virus 35S RNA,[119] was added to chick embryo fibroblast tissue cultures infected with *Rous sarcoma* virus. Inhibition of virus production resulted. This is because the synthetic oligomer hybridizes with the terminal reiterated sequences at the 3'-and 5'-ends of the 35S RNA and interferes with one or more steps involved in viral production and cell transformation.

F. Molecular Structure of Synthetic DNA Fragments

Various self-complementing synthetic deoxyoligomers have been crystallized as a left-hand double helical molecule with Watson–Crick base pairs and an antiparallel organization of the sugar phosphate chains.[120,121] It differs significantly from right-hand B-DNA.

G. Uptake of Synthetic Deoxyribooligonucleotide Sequence in *Haemophilus* Cells

Only certain DNA fragments are taken up efficiently by competent *Haemophilus* cells, which implies that efficient uptake requires the presence of a specific nucleotide sequence on the incoming DNA. Such a fragment has been identified[122] as undecamer base pairs in common 5'-AAGTGCGGTCA-3'. Its synthesis has been accomplished and has been shown to be biologically active in transporting DNA in the *Haemophilus* cell.[123]

6. CONCLUDING REMARKS

Progress during the past decade has resulted in the maturation of the triester methodology of chemical synthesis, sophistication in recombinant DNA techniques, and rapid DNA sequencing methods. It seems safer to predict that during the 1980s the large-scale and economical synthesis of important proteins, such as insulin, human growth hormones, and interferons, directed by synthetic genes will become a reality. It should even be possible to program microorganisms to make proteins that do not occur naturally in any organism. This approach to synthesis can provide a wealth of new molecules when more is known about the relations between the architecture of proteins and their biological properties. It is also predicted that most of the steps to gene synthesis will be automated and our knowledge of gene regulation and expression will increase. Ultimately, this branch of science may find its best application in solving basic problems in the essential fields of health, food, and energy.

ACKNOWLEDGMENT

Original research by S.A.N. and W.L.S. was supported by the National Research Council of Canada.

REFERENCES

1. S. A. Narang and R. H. Wightman, in *Total Synthesis of Natural Products*, Vol. 1 (J. W. ApSimon, (ed.), Wiley, New York, 1973, p. 279.

2. R. I. Zhdanov and S. M. Zhendarova, *Synthesis*, 222 (1975).

3. H. Kössel and H. Seliger, *Fortschr. Chem. Org. Naturst.* **32**, 297 (1975).

4. V. Amarnath and A. D. Broom, *Chem. Rev.*, **77**, 183 (1977).

5. J. H. van Boom, *Heterocycles*, **7**, 1197 (1977).

6. C. B. Reese, *Tetrahedron*, **34**, 3143 (1978).

7. M. Ikehara, E. Ohtsuka, and A. F. Markham, *Adv. Carbohydr. Chem. Biochem.*, **36**, 135 (1979).

8. A. M. Michelson and A. R. Todd, *J. Chem. Soc.*, 2632 (1955).

9. H. G. Khorana, *Science*, **203**, 614 (1979).

10. R. L. Letsinger and V. Mahaderan, *J. Am. Chem. Soc.*, **87**, 3826 (1965).

11. C. B. Reese and R. Saffhill, *Chem. Commun.*, 767 (1968).

12. F. Eckstein and I. Risk, *Chem. Ber.*, **102**, 2362 (1969).

13. K. Itakura, C. P. Bahl, N. Katagiri, J. Michniewicz, R. H. Wightman, and S. A. Narang, *Can. J. Chem.*, **51**, 3469 (1973).

14. J. C. Catlin and F. Cramer, *J. Org. Chem.*, **38**, 245 (1973).

15. C. P. Bahl, R. Wu, K. Itakura, and S. A. Narang, *Proc. Natl. Acad. Sci. USA*, **73**, 91 (1976).

16. *Nucleic Acid Synthesis*, Proceedings of an International Symposium at Hamburg, Germany, May 3–9, 1980, published by Nucleic Acids Research, Special Publication No. 7 (September 1980).

17. P. T. Gilham and H. G. Khorana, *J. Am. Chem. Soc.*, **80**, 6212 (1958).

18. K. L. Agarwal, Y. A. Berlin, H. J. Fritz, M. J. Gait, D. G. Kleid, R. G. Lees, K. E. Norris, B. Ramamoorthy, and H. G. Khorana, *J. Am. Chem. Soc.*, **98**, 1065 (1976).

19. R. A. Jones, H. J. Fritz, and H. G. Khorana, *Biochemistry,* **17,** 1268 (1978).
20. G. J. Powers, R. L. Jones, G. A. Randall, M. H. Caruthers, J. H. van de Sande, and H. G. Khorana, *J. Am. Chem. Soc.,* **97,** 875 (1975).
21. K. Itakura, N. Katagiri, C. P. Bahl, R. H. Wightman, and S. A. Narang, *J. Am. Chem. Soc.,* **97,** 7327 (1975).
22. N. Katagiri, K. Itakura, and S. A. Narang, *J. Am. Chem. Soc.,* **97,** 7332 (1975).
23. A. K. Sood and S. A. Narang, *Nucl. Acids Res.,* **4,** 2527 (1977).
24. R. W. Adamiak, M. Z. Barciszewska, E. Biala, K. Grzeskowiak, R. Kierzek, A. Kraczewski, W. T. Markiewicz, and M. Wiewiorowski, *Nucl. Acids Res.,* **3,** 3397 (1976).
25. W. L. Sung and S. A. Narang, *Can. J. Chem.,* **60,** 111 (1982).
26. G. R. Gough, C. K. Singleton, H. L. Weith, and P. T. Gilham, *Nucl. Acids Res.,* **6,** 155 (1979).
27. S. S. Jones, R. Raynes, C. B. Reese, A. Ubasawa, and M. Ubasawa, *Tetrahedron,* **36,** 3075 (1980).
28. H. Hsiung, R. Brousseau, J. Michniewicz, and S. A. Narang, *Nucl. Acids Res.,* **6,** 1371 (1979).
29. H. Hsiung, W. Sung, R. Brousseau, R. Wu, and S. A. Narang, *Nucl. Acids Res.,* **8,** 5753 (1980).
30. R. L. Letsinger and W. B. Lunsford, *J. Am. Chem. Soc.,* **98,** 3655 (1976).
31. K. K. Ogilvie, N. Theriault, and K. L. Sadana, *J. Am. Chem. Soc.* **99,** 7741 (1977).
32. M. D. Mateucci and M. H. Caruthers, *J. Am. Chem. Soc.,* **103,** 3186 (1981).
33. M. J. Gait and R. C. Sheppard, *J. Am. Chem. Soc.,* **98,** 8514 (1976).
34. M. J. Gait, M. Singh, R. C. Sheppard, M. D. Edge, A. R. Greene, G. R. Heathcliffe, T. C. Atkinson, C. R. Newton, and A. F. Markham, *Nucl. Acids Res.,* **8,** 1081 (1980).
35. K. Miyoshi and K. Itakura, *Tetrahedron Lett.,* **28,** 2449 (1979).
36. K. K. Ogilvie and M. J. Nemer, *Tetrahedron Lett.,* **29,** 4159 (1980).
37. P. Dembek, K. Miyoshi, and K. Itakura, *J. Am. Chem. Soc.,* **103,** 706 (1981).
38. J. Stawinski, T. Hozumi, S. A. Narang, C. P. Bahl, and R. Wu, *Nucl. Acids Res.,* **4,** 353 (1977).
39. J. B. Chattopadhya and C. B. Reese, *Chem. Commun.,* 639 (1978).
40. R. W. Adamiak, R. Arentzen, and C. B. Reese, *Tetrahedron Lett.,* 1431 (1977).
41. J. H. van Boom and P. M. J. Burgers, *Rec. Trav. Chim. Pays-Bas,* **97,** 73 (1976).
42. J. H. van Boom, P. H. van Deusen, J. Meewose, and C. B. Reese, *Chem. Commun.,* **766** (1972).
43. R. W. Adamiak, E. Biala, K. Grzeskowiak, R. Kierzek, A. Kraszewski, W. T. Markiewicz, J. Stawinski and M. Wiewiorowski, *Nucl. Acids Res.,* **4,** 2321 (1977).
44. T. Neilson, K. V. Deugau, T. E. England, and E. S. Werstiuk, *Can. J. Chem.* **53,** 1093 (1975).
45. J. H. van Boom and P. M. J. Burgers, *Tetrahedron Lett.,* 4875 (1976).
46. G. Reitz and W. Pfleiderer, *Chem. Ber.,* **108,** 2878 (1975).
47. E. Ohtsuka, T. Tanaka, and M. Ikehara, *Nucl. Acids Res.,* **1,** 1351 (1974).
48. E. Ohtsuka, T. Tanaka, and M. Ikehara, *Chem. Pharm. Bull.* (Tokyo), **28,** 120 (1980).
49. G. Kumar, L. Celewicz, and S. Chladek, *J. Org. Chem.,* **47,** 614 (1982).
50. K. K. Ogilvie, A. L. Schifman, and C. L. Penney, *Can. J. Chem.,* **57,** 2230 (1979).
51. G. K. Hakimelahi, Z. A. Proba, and K. K. Ogilvie, *Tetrahedron Lett.,* **22,** 4775 (1981).

52. D. G. Bartholomew and A. D. Broom, *Chem. Commun.*, **38** (1975).

53. E. Ohtsuka, S. Tanaka, and M. Ikehara, Synthesis, **453** (1977).

54. Y. Ishido, N. Nakasaki, and N. Sakairi, *J. Chem. Soc., Perkins I*, 2088 (1979).

55. Y. Ishido, N. Sakairi, and I. Hirao, *Nucl. Acids Res.*, **55**, 263 (1978).

56. C. B. Reese, J. C. M. Stewart, J. H. van Boom, H. P. M. de Leeuw, J. Nagel, and J. F. M. de Rooij, *J. Chem. Soc., Perkins I*, 934 (1975).

57. E. Ohtsuka, E. Nakagawa, T. Tanaka, A. F. Markham, and M. Ikehara, *Chem. Pharm. Bull.* (Tokyo), **26**, 2998 (1978).

58. J. Smrt and F. Cramer, *Collect. Czech., Chem. Commun.*, **35**, 1456 (1970).

59. H. Schaller, G. Weimann, B. Lerch, and H. G. Khorana, *J. Amer. Chem. Soc.*, **85**, 384 (1963).

60. R. Charubala and W. Pfleiderer, *Heterocycles*, **15**, 761 (1981).

61. R. W. Barnett and R. L. Letsinger, *Tetrahedron Lett.*, **22**, 991 (1981).

62. E. Ohtsuka, H. Tsuji, T. Miyake, and M. Ikehara, *Chem. Pharm. Bull.* (Tokyo), **25**, 2844 (1977).

63. R. W. Adamiak, E. Biala, D. Grzeskowiak, R. Kierzek, A. Krszewski, W. T. Markiewicz, J. Okupniak, J. Stawinski, and M. Wiewiorowski, *Nucl. Acids Res.*, **5**, 1889 (1978).

64. K. K. Ogilvie and S. L. Beaucarge, *Nucl. Acids Res.*, **7**, 805 (1979).

65. W. S. Zielinski and Z. Lesnikowski, *Synthesis*, 185 (1976).

66. J. H. van Boom, P. M. J. Burgers, P. H. van Deursen, R. Arentzen, and C. B. Reese, *Tetrahedron Lett.*, 3785 (1974).

67. C. B. Reese, R. C. Titmas, and L. Yau, *Tetrahedron Lett.*, 2727 (1978).

68. J. F. M. de Rooij, G. Wille-Hazeleger, P. M. J. Burgers, and J. H. van Boom, *Nucl. Acids Res.*, **6**, 2237 (1979).

69. C. B. Reese and L. Yan, *Tetrahedron Lett.*, 4443 (1978).

70. G. W. Daub and E. E. van Tamelen, *J. Am. Chem. Soc.*, **99**, 3526 (1977).

71. C. B. Reese and Y. T. Yan Kui, *Chem. Commun.*, 802 (1977).

72. Y. Stabinsky, R. T. Sakata, and M. H. Caruthers, *Tetrahedron Lett.*, **23**, 275 (1982).

73. E. Ohtsuka, T. Tanaka, and M. Ikehara, *Nucl. Acids Res.*, **7**, 1283 (1979), also ref. 7, p. 155.

74. J. H. van Boom, P. M. J. Burgers, R. Crea, W. C. M. M. Luyten, A. B. J. Vink, and C. B. Reese, *Tetrahedron*, **31**, 2953 (1975).

75. H. Takaku, M. Kato, and T. Hata, *Chem. Commun.*, 190 (1977).

76. Y. A. Berlin, O. G. Chakhmakhchera, V. A. Efimov, M. N, Kolosov, and V. G. Korobko, *Tetrahedron Lett.*, 1353 (1973).

77. H. A. Staab, *Angew. Chem. Int. Ed.*, **1**, 350 (1962).

78. N. Katagiri, K. Itakura, and S. A. Narang, *Chem. Commun.* 326 (1974).

79. J. Stawinski, T. Hozumi, and S. A. Narang, *Can. J. Chem.*, **54**, 670 (1976).

80. S. S. Jones, C. B. Reese, S. Sibanda, and A. Ubasawa, *Tetrahedron Lett.*, **22**, 4755 (1981).

81. G. van der Marcel, C. van Boeckel, G. Wille, and J. H. van Boom, *Tetrahedron Lett.*, **22**, 3887 (1981).

82. V. A. Efimor, S. V. Reverdatta, and O. G. Chakhmakhchera, *Tetrahedron Lett.*, **23**, 961 (1982).

83. E. Ohtsuka, Z. Tozuka, and M. Ikehara, *Tetrahedron Lett.*, **22**, 4483 (1981).

84. D. G. Knorre and V. F. Zarytova, in *Phosphorus Chemistry Directed towards Biology* (W. J. Stec, Ed.), Pergamon Press, Oxford and New York, 1980, p. 13.

85. C. B. Reese and A. Ubasawa, *Nucl. Acids Res.*, Symposium Series No. 7, 5 (1980).

86. W. L. Sung, *Nucl. Acids Res.*, 9, 6139 (1981).

87. C. J. McNeal, S. A. Narang, R. D. Macfarlane, H. M. Hsiung, and R. Brousseau, *Proc. Natl. Acad. Sci. USA*, 77, 735 (1980).

88. C. J. McNeal and R. D. Macfarlane, *J. Am. Chem. Soc.*, 103, 1609 (1981).

89. F. Sanger, J. E. Donelson, A. R. Coulson, H. Kössel, and D. Fischer, *Proc. Natl. Acad. Sci. USA*, 70, 1209 (1973).

90. E. Jay, A. Bambara, R. Padmanabhan, and R. Wu, *Nucl. Acids Res.*, 1, 331 (1974).

91. A. M. Maxam and W. Gilbert, *Proc. Natl. Acad. Sci. USA*, 74, 560 (1977).

92. F. Sanger and A. R. Coulson, *J. Mol. Biol.*, 94, 441 (1975).

93. H. G. Khorana, K. L. Agarwal, P. Besmer, H. Büchi, M. H. Caruthers, P. J. Cashion, M. Fridkin, E. Jay, K. Kleppe, R. Kleppe, A. Kumar, P. C. Loewen, R. C. Miller, K. Minamoto, A. Panet, U. L. RajBhandary, B. Ramamoorthy, T. Sekiya, T. Takeya, and J. H. van de Sande, *J. Biol. Chem.*, 251, 565 (1976).

94. M. Gellert, *Proc. Natl. Acad. Sci. USA*, 57, 148 (1967); B. Weiss and C. C. Richardson, *Proc. Natl. Acad. Sci. USA*, 57, 1021 (1967); B. M. Olivera and I. R. Lehman, *Proc. Natl. Acad. Sci. USA*, 57, 1426 (1967).

95. C. M. Tsiapalis and S. A. Narang, *Biochem. Biophys. Res. Commun.*, 39, 631 (1970).

96. K. Itakura, T. Hirose, R. Crea, A. D. Riggs, H. L. Heynekar, F. Bolivar, and H. W. Boyer, *Science*, 198, 1056 (1977).

97. D. V. Goeddel, D. G. Kleid, F. Bolivar, H. L. Heynekar, D. C. Yansura, R. Crea, T. Hirose, A. Krazewski, K. Itakura, and A. D. Riggs, *Proc. Natl. Acad. Sci. USA*, 76, 106 (1979).

98. R. C. Scarpulla, S. A. Narang, and R. Wu, *Anal. Biochem.*, 121, 356 (1982).

99. R. Brousseau, R. Scarpulla, W. Sung, H. M. Hsiung, S. A. Narang, and R. Wu, *Gene*, 17, 279 (1982).

100. M. D. Edge, A. R. Greene, G. R. Heathcliffe, P. A. Meacock, W. Schuch, D. B. Scanlon, T. C. Atkinson, C. R. Newton, and A. F. Markham, *Nature*, 292, 756 (1981).

101. D. V. Goeddel, H. L. Heynekar, R. Arentzen, K. Itakura, D. G. Yansura, M. J. Ross, G. Miozzari, R. Crea, and P. H. Seeburg, *Nature*, 281, 544 (1979).

102. R. J. Roberts, *C.R.C. Crit. Rev. Biochem.*, 4, 123 (1976).

103. S. N. Cohen, C. Y. Chang, H. W. Boyer, and R. B. Helling, *Proc. Natl. Acad. Sci. USA*, 70, 3240 (1973).

104. K. J. Marians, R. Wu, J. Stawinski, and S. A. Narang, *Nature*, 263, 744 (1976).

105. C. P. Bahl, K. J. Marians, R. Wu, J. Stawinski, and S. A. Narang, *Gene*, 1, 1772 (1979).

106. H. L. Heynekar, J. Shine, H. M. Goodman, H. W. Boyer, J. Rosenberg, R. E. Dickerson, S. A. Narang, K. Itakura, S. Lin, and A. D. Riggs, *Nature*, 263, 748 (1976).

107. C. P. Bahl, R. Wu, R. Brousseau, A. K. Sood, H. M. Hsiung, and S. A. Narang, *Biochem. Biophys. Res. Commun.*, 81, 695 (1978).

108. R. Wu, R. Tu, and R. Padmanabhan, *Biochem. Biophys. Res. Commun.*, 55, 1092 (1973).

109. J. W. Szostak, J. E. Stiles, C. P. Bahl, and R. Wu, *Nature*, 265, 61 (1977).

110. B. E. Noyes, M. Mevarech, R. Stein, and K. L. Agarwal, *Proc. Natl. Acad. Sci. USA*, 76, 5036 (1979).

111. S. J. Chan, B. F. Noyes, K. A. Agarwal, and D. F. Steiner, *Proc. Natl. Acad. Sci. USA*, **76**, 5036 (1979).

112. M. Houghton, A. G. Stewart, S. M. Doel, J. S. Emtage, M. A. Eaton, J. C. Smith, T. P. Patel. H. M. Lewis, A. G. Porter, A. J. Birch, T. Cartwright, and N. H. Carey, *Nucl. Acids Res.*, **8**, 1813 (1980).

113. S. V. Suggs, R. B. Wallace, T. Hirose, E. H. Kawashima, and K. Itakura, *Proc. Natl. Acad. Sci. USA*, **78**, 6613 (1981).

114. M. Noda, Y. Furutani, H. Takahashi, M. Toyosato, T. Hirose, S. Inayama, S. Nakaanishi, and S. Numa, *Nature*, **295**, 202 (1982); U. Gubler, P. Seeburg, B. J. Hoffman, L. P. Gage, and S. Udenfriend, *ibid*, **295**, 206 (1982).

115. A. Razin, T. Hirose, K. Itakura, and A. D. Riggs, *Proc. Natl. Acad. Sci. USA*, **75**, 4268 (1978).

116. S. Gillam and M. Smith, *Gene*, **8**, 99 (1979).

117. O. S. Bhanot, S. A. Khan, and R. W. Chambers, *J. Biol. Chem.*, **254**, 12,684 (1979).

118. G. F. Temple, A. M. Dozy, K. L. Roy, and Y. W. Khan, *Nature*, **296**, 537 (1982).

119. P. C. Zemecnik and M. L. Stephenson, *Proc. Natl. Acad. Sci. USA*, **75**, 280 (1978).

120. A. H. J. Wang, G. J. Quigley, F. J. Kolpak, J. L. Crawford, J. H. van Boom, G. van der Marel, and A. Rich, *Nature*, **282**, 680 (1979).

121. H. R. Drew, R. M. Wing, T. Takano, C. Broka, S. Tanaka, K. Itakura, and R. E. Dickerson, *Proc. Natl. Acad. Sci. USA*, **78**, 2179 (1981).

122. D. B. Danner, R. A. Deich, K. L. Sisco, and H. O. Smith, *Gene*, **11**, 311 (1980).

123. D. B. Danner, H. O. Smith, and S. A. Narang, *Proc. Natl. Acad. Sci. USA*, **79**, 2393 (1982).

SUGGESTED READING

T. Tanaka and R. L. Letsinger, "Syringe method for stepwise chemical synthesis of oligonucleotide," *Nucl. Acids Res.*, **10**, 3249 (1982).

E. M. Ivanova, L. M. Khalimshaya, V. P. Romanenko, and V. F. Zarytova, "Pyrophosphate tetraester intermediate of coupling reactions in the phosphotriester approach to the synthesis of deoxyoligoribonucleotides," *Tetrahedron Lett.*, **23**, 5447 (1982).

Shanghai Institute of Biochemistry, Academia Sinicas, "Total synthesis of yeast alanine transfer ribonucleic Acid," *Kexue Tongbao*, **27**, 216 (1982).

The Total Synthesis of Triterpenes 1973–1981

JOHN W. APSIMON, KIM E. FYFE, and AUSTIN M. GREAVES

Department of Chemistry, Carleton University, Ottawa, Canada

Although the triterpene field has not seen anything approaching the synthetic activity devoted to the lower terpenes, several notable synthetic achievements have been reported since publication of the last review in 1973.[1] Ireland's group has succeeded in their second total synthesis of alnusenone[66] as well as the only syntheses reported to date of shionone[56] and friedelin.[71] Kametani and co-workers[73] have obtained formal total syntheses of alnusenone and friedelin by developing new routes to Ireland's key intermediates. Utilizing the biogenetic type of polyene cyclization investigated extensively by W.S. Johnson,[52] Prestwick and Labovitz obtained the modified hopane type of triterpene serratenediol.[108]

The total syntheses, as well as the many reports on approaches to the synthesis of triterpenes, have added considerably to knowledge of the chemistry of these complex molecules.

1. SQUALENE EPOXIDE

Optically active epoxy-terpenes have aroused considerable interest in recent years because of their intermediacy in biosynthetic pathways. Several reports on the syntheses of chiral terpenes have appeared,[2,3] among which are the syntheses of R-(+)- and S-(−)-squalene-2,3-oxide from L-glutamic acid (Scheme 1).

The (S)-lactone acid 1, obtained from L-glutamic acid by nitrous acid deamination, was converted to the acid chloride, then treated with excess diazomethane followed by hydrogen iodide to yield the keto-lactone 2. Amidation occurred quantitatively to give the partially racemized amide 3, which was purified by repeated recrystallizations. The vicinal diol resulting from reaction with excess methylmagnesium iodide was protected as the acetonide 4. An isomeric mixture of olefins ($Z:E$, 26:74) was obtained from the subsequent Wittig reaction. Reduction followed by separation on silver nitrate coated silica gel gave the (Z)- and (E)-alcohols in 20% (6) and 61% (5) yield, respectively. Conversion of the (S)-(E)-alcohol (5) to the chloride then afforded the thioether (7) on reaction with sodium phenylsulfide. The thio ether anion was formed by treatment with n-butyllithium. Alkylation with the allylic chloride[4] (8), followed by removal of sulfur, then yielded the diene 9, which was converted in several steps to (R)-(+)-10,11-epoxyfarnesol.

The allylic alcohol of 9 was converted to the chloride 10 and coupled with the thioether anion of previously synthesized triene 11[5] (Scheme 2). The compound 12 now has all 30 carbons, and five double bonds, of squalene epoxide. Reduction with lithium in ethylamine removed the thioether activating group (13), and hydrolysis of the acetonide produced the (S)-diol 14. Epoxidation of 14 via mesylation of the secondary alcohol occurred with inversion at the chiral center, yielding (R)-squalene 2,3-oxide (16). Enantiomeric conversion was car-

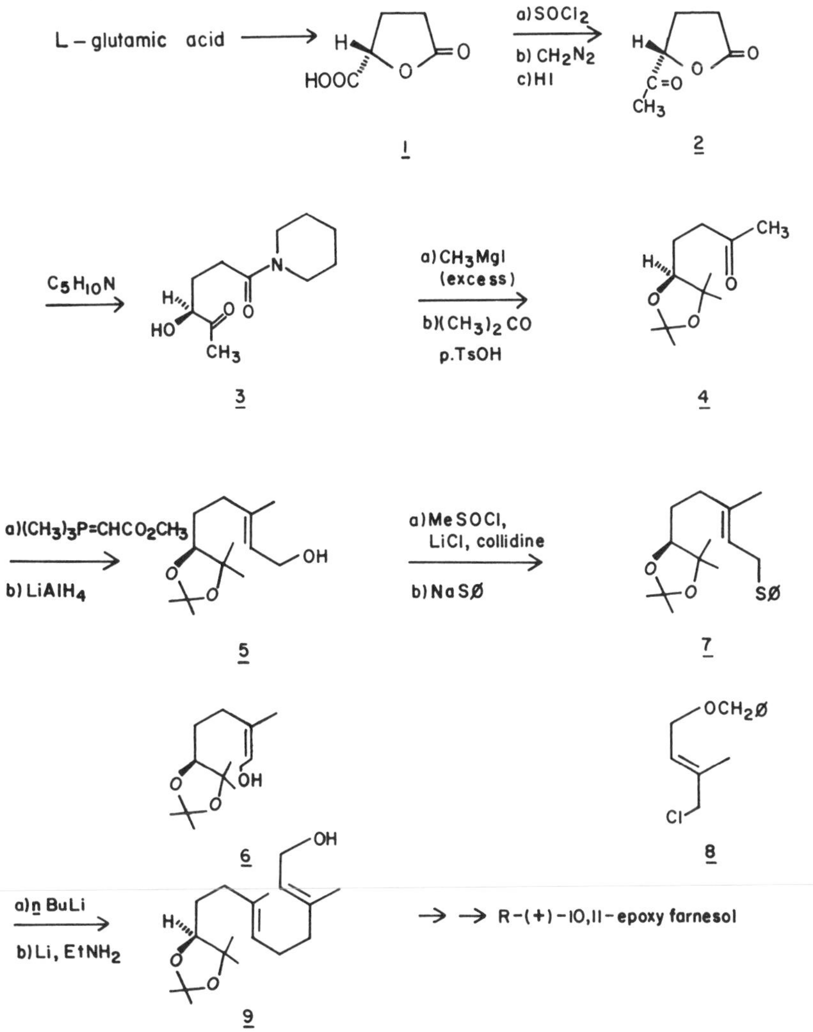

Scheme 1

ried out by hydrolysis in perchloric acid-dimethoxyethane to the diol **15**, which proceeded with retention at the chiral center, followed by re-epoxidation to **17** under the previously employed conditions. Both (R)- and (S)-enantiomers showed optical rotations $\{[\alpha]_D + 2.0$ and -1.8 (MeOH), respectively$\}$ in close agreement with those obtained by Boar and Damps.[6]

Boar and Damps obtained **16** and **17** by resolution of 2,3-dihydrosqualene-

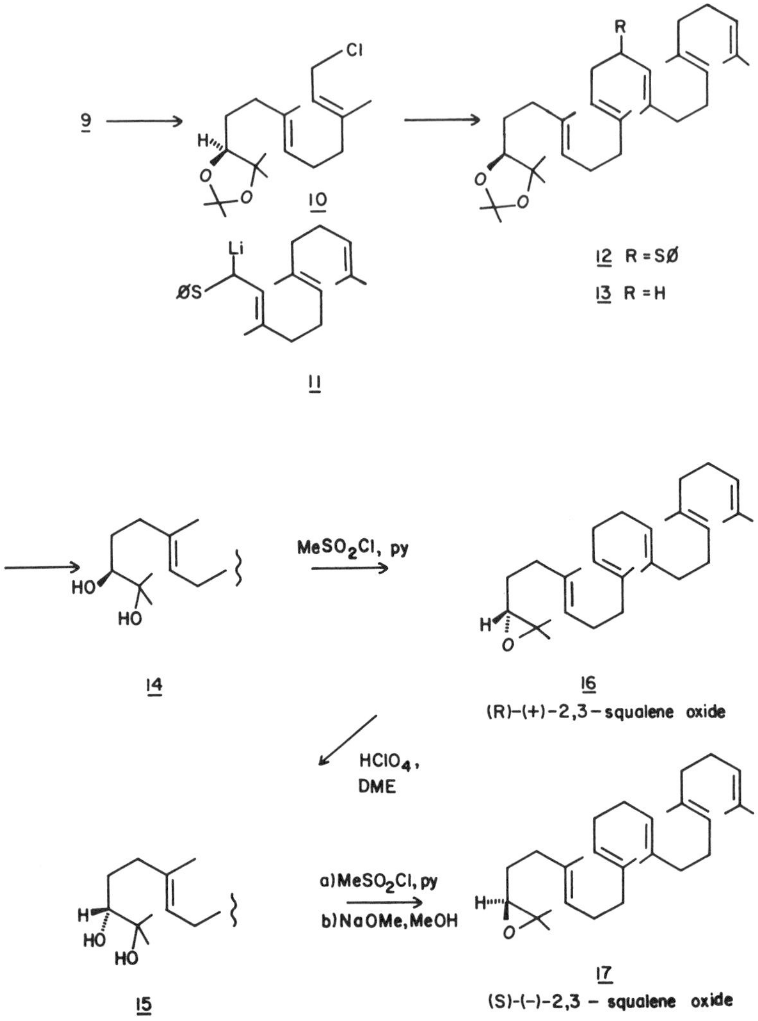

Scheme 2

2,3-diol. The enantiomeric mixture of diols was esterified with 3β-acetoxy-17β-chloroformylandrost-8-ene[7] to give the mixture of diastereomers. The esters were separated by chromatography, hydrolyzed, and epoxidized by treatment of the respective monotosylates with sodium hydroxide in ethanol.

Damps and Boar, in collaboration with Barton and Widdowson, tested radiochemically labeled samples of (3R)- and (3S)-squalene epoxides as precursors to a range of 3β-hydroxy-triterpenes (**18**).[8] In all cases the (3*S*)-epoxide was shown to act as an essentially exclusive precursor. Although this had been assumed to be the case, only indirect experimental support had been previously obtained.[9]

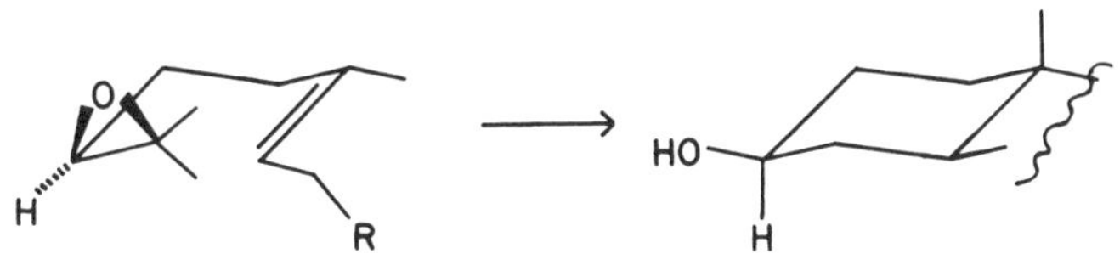

(17)   (18)

2. SQUALANE AND SQUALENES

Selective linear trimerization of isoprene has been effected[10] using a nickel alkoxide catalyst, $Ni(OR)(\eta^3\text{-}C_3H_5)$ $PPh(NEt_2)_2$, the major component of the product mixture being *trans*-β-farnesene (**19**) (70%). The alkoxide catalyst is the only nickel system so far developed that yields natural β-farnesene as the major oligomerization product. The effect of the anionic ligand is most significant, the selectivity for trimers increasing in the order $R = CH_3O < n - C_4H_9O < n - C_{15}H_{31}O$.

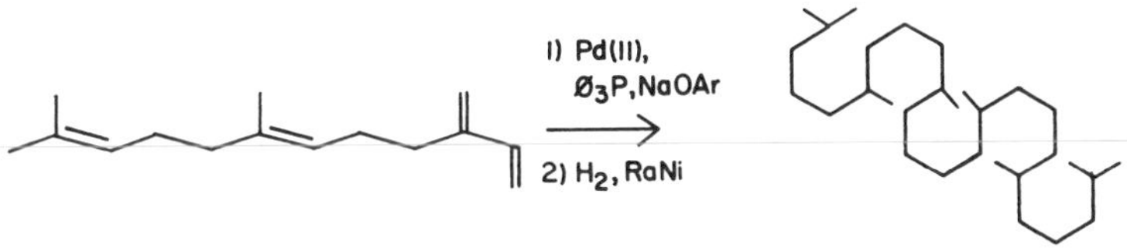

1) Pd(II),
Ø₃P,NaOAr

2) H₂, RaNi

(19)   (20)

Regioselective tail-to-tail coupling of β-farnesene (**19**) was achieved using a modified palladium catalyst[10] consisting of palladium nitrate, triphenylphosphine, and sodium *o*-methoxyphenoxide. Catalytic hydrogenation gave squalene **20** in 85% yield.

A total synthesis of squalene and lower terpenes[11] using the functionalized isoprene 2-hydroxymethyl-4-phenylthio-1-butene[12] (**21**) has been recently reported. The starting material (**21**) has three functionalities: an allylic alcohol allowing two-carbon elongation by Claisen rearrangement; a homoallyl phenyl

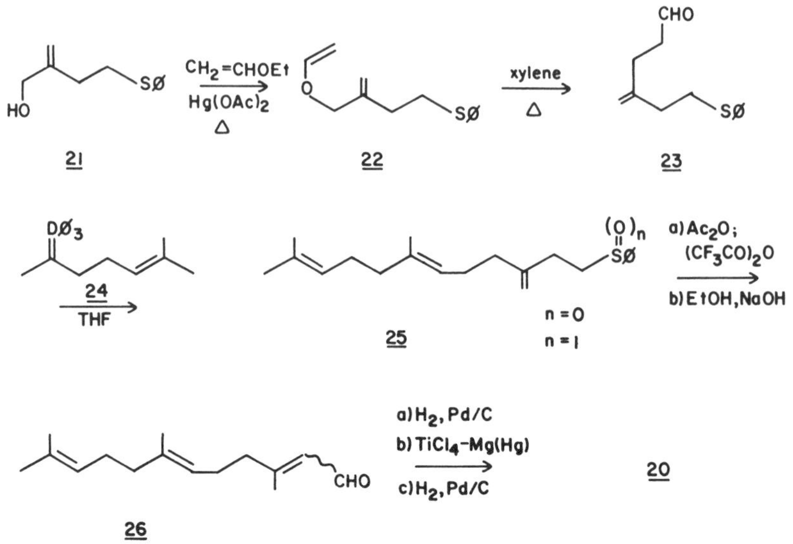

Scheme 3

sulfide, synthetically equivalent to an α,β-unsaturated aldehyde via the Pummerer reaction; and a thiophenyl function allowing α-alkylation and thermolytic generation of a double bond.

Conversion of **21** to the vinyl ether (**22**) followed by Claisen rearrangement gave the aldehyde (**23**) (82%) (Scheme 3). Reaction with the ylide (**24**)[13] gave the triene (**25**) as a mixture of geometric isomers, (*E:Z*, 6:4). Pummerer rearrangement of the derived sulfoxide on treatment with acetic anhydride, followed by hydrolysis of the intermediate acetate, provided **26** as a mixture of three geometric isomers. Catalytic hydrogenation and coupling of the saturated aldehyde with low valent titanium [TiCl$_4$-Mg(Hg)][14] afforded the symmetrical olefin. A second hydrogenation yielded squalene (**20**).

All *E*-isomers 11,14-dimethylsqualene and 11-methylsqualene were synthesized[15] as reference compounds for biosynthetic studies of squalene synthetase binding sites. 11,14-Dimethylsqualene, obtained in 18% yield by nickel tetracarbonyl coupling[16] of 2-methoxyfarnesyl bromide, was isolated from a mixture of geometric and positional isomers as a thiourea clathrate. Condensation between 2-methylfarnesyl bromide and the anion of 2-(1-farnesylthio)-1-methyl-imidazole followed by Raney nickel desulfuration afforded 11-methylsqualene in 40% yield.

3. TRITERPENES WITH STEROIDAL RING SYSTEMS

A. Lanostanes

As part of an investigation into the synthesis of tetracyclic triterpenes of the lanostane (**27**)-cycloartane (**28**) group, Packer and Whitehurst[17] described the synthesis of the indan-2-one **29**. Compound **29** has potential as a key intermediate in the synthesis of tetracyclic triterpenes, possessing the *trans*-CD ring junction and a carbonyl suitably positioned for introduction of the side chain at C-17.

Reaction of 6-methoxy-1-tetralone (**30**) with methylmagnesium iodide gave the dihydronaphthalene (**31**) in high yield (Scheme 4). However, the transformation of **31** to the tetralone **33**, via perbenzoic acid epoxidation followed by acid workup,[18,19] was capricious and resulted in low yields. A better route was developed, involving hydroboration (hydrogen peroxide oxidation) to the alcohol **32**, followed by Pfitzner–Moffatt oxidation[20] to the tetralone **33** in an overall yield of 61%. Other oxidation methods were tried but with varied and poor results. Alkylation of the tetralone **33** with 3-bromopropyne yielded **34**, which underwent hydration [mercury (II) acetate in acetic acid-formic acid] to the diketone (**35**). The enone (**36**) obtained by base catalyzed cyclization was stereospecifically reduced with lithium aluminum hydride[21] to the allylic alcohol (**37**). Simmons–Smith cyclopropanation[22] yielded **38**, in which the cyclopropane ring and the hydroxyl group were expected to be *cis* oriented, thereby resulting in the desired *trans* ring junction. Lithium-ammonia reduction of the corresponding ketone (**39**) gave the ketone **29**.

Since the nmr spectroscopic study of **29** did not provide sufficient evidence

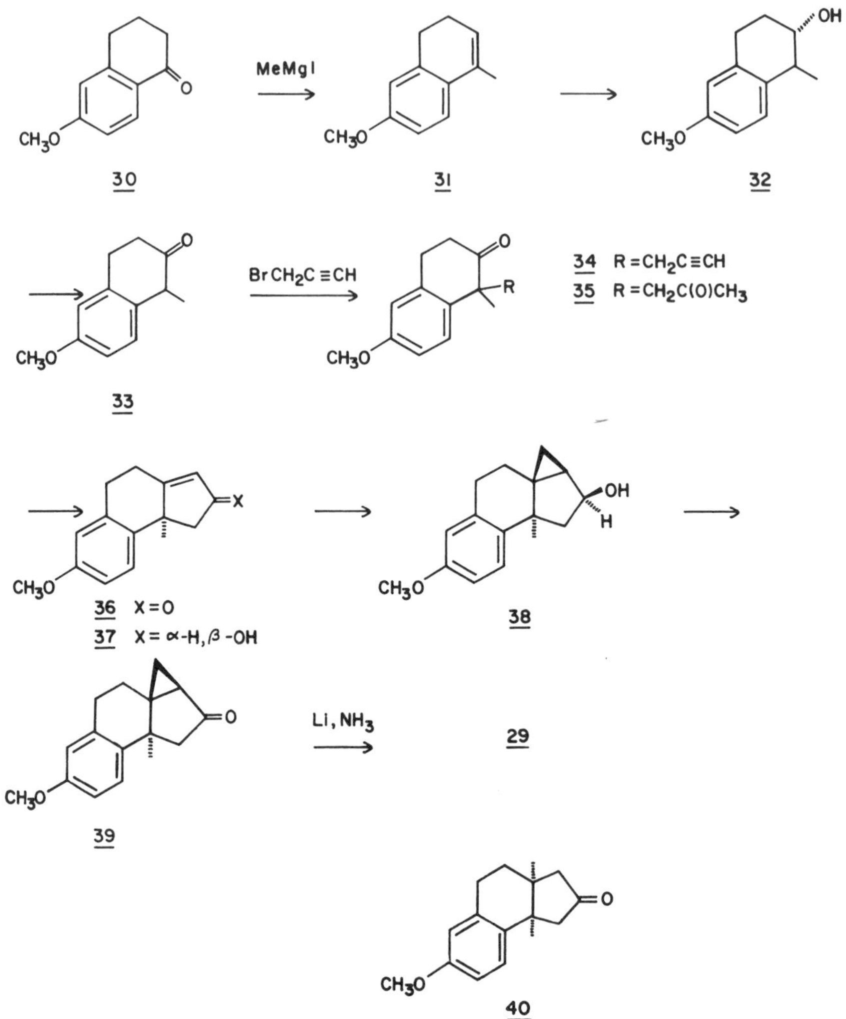

Scheme 4

for the unequivocal assignment of the *trans* stereochemistry, the ketone (**40**), of *cis* configuration, was prepared. A comparative study of the ¹H and ¹³C nmr spectral features[23] of the isomeric ketones allowed establishment of the ring fusion configurations.

The perhydroindene (**43**) is of interest as an intermediate in natural product synthesis since the skeleton is found in a variety of triterpenes such as lanosterol, euphol, and the cucurbitanes. An efficient synthesis of **43** from the Wieland–Mieschler ketone (**41**), by reductive cyclization[24] followed by opening of the cyclopropanol ring,[25] of **42** was reported. In order for **43** to be a useful CD synthon in the synthesis of lanostane and euphane triterpenes, it would be necessary to conduct selective annulation reactions at the six-membered ketone and introduce a branched alkyl at the five-membered ketone.

The perhydroindane (**43**) was found to react selectively at the six-membered ring ketone, without protection of the other carbonyl.[26] Unfortunately, the unreactivity of the cyclopentanone was found to be so great that many organometallic reagents, ylides, and enolate salts were ineffective. This was attributed in part to facile enolization of the carbonyl by strong base. Alkylation was finally achieved (**44**) by the action of allylic Grignard reagents.[27] Selective enolate formation was also studied, since the usefulness of the diketone as a synthetic intermediate would be enhanced if selective enolate formation could be achieved. Both carbonyl functions may be trapped as the silyl enol ether derivatives. One equivalent of base generates the six-membered ring enolate. However, condensation of **43** with benzaldehyde yielded the C-8 benzylidene product (**45**) (and dialkylated material) under equilibrating conditions.

Previous attempted syntheses of the cucurbitane skeleton from the lanostane skeleton by migration of the 19-methyl group have failed[28] or resulted in low yields.[29] An entry into the cucurbitane series via rearrangement of the epoxide from 3β-acetoxylanostane-9,11-ene-7-one has been reported ($46 \rightarrow 47$).[30] The

46   47

C-19 methyl migration occurs in the presence of acetic anhydride-boron trifluoride etherate. Prolonged hydrolysis, followed by selective reacetylation of the C-3 hydroxyl and Jones oxidation, yielded the cucurbitone (48).

48

B. Approaches to the Fuisidic Acid Nucleus

The steroidal antibiotic fusidic acid (49)[31] is characterized by the highly strained *trans-syn-trans* perhydrophenanthrene system of the A, B, and C rings. The B ring is fused in a full boat conformation, and isomerization at either of the synthetically accessible C-5 or C-9 positions leads to a more stable structure.

Recently, Ireland et al.[32] described a multistep synthesis of the *trans-syn-trans* diketone 53 from the readily available enedione (50)[33] (Scheme 5). Direct saturation of the enone by either metal-ammonia or catalytic reduction had previously proven to yield only AB *cis*-fused material.[33] The tendency for β-face attack is understandable from models, in which it appears that the α-C-8 methyl group shields the α-side of the C-5 carbon from approach by an external reagent.

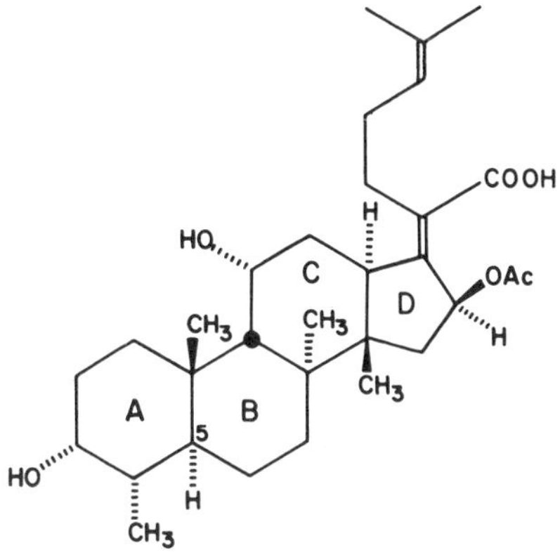

49

Therefore, a more circuitous route was pursued, in which the desired C-5 α-hydrogen was introduced by intramolecular rearrangement (51 → 55).

Since the synthesis of fusidic acid 49 was the ultimate objective,[32] the synthesis of a tetracyclic analog of 50 was pursued. Two synthetic schemes were outlined,

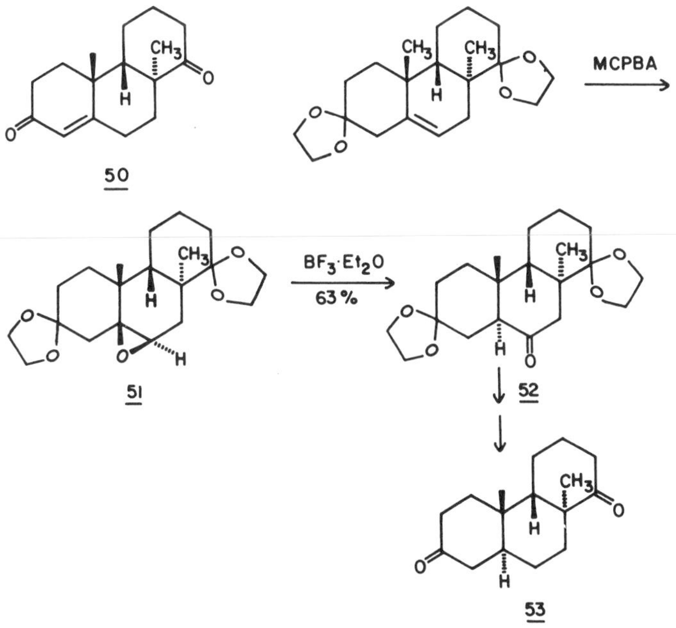

Scheme 5

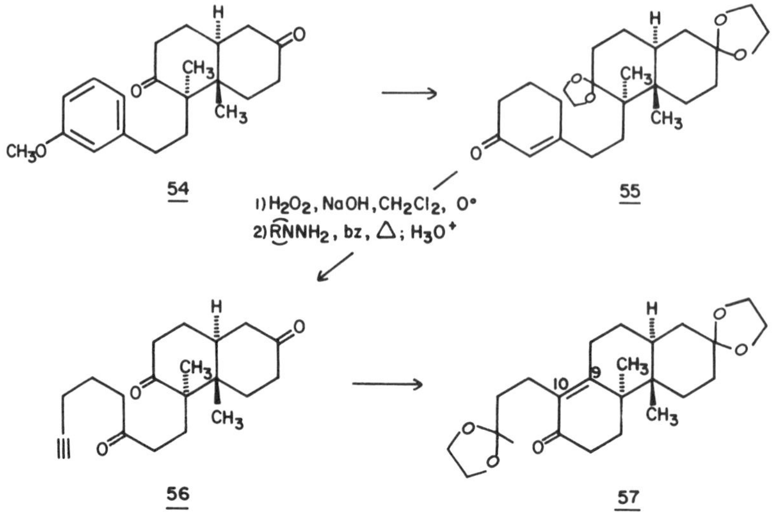

1) H_2O_2, NaOH, CH_2Cl_2, 0°
2) $\widehat{R}NNH_2$, bz, $\triangle$; H_3O^+

Scheme 6

both of which utilize the tricyclic bisketal (50) as the key intermediate. The first approach (Scheme 6), the conversion of the aromatic ring of 54[34] to a saturated chain without oxidative loss of carbon, was undertaken as follows.

Birch reduction of the bisketal of 54 to the enone 55, followed by Eschenmoser cleavage[35] of the derived epoxyketone with aminodiphenylaziridine,[36] yielded the acetylide 56. Ring B was formed by aldol condensation, hydrolysis of the acetylene[37] to the methyl ketone, and selective ketalization to give 57. The extensive manipulation of protecting groups resulted in a somewhat unsatisfactory overall yield of 14%.

The second route (Scheme 7), involving the Diels–Alder condensation between the α-methylene ketone 58[34] and the readily available dienophile 59,[32] afforded 57 in much improved overall yield.

A noteworthy aspect of this reaction is that although the dienophile is best used in fivefold excess over the ketone, unreacted dienophile may be recovered and reused. Therefore, two valuable reagents may be used economically and efficiently in the key reaction of a convergent synthesis.

Cleavage of the dihydropyran ring of 60 proved more difficult than expected, due to two main constraints on the conditions: that protonation of the enolate, resulting in re-enolization to the more stable isomer, not be favored, and that cleavage of the ring be so arranged that the carbomethoxy-bearing carbon of 59 could ultimately become a ketone.

The method chosen involved the recently developed phosphorodiamidate

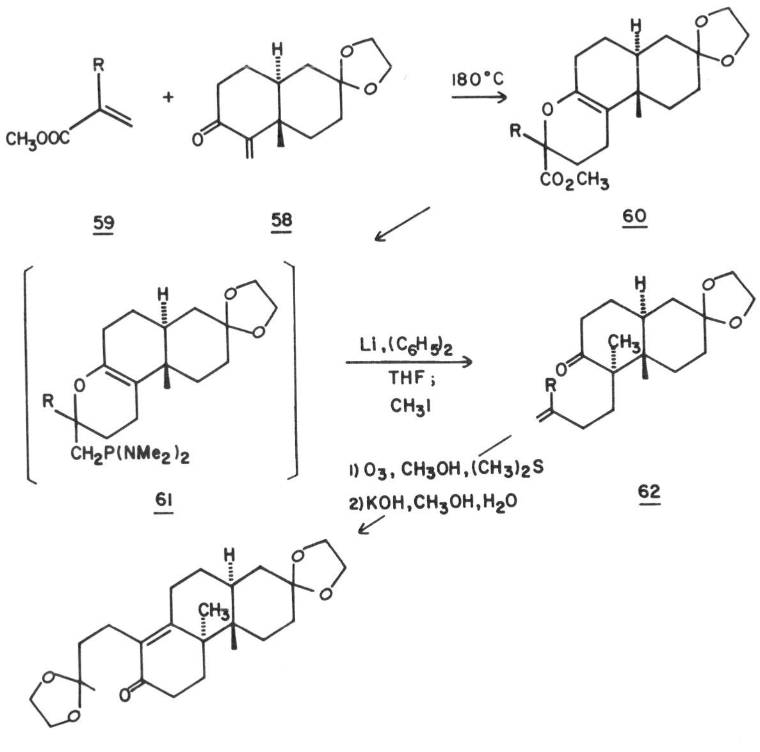

Scheme 7

method,[38] which should undergo reductive alkylation to **62**. However, reduction of the phosphorodiamidate (**61**) proved unsuccessful in ammonia; ethylamine caused overreduction of the methylene; and methylamine, although preventing overreduction, was too basic to allow direct methylation of the generated enolate. Reductive fragmentation and direct methylation were achieved with biphenyl radical anion in tetrahydrofuran (THF) followed by methyl iodide trapping, but small amounts of the *cis*-4,5-dimethyl ketone (5–14%) and the methyl enol ether (12–14%) were also formed.

Ozonolysis followed by internal aldol condensation led to the desired tricyclic enone **57** in 45% yield from **58**,[34] significantly improved over the first approach (Scheme 6).

Formation of the tetracyclic analog of **50** was effected in good yield by reductive methylation followed by aldol condensation (Scheme 8, **57 → 63**).

The tetracyclic enone **63** lacks the C-11 hydroxyl function of fusidic acid,

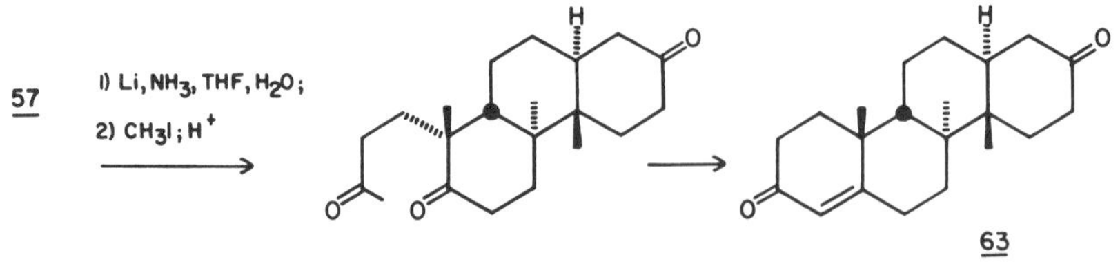

Scheme 8

although it is a suitable intermediate in the synthesis of fusidane or helvolic acid.[39] The acid function at C-11 was introduced by peracid oxidation of the enol acetate[40] derived from **57**, furnishing the hydroxyenone **64** (Scheme 9). Direct lithium-ammonia reduction led only to hydrogenolysis, as expected.[41] Conversion of the alcohol to the lithium salt before reductive alkylation allowed formation of the saturated ketone **65** in 54% yield. However, reductive meth-

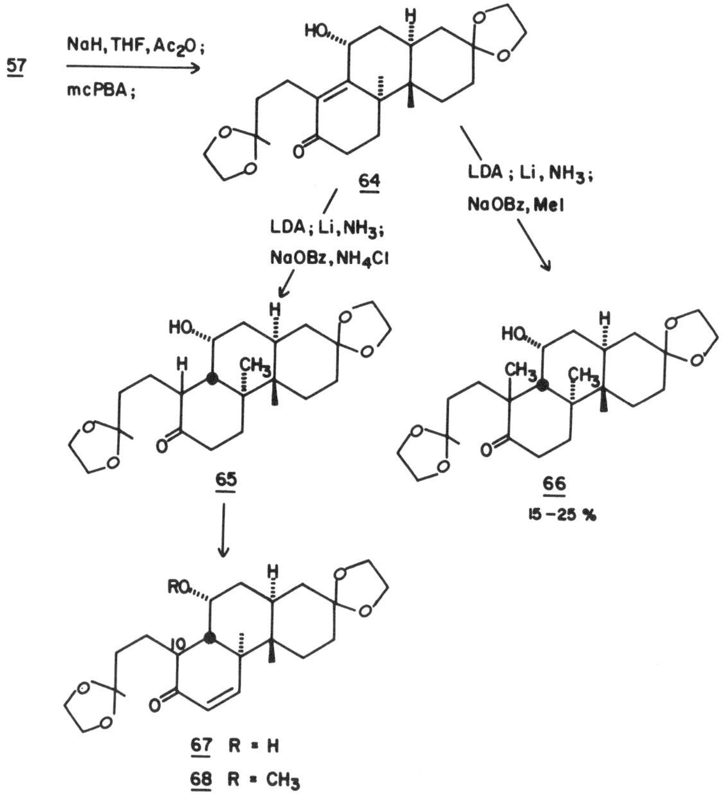

Scheme 9

ylation by a similar process led to a crude product mixture with at least seven components, the desired ketone (**66**) being formed in only 15–25% yield. A final attempt was made to effect C-10 methylation through the enone **67**, available from **65** by the procedures of Sharpless and Reich.[42] Several methylation procedures yielded only the O-methyl ether (**68**). These difficulties attest to the problems attending the formation of the *trans-syn-anti-trans* backbone of these systems. The C-11-deoxygenated enone **63** was used in studies on the tetracyclic degradation products of fusidic acid.[43]

4. TRITERPENES WITH NONSTEROIDAL SKELETONS

A. Tetracyclic

In their continuing work on the synthesis of pentacyclic triterpenes, Ireland's group proposed the tetracyclic ketone **69** as a key intermediate. However, such a structure could also be elaborated to the tetracyclic triterpene, (±)-shionone (**70**).

69   70

Three approaches to the synthesis of **69** have been described.[44] The first approach entailed the stereoselective introduction of the C-8a angular methyl group. The required starting material was the methoxy enone **71**, prepared in 24% overall yield by a five-step sequence from 2-methyl-1,3-cyclohexanedione and 1,4-dimethoxy-2-butenone,[45] via annelation[46] and reductive removal of the allylic oxygen.[47]

Selective reduction of **71** with lithium 9b-boraperhydrophenalyl hydride[48] afforded a 70:30 mixture of α-(axial) **72** and β-(equatorial) alcohols. Due to the lability of the α-alcohol to ketal formation (**73**), the crude mixture was transformed directly[49] into the methoxycyclopropane (**74**). Protolysis[46] yielded the ketoalcohol (**75**), which was transformed into the decalone **76** by Wolff–Kishner reduction of the ketone, followed by oxidation of the alcohol. The stereochemistry at the ring junction was unambiguously established by X-ray single crystal structural analysis.

Having established a method for the preparation of a *trans*-fused, diangularly methylated 1-decalone, attention was turned to the desired tetracyclic ketone (**69**). An initial plan to transform the ketone **77** to the α-methylene ketone via photooxygenation of the ketal was unsuccessful, the intermediate allylic alcohol (**78**) being obtained in only 32% yield.

Attempted synthesis of the ketone **80** from **75** also met with the insurmountable problems of steric congestion. The overall yield of **79** from **75**, after alkylation and dehydration, was only 38%, and the desired secondary alcohol was obtained in 38% yield by hydroboration. Oxidation to **80** proceeded in excellent yield, but the ketone proved unreactive to a variety of organometallic nucleophiles. Steric hindrance from the two angular methyl groups and the C-1 ketal precluded further investigation of this route.

75

1) $(CH_2OH)_2$, H^+
2) H_2CrO_4
3) m $CH_3O-C_6H_4C\equiv CLi$, 10% Pd-C, H_2
4) Py, $SOCl_2$, 0°

79
38%

1) BH_3·THF OH^-, H_2O_2
2) H_2CrO_4

80

Attention was turned to an approach (Scheme 10) that avoided tetrahedral substitution at positions 1 and 8a, utilizing the stable methoxycyclopropane moiety. Alcohol **74**, after Jones oxidation, reductive alkylation with lithium *m*-methoxyphenylacetylide, and catalytic hydrogenation yielded the α-alcohol **81** in 44% yield, together with a 46% yield of the epimeric β-alcohol. Unfortunately, only the α-alcohol could be dehydrated to the desired endocyclic olefin (**82**), the β-alcohol forming only the exocyclic olefin, in poor yield. The remainder of the transformations were carried out on olefin **82** to give the alcohol **83**. As expected from the model studies,[34] protolysis followed by acid catalyzed cyclization yielded the desired BC *trans* isomer, providing the first sample of the tetracyclic ketone **69**. The overall yield was 3.3% in 15 steps.

The second and ultimately most practical sequence involved the introduction of the C-4a angular methyl group via the Nagata hydrocyanation reaction[50] on the enone (**85**) (Scheme 11). Robinson annelation between the enone (**84**) and 2-methylcyclohexane-1,3-dione, followed by selective reduction, led to the hydroxyenone (**85**). Both preformed diethylaluminium cyanide,[51] which should exercise thermodynamic control to give the *cis* product, and diethylaluminum-hydrogen cyanide,[51] which should exercise kinetic control, yielded the same *trans*-fused product (**86**). It was proposed that the lack of selectivity was due to

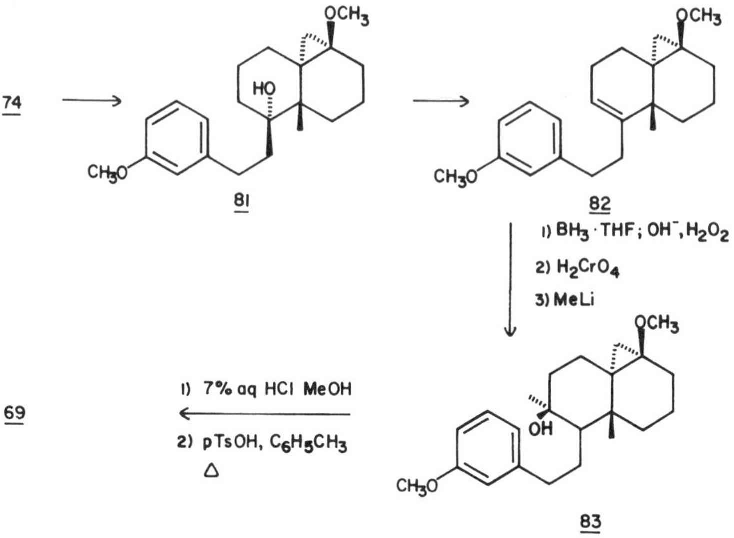

Scheme 10

in situ protonation by the hydroxyl group. In support of this possibility, it was found that the acetate or dimethyl silyl ether derivative yields a mixture of products on hydrocyanation, but the *cis*-cyano ketone has not been isolated.

The tetracyclic ketone **69** was ultimately obtained in 20.4% yield in 10 steps from **84** via **87**. Single-crystal X-ray structural analysis established the structure of **69**.

Johnson, Corey, and van Tamelen have successfully utilized the biomimetic acid-catalyzed cyclization of polyolefinic substrates for the construction of steroids and some terpenes.[52] Based on the previous work, Ireland and co-workers designed a route to the desired tetracycle **69** utilizing polyene cyclization (Scheme 12), which, although preparatively uncompetitive with the hydrocyanation (Scheme 11), was of some interest. Based on a model study, the symmetric dibromide **88** was chosen as starting material to form the crucial intermediate aldehyde **95**.

Attempted coupling with lithio-1-trimethylsilyl propyne[53] to give the enediyne **89** was unsuccessful, but reaction with propargylmagnesium bromide followed by trimethylsilylation of the crude acetylene product produced **89** in good yield. Cleavage of the trimethylsilyl groups[54] to **90**, followed by monoreduction of one acetylenic function, yielded the dienyne **91**. The remaining acetylenic function was utilized for stereoselective synthesis of the trisubstituted double bond,[55] and the terminal double bond served to mask the aldehyde as illustrated (**92, 93**, and **94**).

The cationic cyclization of the aldehyde **95** was investigated under several sets of conditions suggested by the work of Johnson's group,[52] but all of the

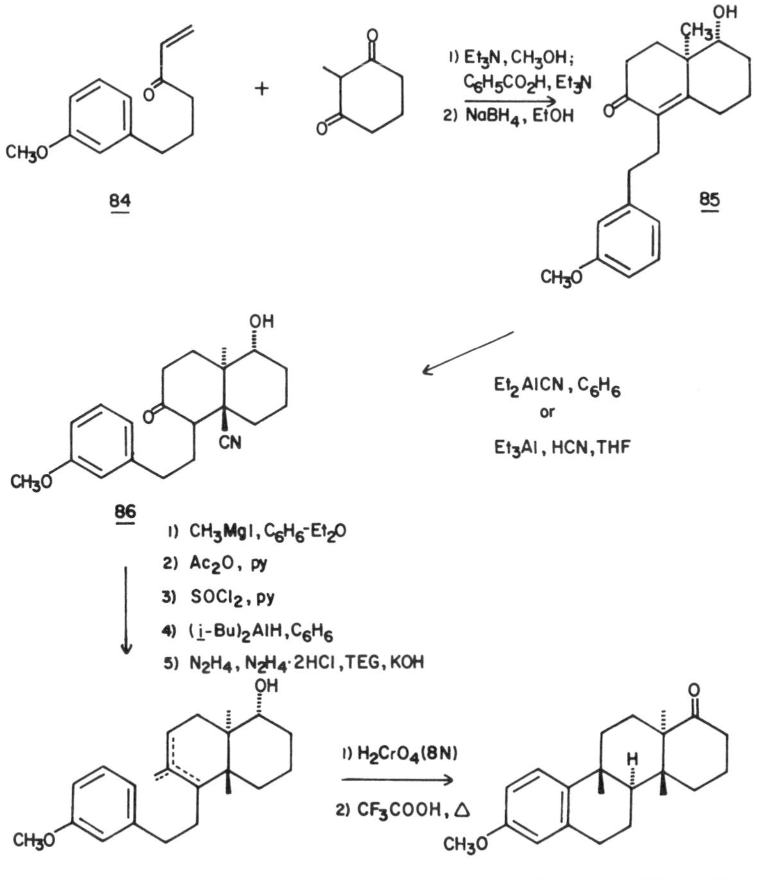

Scheme 11

reactions yielded polymeric material (42%) and a product mixture of four or more compounds (58%). The desired tetracyclic ketone **69** and the octalinone **96** represented 23% and 44% of the nonpolymeric reaction mixture respectively. Strong protic acid catalyzed cyclization of **96** completes formation of the desired tetracycle, but even after attempted optimization of formation of **96**, the ketone **69** was obtained in only 29% yield.

Although the overall yield was modest, an interesting point was elucidated concerning this type of reaction. The polyene **95**, although deviating substantially from the pattern of squalene-like disposition of methyl substituents previously studied[52] and in no way simulating a natural intermediate, still cyclizes in significant amounts to a tetracyclic system.

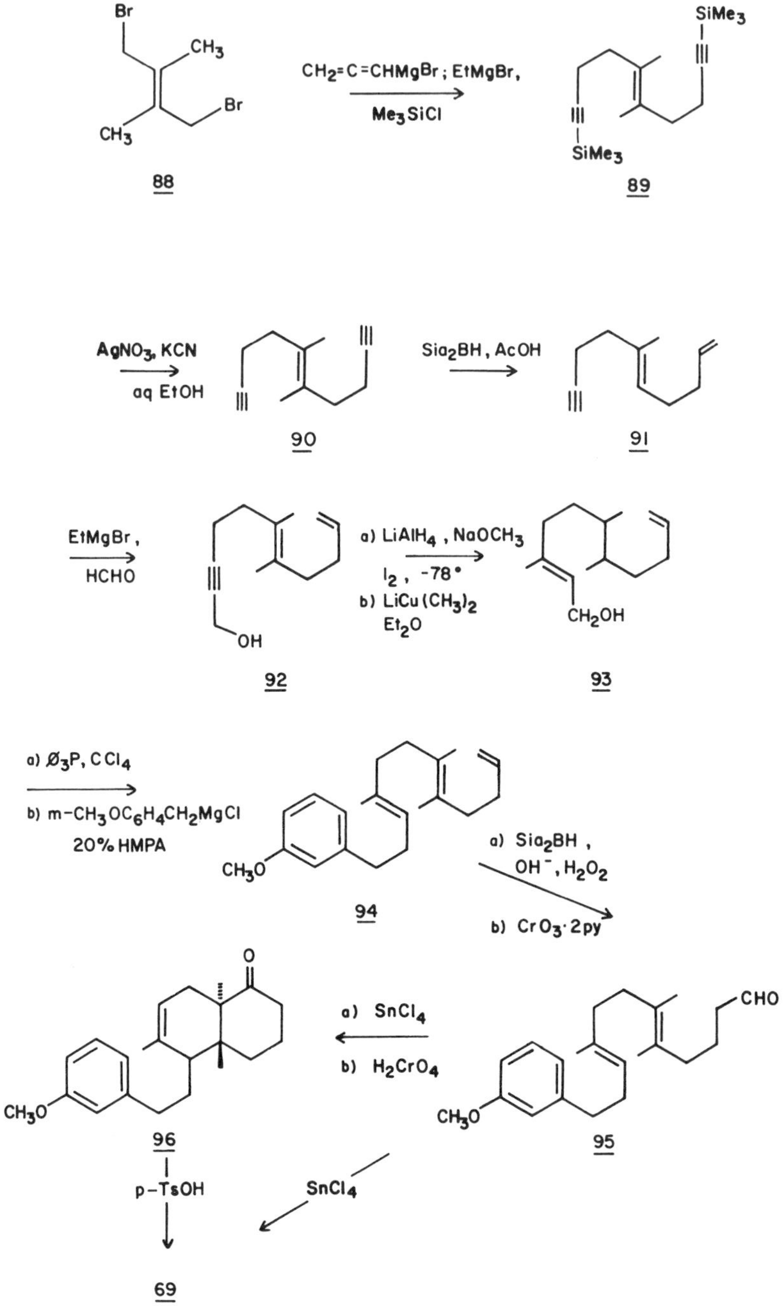

Scheme 12

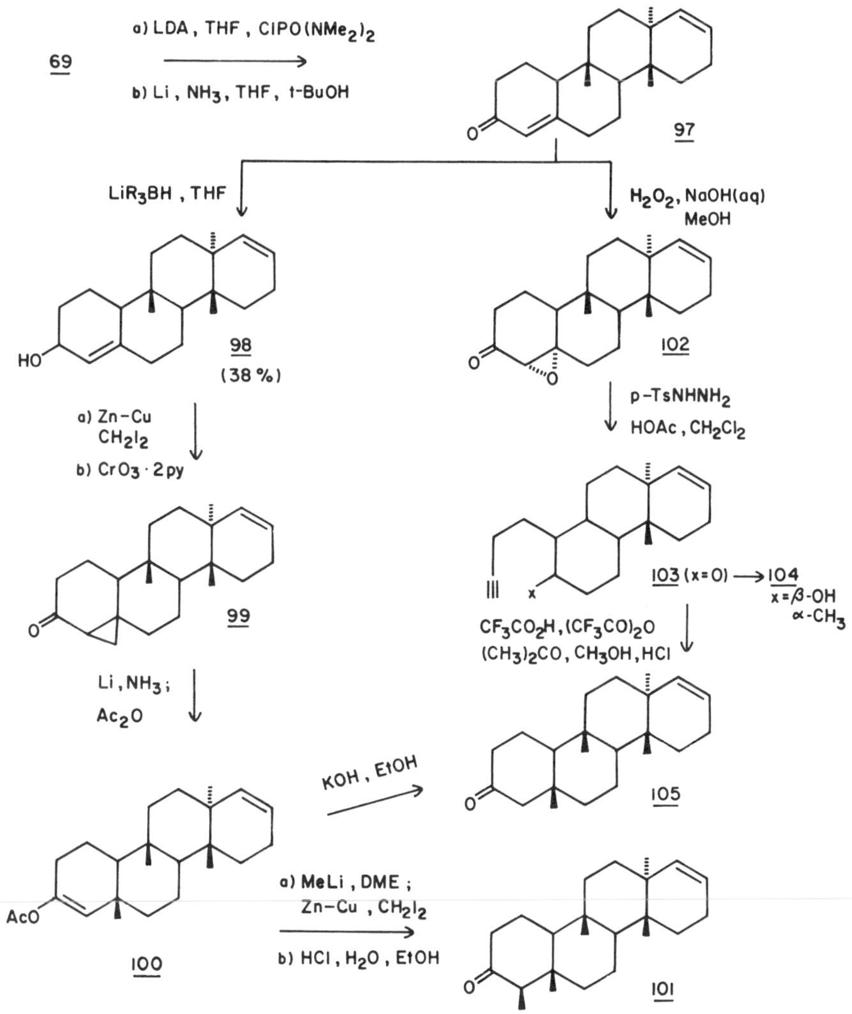

Scheme 13

In order to convert the ketone **69** to shionone **70**,[56] it was necessary to introduce the side chain in ring D and modify the aromatic A ring.

The ring A modification was undertaken on the enone **97**, obtained in two steps from **69** (Scheme 13). The first route studied, shown on the left of the scheme, suffered from poor yields in formation of the allylic alcohol **98**. Therefore, an alternative sequence was investigated in which the stereochemistry of carbon–carbon bond formation at C-12a was controlled by the C-4bβ methyl group.

The sequence developed by Eschenmoser et al.[35,57] provided a means for cleavage of the A ring (**97** → **103**) without loss of carbon. Addition of meth-

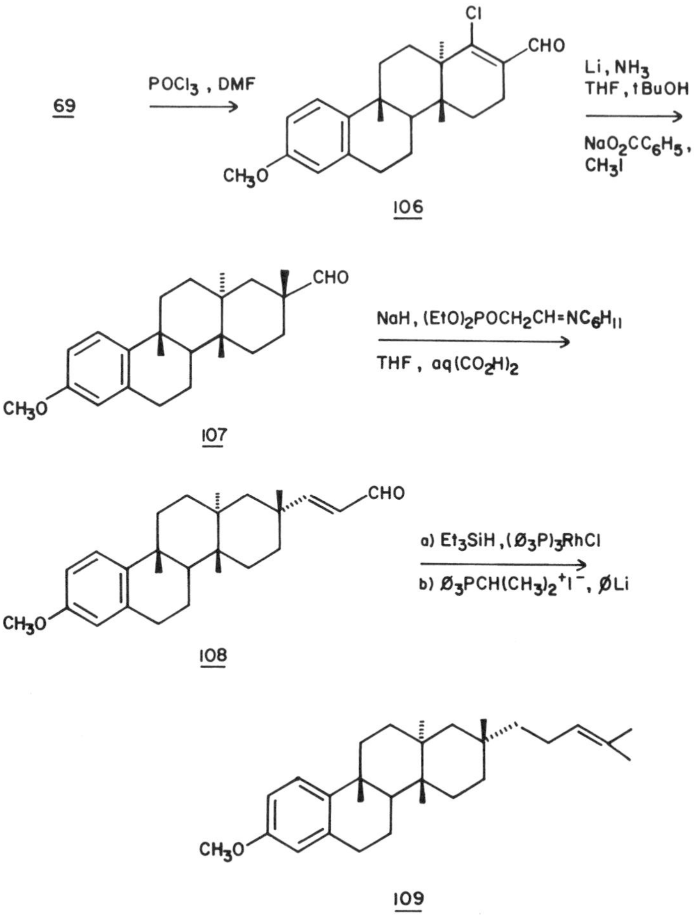

Scheme 14

yllithium to **103** incorporated the potential C-12a angular methyl group. The stereochemical control provided by the axial C-4bβ methyl group during similar cationic ring closures was known from previous experience.[34] Conditions for the reaction were provided by the work of Peterson and Kamat[58] and of the groups headed by Johnson and Lansbury.[59] Indeed, cyclization of the acetylenic alcohol **104** in trifluoroacetic acid, followed by hydrolysis of the enoltrifluoroacetate formed, yielded the ketone **105**, which was identical to the hydrolysis product of **100**.

A method for introduction of the ring D side chain was then investigated, starting with the tetracycle **69**. Although the ketone functionality in the D ring could serve to introduce the alkyl group in the α-position, the stereochemical outcome and efficiency of this alkylation was doubtful and would require subsequent removal of a very hindered ketone. It was decided to remove the D ring

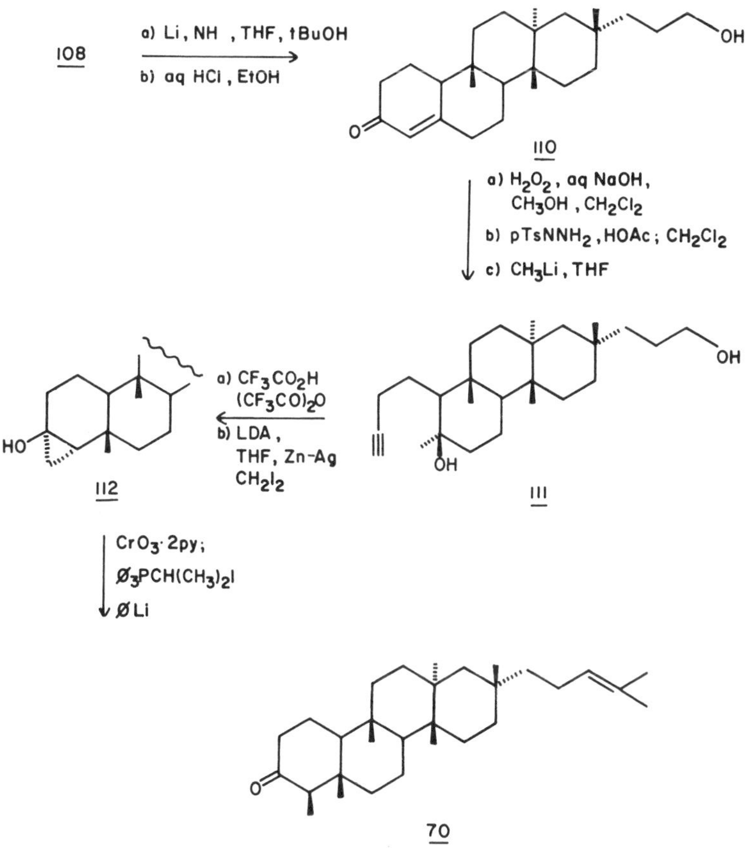

Scheme 15

ketone, at the same time introducing an activating functionality external to the ring.

Vilsmeier reaction[60] to the chloroaldehyde (106), followed by reductive methylation, yielded the aldehyde 107 (Scheme 14) with the precedented stereochemical outcome.[61] Attempted halogenation of the derived neopentyl-type alcohol was unsuccessful under a variety of conditions, so an alternative scheme involving two successive Wittig-type reactions was developed. The aldehyde 107 was converted to the unsaturated aldehyde 108 by the method of Nagata and Hayase.[62] Reduction with triethylsilane and tris (triphenylphosphine) rhodium chloride[63] was followed by condensation with isopropylidene phosphorane to give the desired product 109.

When the previously developed route to convert 109 to shionone was undertaken, however, it was found that the conditions for reformation of the A ring invariably led to hydration of the trisubstituted side chain double bond. The problem was readily circumvented by a small rearrangement of the final sequence of steps. The unsaturated aldehyde **108** was completely reduced to **110**, the usual reaction sequence carried out on ring A, and the alcohol oxidized back to the aldehyde previous to incorporation of the isopropylidene moiety (Scheme 15) completing the shionone synthesis (**111, 112 → 70**).

B. Stereocontrolled Side Chain Synthesis

Trost and co-workers[64] have directed some attention to the stereocontrolled creation of the acyclic C-7 side chain found in a variety of natural products, including the tetracyclic triterpenes such as lanosterol, cycloartenol, and euphol as well as steroids and vitamin D metabolites. The selective control of the stereochemical relationship between the D ring and the carbon bearing methyl is required.

The target compound **113** was envisioned as an intermediate in the total synthesis of the above-mentioned types of compounds. The authors based their approach on the opening of a cyclopropane ring with an organocopper reagent, as outlined in Scheme 16.

113

7-Methyloct-Z-2-en-1-ol, available as previously described,[65] was converted to the bromide with phosphorus tribromide and calcium hydride. Alkylation with methylacetoacetonate dianion gave the unsaturated keto ester **114**. Diazo transfer from tosylazide was followed by direct intramolecular cyclization of the crude product (**115**) in the presence of copper-bronze powder, to give the desired cyclopropane (**116**) which on treatment with lithium dimethylcuprate resulted in formation of the synthon **113**, the stereochemistry at C-7 relative to the ring junction being determined by the *cis* stereochemistry of the original olefin and the inversion of configuration in the cuprate ring opening. The E olefin should give the opposite stereochemistry (**116a → 113a**).

C. Nonsteroidal Polycyclic Triterpenes

(a) dl-Alnusenone

The polyene cyclization used in Ireland's investigation of the total synthesis of shionone (Scheme 13)[44] also has potential in the synthesis of pentacyclic triter-

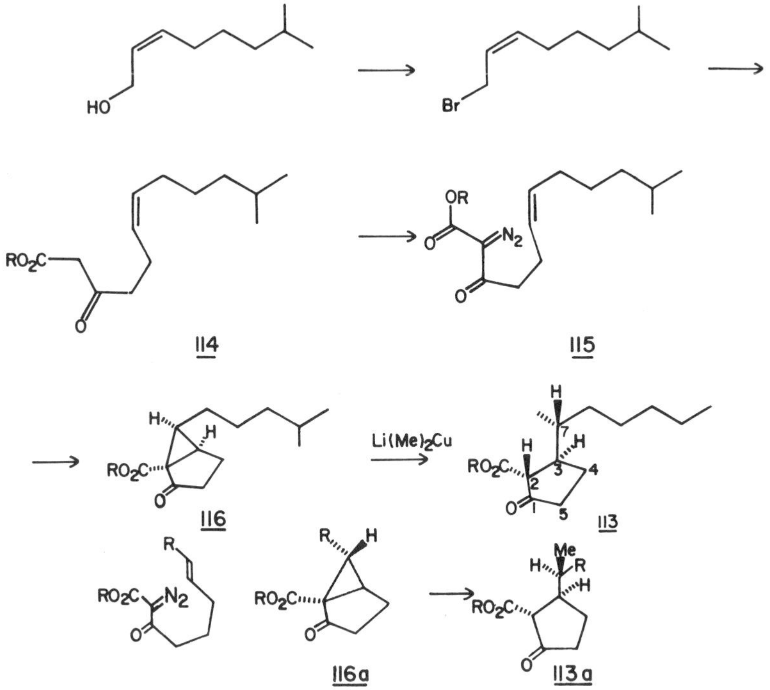

Scheme 16

penes such as alnusenone **117**. In order to pursue the proposed route[66] it was desirable to develop systems that could be synthesized from aldehyde **95**,[44] with the modification of incorporating a less labile but still effective source of cationic character in the aldehyde position to initiate cyclization. Finally, a system that would result directly in formation of a nonaromatic E ring was sought to avoid the problems of selective reduction of two dissimilar A and E rings associated with the previously reported conjugate hydrocyanation route.[67]

From the work of Johnson and co-workers[68] it was proposed that a cyclic

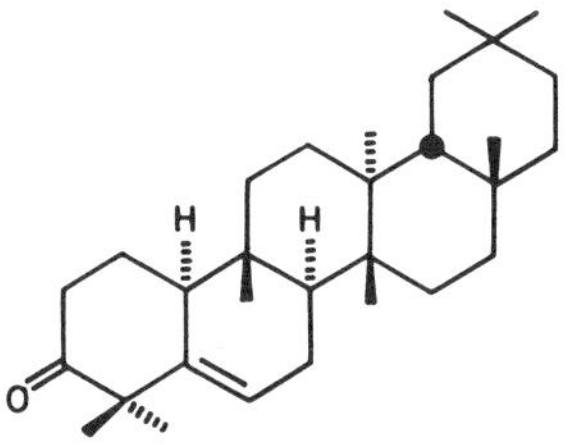

117

(±)-alnusenone

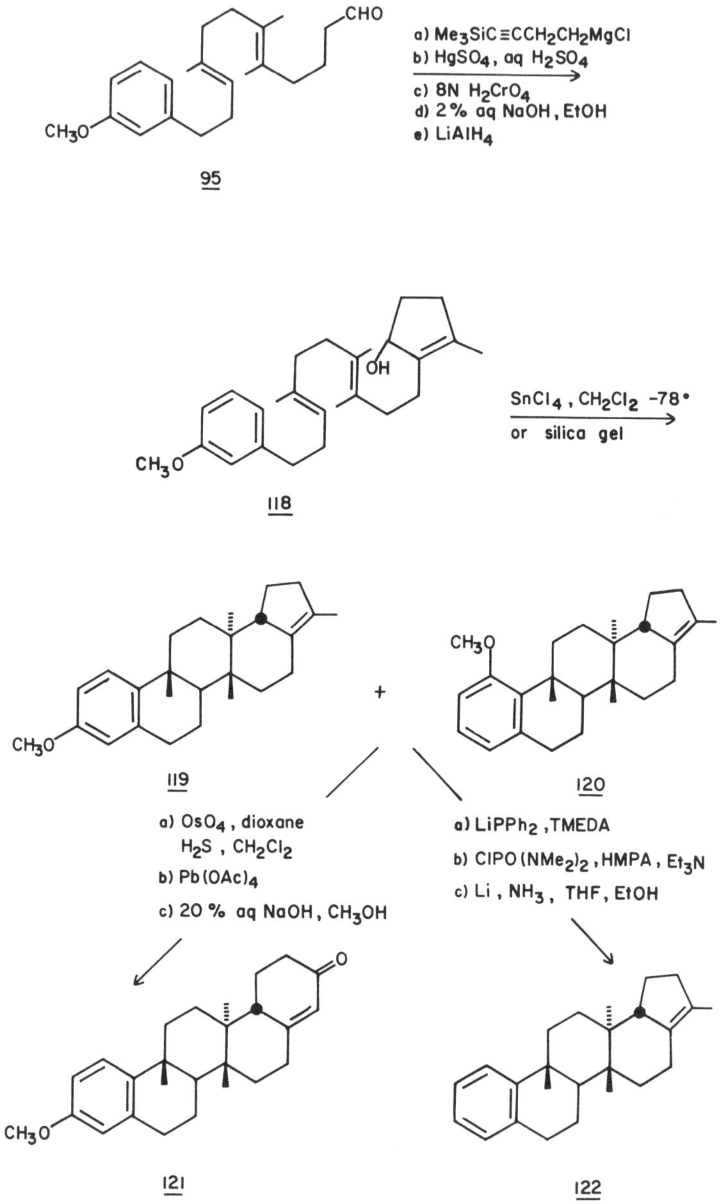

a) Me₃SiC≡CCH₂CH₂MgCl
b) HgSO₄, aq H₂SO₄
c) 8N H₂CrO₄
d) 2% aq NaOH, EtOH
e) LiAlH₄

95

SnCl₄, CH₂Cl₂ −78°
or silica gel

118

119 + 120

a) OsO₄, dioxane
   H₂S, CH₂Cl₂
b) Pb(OAc)₄
c) 20% aq NaOH, CH₃OH

a) LiPPh₂, TMEDA
b) ClPO(NMe₂)₂, HMPA, Et₃N
c) Li, NH₃, THF, EtOH

121

122

Scheme 17

110

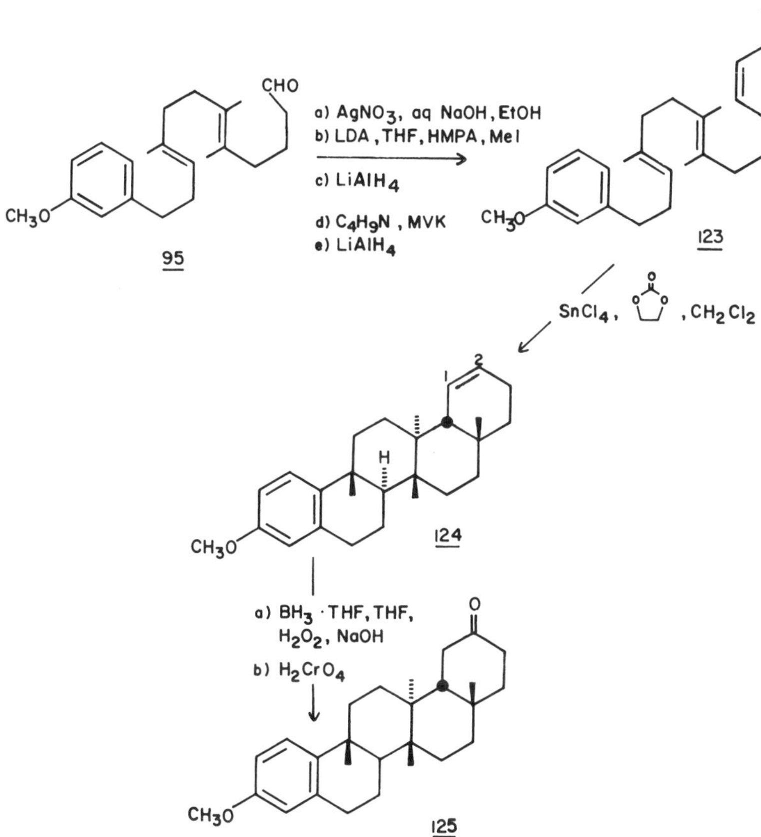

Scheme 18

allylic alcohol (specifically the 3-methyl-2-cyclopentenol unit) would be particularly suited.

After development of the procedure for synthesis and cyclization of the 3-methyl-2-cyclopentenol in a model series, the aldehyde 95[44] was converted to the required cyclopentenol, 118 (Scheme 17).

4-Trimethylsilylhomopropargylmagnesium chloride was used as a masked 2-butanone synthon, on reaction with 95. Hydration with concomitant desilylation, followed by oxidation of the resultant hydroxy ketone and base cyclization yielded the cyclopentenol (118) after reduction. Despite success on the model system, cyclization of the cyclopentenol (118) proceeded only in low yield, to give the olefins 119 and 120. Oxidative ring enlargement[68a] of the major isomer 119 formed the target enone 121, completing the alternate formal total synthesis of alnusenone.[67] Both 119 and 120 provided 122 on demethoxylation.

At the inception of the above-described synthetic scheme for the synthesis of alnusenone, two possible cyclic allylic alcohols were considered for initiation of the cyclization step. As well as the cyclopentenol 118, the 4-methyl-2-cyclohexenol system **123** was considered to have potential. Cyclization of this molecule would give a pentacyclic intermediate with a six-membered E ring, and the C-17 methyl group at a *cis* DE ring fusion. This route would avoid both the selective reduction of the AE-diaromatic intermediate used originally[67] and the necessity for subsequent incorporation of the C-17 methyl group.[66,67]

Therefore, the cyclohexenol **123** was synthesized[69] and subjected to acid catalysed cyclization (Scheme 18).

A complex mixture resulted from cyclization of **123** with stannic chloride in dichloromethane-ethylene carbonate.[68a] The desired pentacyclic olefin **124** could be isolated in only 12% yield, and several alternate reaction conditions were no improvement.

The only remaining step to link this scheme with the previously successful synthesis of alunesenone was the introduction of a gem-dimethyl group at C-2 of **124**. However, while hydroboration[70] and oxidation yielded the ketone **125** readily enough, steric congestion around C-2 resulting from the *cis* DE ring fusion and the proximity of the C-14aα methyl group prevented several attempts to affect further transformation, and the route halted at this step.

(b) Friedelin

In 1976, Ireland and Walba reported the first total synthesis of (±)-friedelin (**126**).[71] The synthetic route was based on results obtained from previous successful syntheses of (±)-shionone[56] and (±)-alnusenone.[67,66]

126

Some modifications of the alnusenone synthesis[67] were introduced, in order to avoid problems encountered in the earlier route. The aromatic ring elaborations were reversed (i.e., A before E rather than E before A) since the overall yield observed for modification of the E ring was only 14% in the original work.[67] Furthermore, cationic cyclization was envisaged for modification of the aromatic ring (e.g., **69** → **103** → **105** and **108** → **112**). If this reaction had been performed on a precursor with a modified E ring, in which the severe steric strain in the CDE ring system had already been introduced, backbone rearrangement[72] might compete with cyclization.

The required diether **127**, in which the positions of the ethoxy and methoxy groups were interchanged relative to those in the (±)-alnusenone precursor, was synthesized by the previously reported procedures[67] (Scheme 19).

Selective reduction to **128** was achieved in much improved yield (56%). The A ring modification to **129** was then accomplished by Eschenmoser cleavage[59] of the derived epoxide, addition of methyllithium, and trifluoracetic acid mediated cyclization to **130**.

The synthesis of (±)-friedelin (**126**) was completed by modification of the A ring (**131**) and of the E ring (**132, 133,** and **134**), following the previously reported procedure for the E ring of (±)-alnusenone.[67] Friedelin was obtained in 0.3% overall yield, after 31 steps.

Kametani and co-workers[73] reported formal total syntheses of both (±)-alnusenone and (±)-friedelin via the stereoselective synthesis of the two diethers **135** and **127**, both intermediates in previously reported total syntheses of (±) alnusenone[66,67] and (±)-friedelin,[71] respectively.

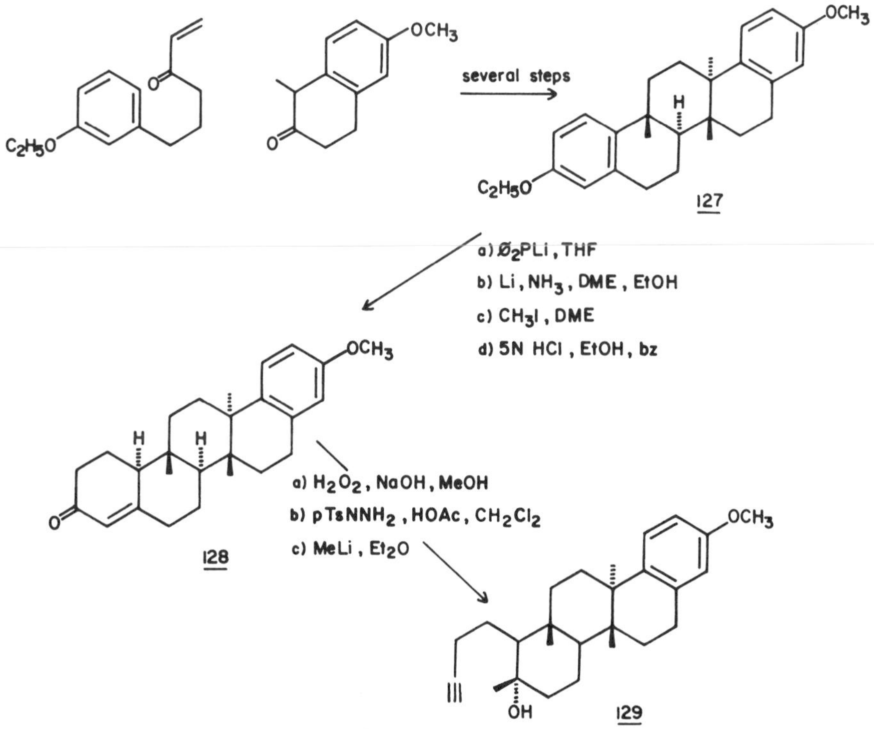

Scheme 19

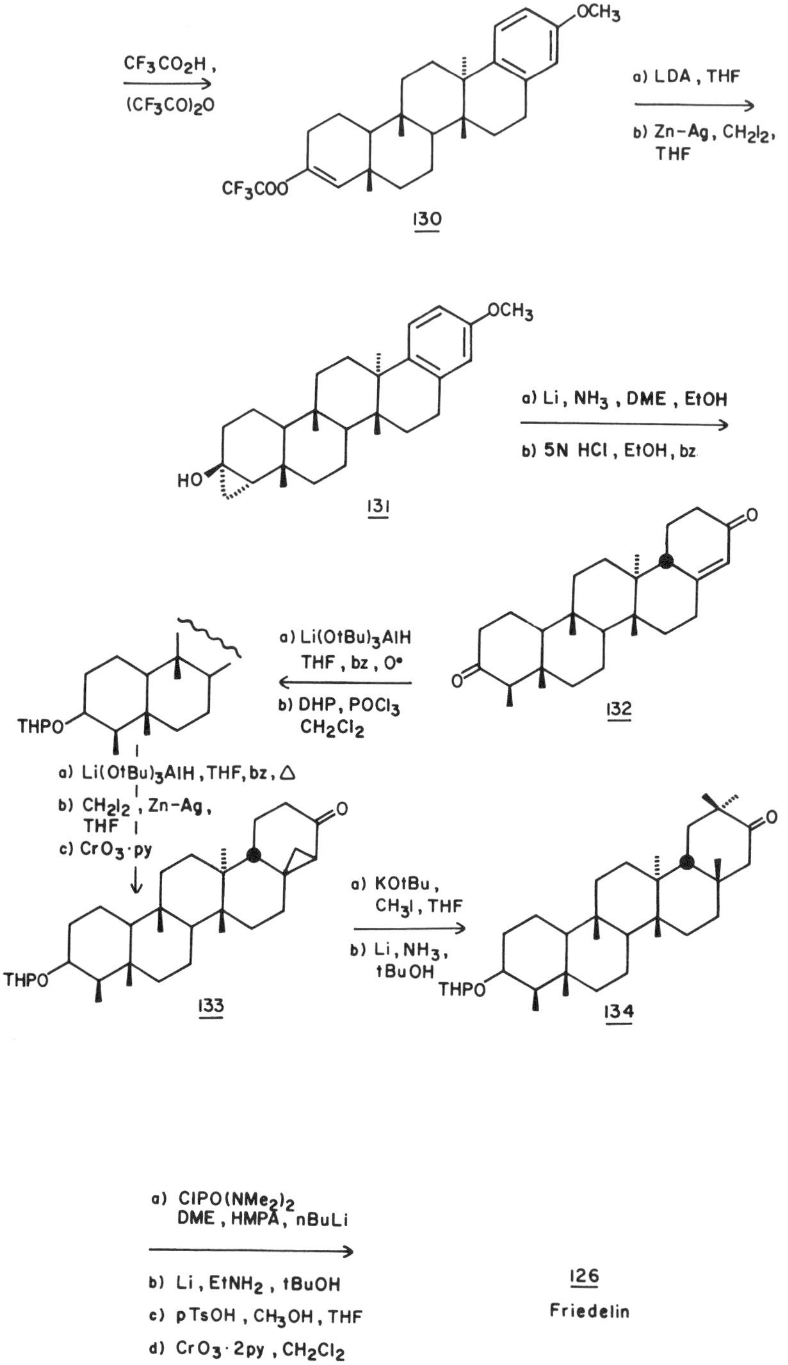

$$\xrightarrow[\text{(CF}_3\text{CO)}_2\text{O}]{\text{CF}_3\text{CO}_2\text{H},}$$

130

$$\xrightarrow[\substack{\text{b) Zn–Ag, CH}_2\text{I}_2,\\ \text{THF}}]{\text{a) LDA , THF}}$$

131

$$\xrightarrow[\text{b) 5N HCl , EtOH, bz}]{\text{a) Li, NH}_3\text{ , DME , EtOH}}$$

132

$$\xleftarrow[\substack{\text{b) DHP, POCl}_3\\ \text{CH}_2\text{Cl}_2}]{\substack{\text{a) Li(OtBu)}_3\text{AlH}\\ \text{THF , bz , O}^\circ}}$$

a) Li(OtBu)$_3$AlH, THF, bz, Δ

b) CH$_2$I$_2$, Zn–Ag, THF

c) CrO$_3$·py

133

$$\xrightarrow[\substack{\text{b) Li, NH}_3,\\ \text{tBuOH}}]{\substack{\text{a) KOtBu,}\\ \text{CH}_3\text{I, THF}}}$$

134

a) ClPO(NMe$_2$)$_2$
   DME , HMPA , nBuLi

$$\xrightarrow{}$$

b) Li , EtNH$_2$, tBuOH

c) pTsOH , CH$_3$OH , THF

d) CrO$_3$·2py , CH$_2$Cl$_2$

126

Friedelin

114

The retrosynthetic logic was that since the BCD ring portion in **136** corresponds to an isoprene unit, the pentacyclic ring system could be formed in one step by intermolecular double cycloaddition between the bis-*o*-quinodimethane (**137**) and isoprene. The second approach involved the intramolecular cycloaddition of the *o*-quinodimethane **138**. The authors had previously found that intramolecular cycloaddition of *o*-quinodimethanes[74] derived from benzocyclobutenes[75] proceeds with stereo- and regioselectivity. The reaction was used to advantage in the synthesis of the tetracycle D-homoestrone[76] and an intermediate in the syntheses of atisine, veatchine, garryine, and gibberellin A_{15}.[77]

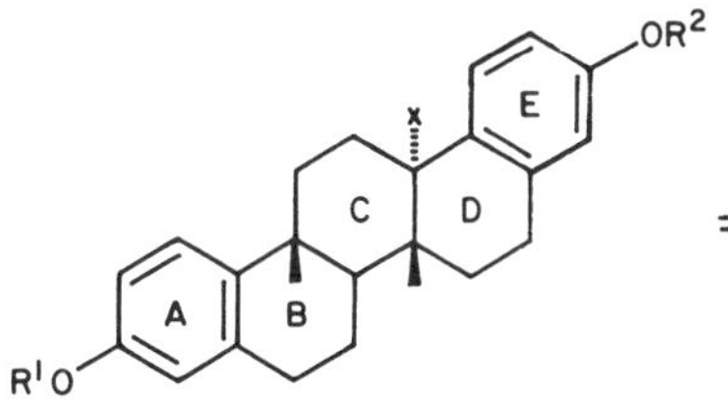

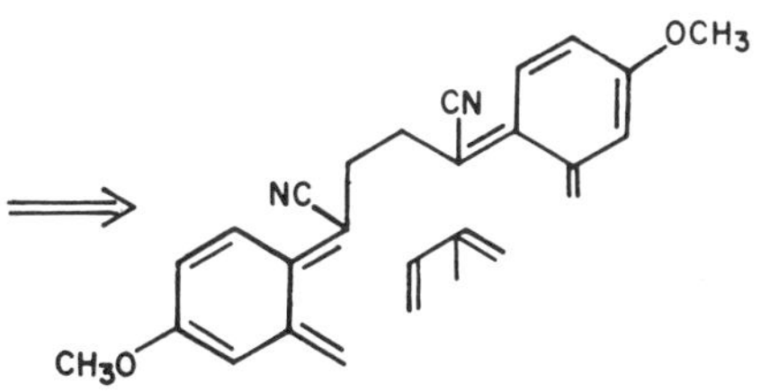

136 x = CN
135 R^1= x = Me , R^2 = Et
127 R^1= Et , R^2 = x = Me

137

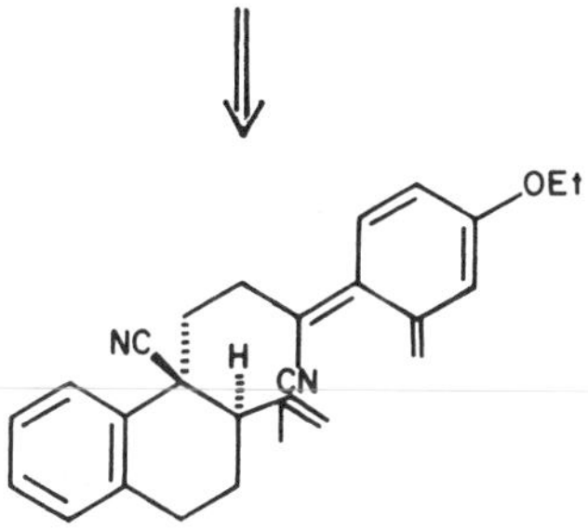

138

The first approach, a one-step intermolecular double cycloaddition (**137** → **136**), required synthesis of the 1,2-di(benzocyclobutenyl) ethane **142**, which proceeded in a straightforward manner (Scheme 20).[78,79] Thus, Knoevenagel reaction of 2-bromo-2-ethoxybenzaldehyde with cyanoacetic acid in the presence of pyridine and ammonium acetate yielded the α-cyanocinnamic acid. Decarboxylation in *N,N*-dimethyl acetamide was followed by cyclization of the resulting nitrile in LDA to yield the benzocyclobutene **139**. The cyanocyclobutene **139** was condensed with the tetrahydropyranyl ether of bromoethanol, followed by hydrolysis to the alcohol **140b** and conversion to the iodide **141** via the tosylate **140c**. A second condensation with **139** yielded the desired 1,2-di(cyano-5-methoxyben-

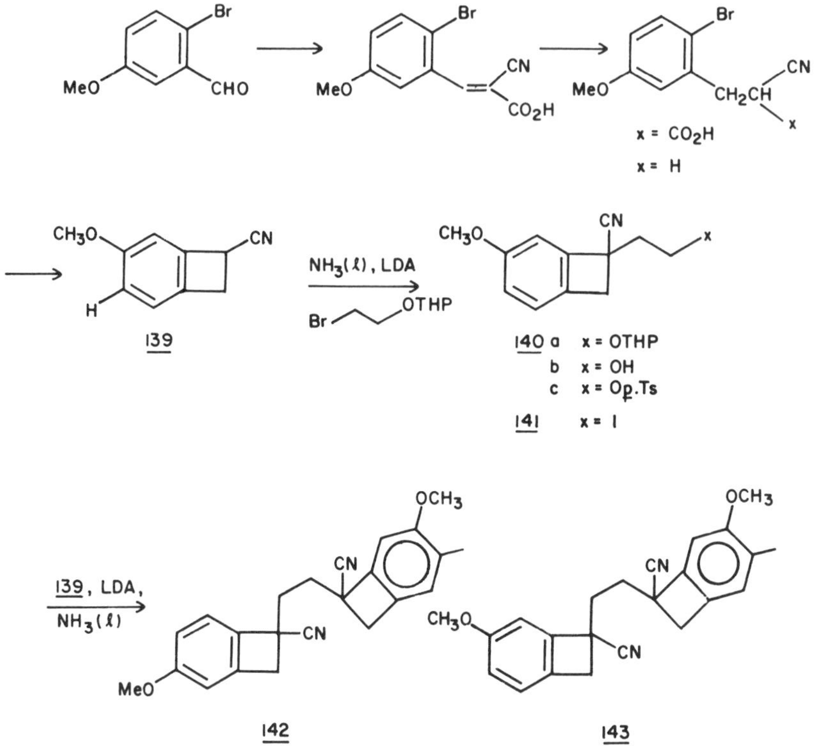

Scheme 20

zocyclobutenyl) ethane (**142**). Similarly, **143** was synthesized from the appropriate 1-cyano-4-methoxybenzocyano-4-methoxybenzocyclobutene.

Attempted conversion of **142** and **143** into the desired pentacyclic compounds by heating in a sealed tube with isoprene yielded not the desired material from

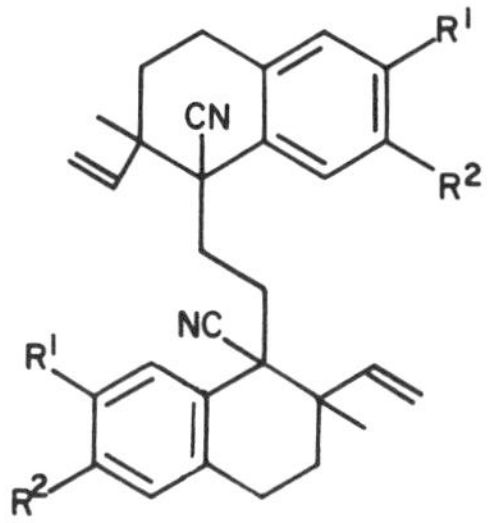

144 a $R^1 = OMe$, $R^2 = H$
b $R^1 = H$, $R^2 = OMe$

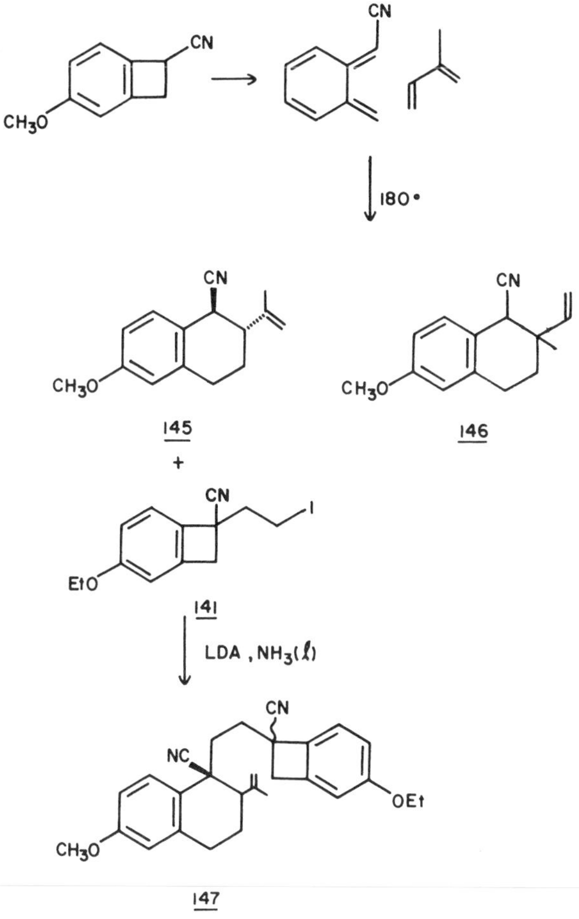

Scheme 21

intermolecular cycloaddition of equimolar isoprene but the tetralin derivatives **144** formed by cycloaddition of two moles of isoprene.

The second approach, formation of the A and B rings by cycloaddition of benzocyclobutene to isoprene, introduction of a second benzocyclobutene moiety forming the E ring, and intramolecular cycloaddition to yield the pentacycle, proved to be more effective.

The benzocyclobutene derivative **141** was prepared as previously described.[78,79] The AB ring system was synthesized from 1-cyano-4-methoxybenzocyclobutene by heating with an excess of isoprene (180°, 2 hours), yielding a mixture of isopropenyl and vinyl methyl tetralins **145** and **146**, respectively (Scheme 21). The desired material **145** was obtained by chromatography in 43%

yield. Condensation with **141** in LDA-liquid ammonia yielded the key inter-mediate **147** in 96% yield. Alkylation proceeded from the least-hindered side, so that the required *cis* configuration between the 1-cyano and 2-isopropenyl was obtained.

Heating **147** in dry toluene at 210°C for 3 hours[77] provided the pentacyclic dinitrile **136** stereoselective in 58% yield. Conrotatory ring opening of cyclobutene forms the sterically favored *o*-quinodimethane (**148**). Intramolecular cy-

cloaddition proceeds through the *exo*-chair conformation, giving **136** with the required stereochemical arrangement at the BC and CD ring junctions.

Reduction with diisobutylaluminum hydride to the diimine, followed by further Wolff–Kishner reduction with hydrazine hydrate,[80] yielded the intermediate **135**, identical to that previously used by Ireland and co-workers in the total synthesis of (±)-alnusenone.[67]

Similarly, the key intermediate **149** was synthesized by modification of the above-described route from the appropriate precursors. Intramolecular cycloaddition by thermolysis of **149** gave **150** in 60% yield. The pentacycle was readily transformed to the desired intermediate **127**, intersecting Kametani's route with Ireland's for the synthesis of (±)-friedelin[71] (Scheme 19).

149   150

127

The vinyl tetralins **146** obtained from the intermolecular Diels–Alder reaction with isoprene were incorporated into a route for nonstereoselective synthesis of the desired pentacycles. Alkylation of **146** with **141** under the usual conditions gave the 1,1,2,2,-tetrasubstituted tetralin **151** as a stereoisomeric mixture. Thermolysis of the mixture yielded the pentacycle (**136** or **150**) and its stereoisomer (**151**) in a ratio of 1:2. The isomer was tentatively assigned the C_{6b}-methyl stereochemistry shown.

Heathcock and co-workers[81] recently initiated a convergent synthetic approach to pentacyclic triterpenes, in which preformed AB and DE synthons would be coupled and then the C ring closed. Since in most of the pentacyclic triterpenes the A and B rings are the same, it was proposed that one AB synthon might be used for the synthesis of a variety of triterpenes. The bicyclic enone **153** was chosen as the common AB synthon.

The precursor to enones **152** and **153** was the ketoalcohol **155** previously prepared by Sondheimer and Elad[82] using a longer and less reliable route. Robin-

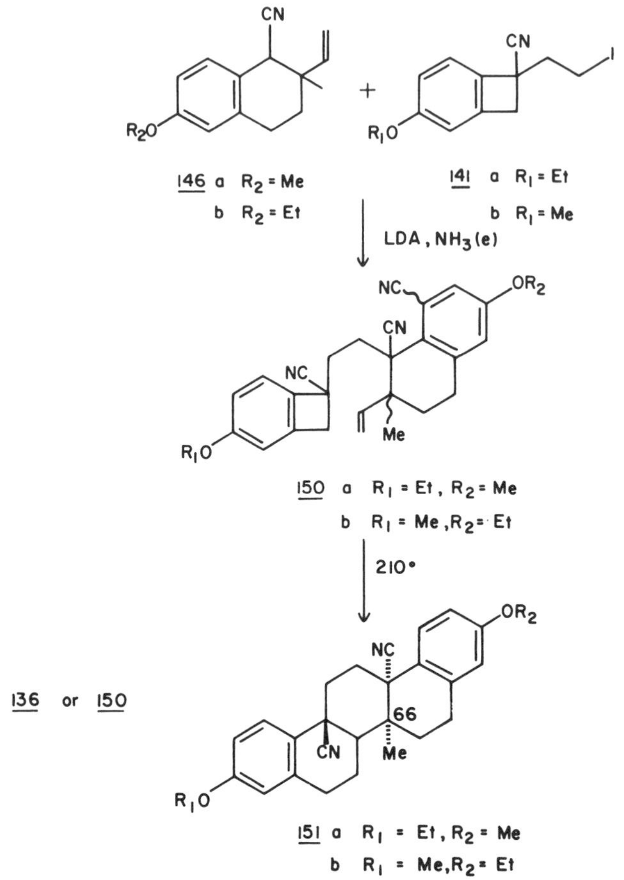
CN
R_2O
+
CN
R_1O
146 a R_2 = Me
b R_2 = Et
141 a R_1 = Et
b R_1 = Me
LDA, NH_3(e)
NC
OR_2
CN
NC
Me
R_1O
150 a R_1 = Et, R_2 = Me
b R_1 = Me, R_2 = Et
210°
136 or 150
OR_2
NC
66
CN
Me
R_1O
151 a R_1 = Et, R_2 = Me
b R_1 = Me, R_2 = Et

son annelation between 2-methyl-1,3-cyclohexanedione and ethyl vinyl ketone yielded the bicyclic dione **154.** Reduction of the saturated carbonyl followed by reductive methylation[83] gave ketoalcohol **155**, as well as reduced unalkylated material (8%) and dialkylated material (8%). After ketalization and oxidation, treatment with methyllithium at -78°C yielded the mixture of tertiary alcohols **156a** and **156b** in a ratio of 3:2. Dehydration afforded a mixture of endocyclic olefin **157**, exocyclic olefin (**157a**), and rearrangement product (**157b**). The yield of **157** was optimized with the use of 25% $H_2SO_4 \cdot H_2O$, stronger acid giving both more rearrangement product **157b** from alcohol (**156**) and subsequent rearrangement of endocyclic olefin (**157**) to **157b**.

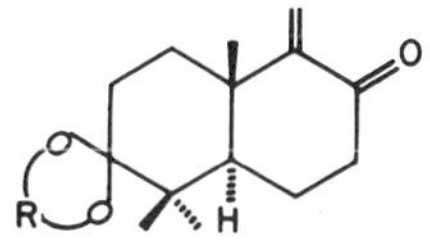

152   R = CH₂CH₂
153   R = CH₂C(CH₃)₂CH₂

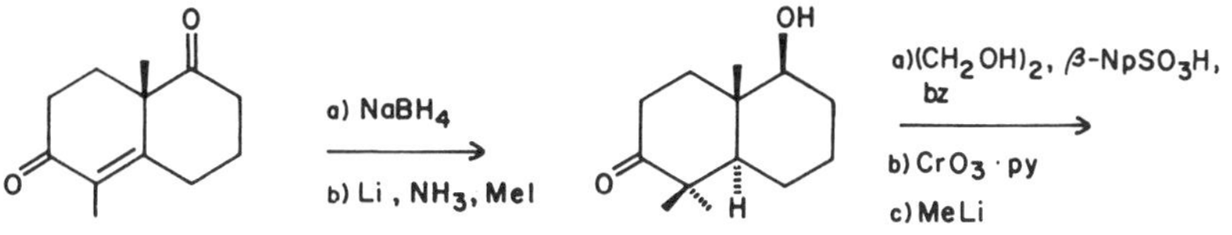

154

a) NaBH₄
b) Li , NH₃, MeI

155

a)(CH₂OH)₂, β-NpSO₃H, bz
b) CrO₃·py
c) MeLi

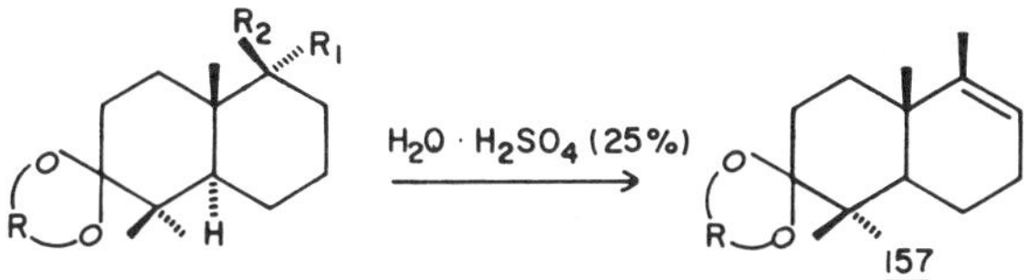

156 a R₁ = OH, R₂ = Me
    b R₁ = Me, R₂ = OH

H₂O · H₂SO₄ (25%)

157
157 a
157 b

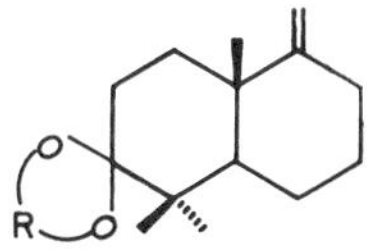

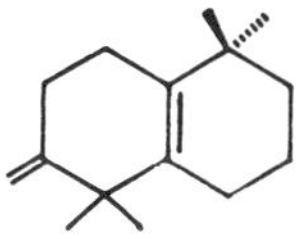

The olefin **157** was epoxidized to an equimolar mixture (**158a,b**). Both epoxides opened with lithium di-*n*-propylamide to the desired allylic alcohols (**159**). Expoxide **158a** also yielded unwanted tertiary alcohol **160** (40%). Reaction of the crude epoxide mixture with lithium di-*n*-propylamide followed by oxidation (CrO₃-py₂) and purification resulted in a 71:29 mixture of enones **152** and **161**, with an overall yield of **152** from **155** of 16%. The oxidation of **160** to **161** occurs by allylic rearrangement of the intermediate chromate ester. The reduction product from **154** was resolved into its enantiomers via the brucine salt of the hydrogen phthalate **162**. The levorotatory enantiomer was found to have the absolute configuration corresponding to that of the pentacyclic triterpenes.

Having completed the synthesis of the optically pure AB ring synthon (−)-**153**, the authors turned their attention to the allylic bromide **163**,[84] a synthon for rings D and E of β-amyrin.

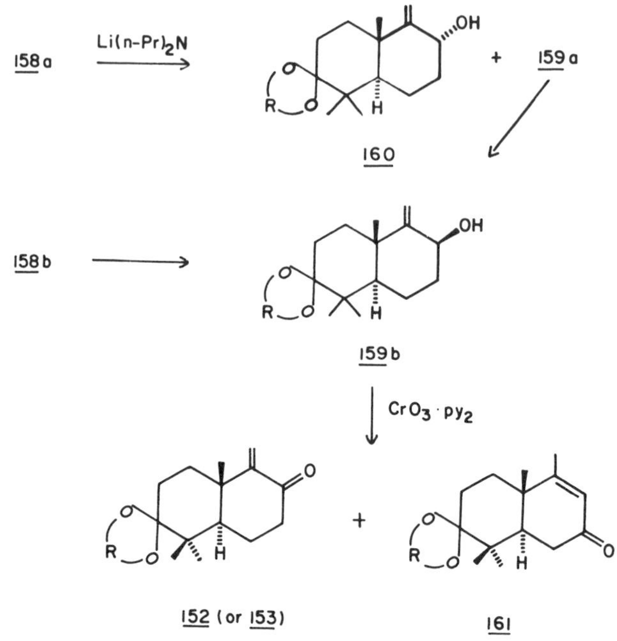

158a $\xrightarrow{\text{Li}(n\text{-Pr})_2\text{N}}$ 160   +   159a

158b $\longrightarrow$ 159b

$\downarrow$ CrO$_3$·py$_2$

152 (or 153)   +   161

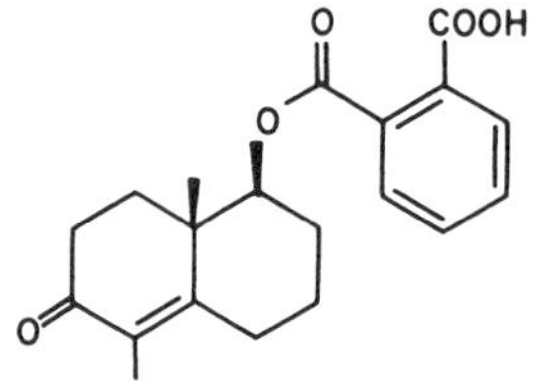

162 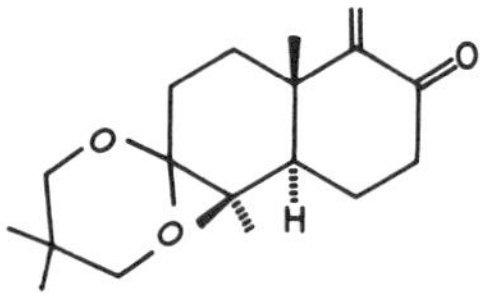   (−) 153 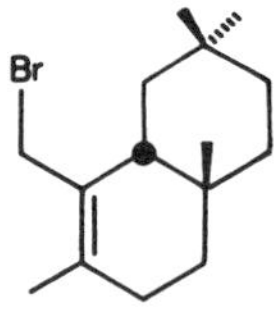   163

The octalone **165** was chosen as starting material. The compound had been previously synthesized by Halsall and Thomas,[85] but the method was inconvenient for large-scale preparation. The alternate route shown in Scheme 22 was developed.

The decalone **166** undergoes preferential enolization to C-3,[85] and it was proposed to first block the C-3 position by introducing a double bond. Bromination-dehydrobromination proceeded readily to give **167**, which proved intractable in all attempts to functionalize C-1 (**167** → **168**). Direct functionalization

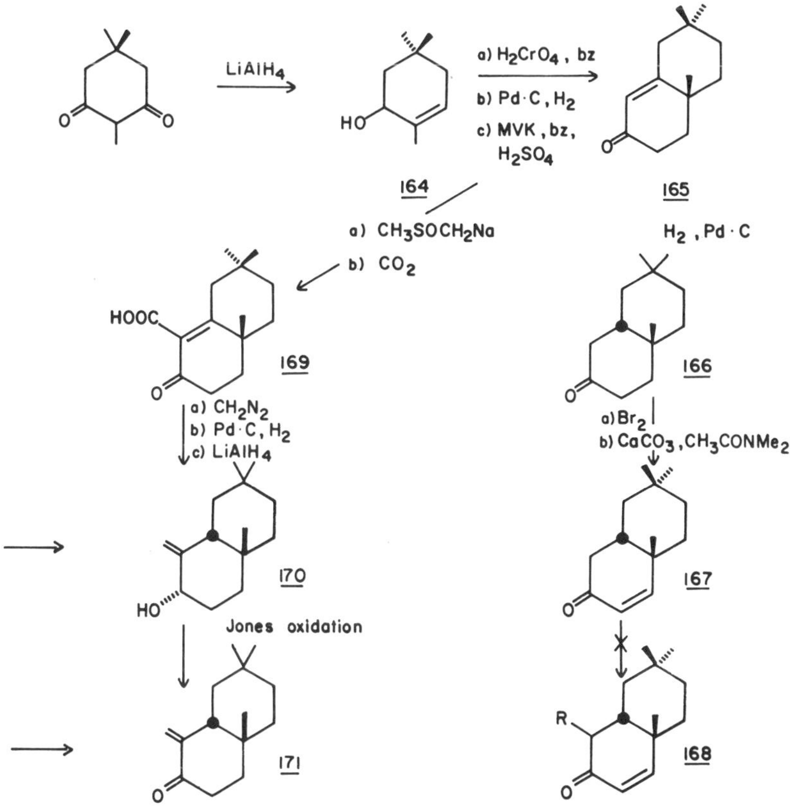

Scheme 22

of C-1 of the enone **165** was then attempted. Reaction of **165** with methylsulfinyl methylide in dimethylsulfoxide followed by carbonation yielded the keto-acid **169**. Esterification with diazomethane followed by catalytic hydrogenation and reduction gave the allylic alcohol **170**. This material was oxidized to the enone **171** in 67% yield. The proposed two-step conversion of **171** to the bromide **163** was thwarted by the predominance of 1,4- rather than 1,2-addition of methyl-lithium to the enone. As a result, the tedious stepwise sequence from **170** through the epoxides **172a,b** via **173** and **174** to the target allyl bromide **163** was required. An alternative synthesis of **163** was previously reported.[84]

The coupling of the AB synthon **153** with the allyl bromide **163** has yet to be undertaken, although preliminary studies on elaboration of the allylic bromide with phenylcrotylsulfide anion have been successfully attempted.

Recently, Sircar and Mukharji[86] also reported synthesis of a DE synthon. Attempted alkylation of the mixture of *cis* and *trans* octalinones corresponding to **167** proved as unsuccessful as those described above. The resistance of the system to alkylation at C-1 was further evidenced by the O-methylation resulting from attempted alkylation of **175**.

The series of closely related triterpenoids alnusenol, multiflorenol, taraxerol, and friedelin all possess the same DE ring system. ApSimon and co-workers have reported the synthesis[87] of unit **176** as a synthon for triterpenoids of these types, having a *cis*-fused decalin system and a C-13 α-methyl group. The work centered on the use of a substrate in which the future C-28 methyl group is functionalized both for its activating effect in the synthesis and for elaboration of pentacyclic triterpenes in which methyl groups have been oxidized.[88]

The octalone **177** was chosen as precursor (Scheme 23). Carboxymethylation

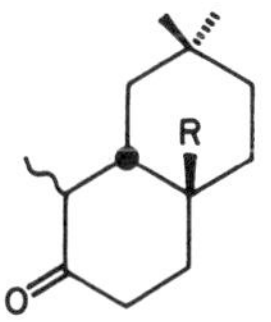

176 a   R = CH3
     b   R = COOMe
     c   R = CH2OH

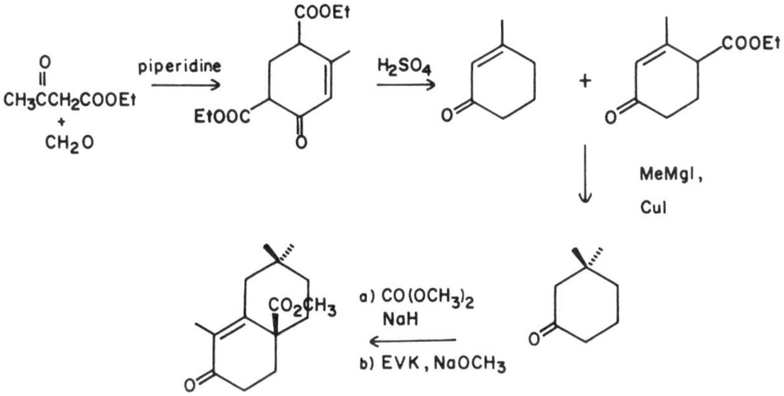

Scheme 23

of 3,3-dimethylcyclohexanone[89] by refluxing in excess dimethylcarbonate with sodium hydride provided 6-carbomethoxy-3,3-dimethyl cyclohexanone in good yield (70%). Robinson annelation with ethyl vinyl ketone furnished the desired octalone **177** in 30% yield from ethylacetoacetate.

Earlier work by Heathcock and co-workers[84a] and Halsall and Thomas[85] suggested that catalytic hydrogenation should yield a *cis*-fused product (**165** → **166**, Scheme 22). On the other hand, the ester **178** hydrogenates to a mixture of *trans*-decalone (**179**), lactone (**180**), and a hydrogenolysis product.[90]

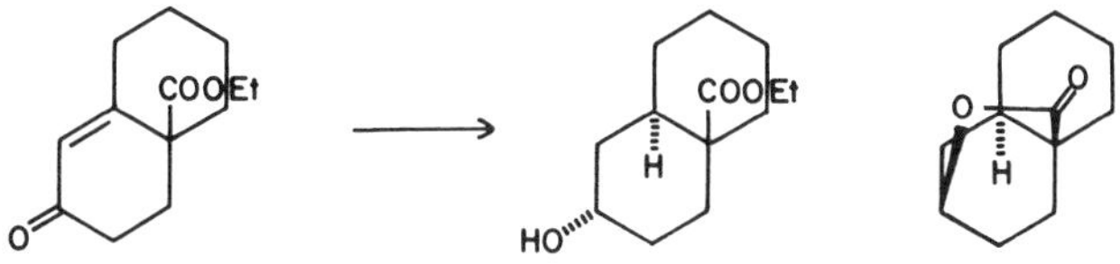

178   179   180

Catalytic hydrogenation of **177** provided a variety of products, depending on the reaction conditions. The stereochemistry of the saturated ketoester **183** was not immediately known, but formation of **181** and lactone **182** were reminiscent of Dauben's work.[90,91] Indeed, X-ray crystallography of **183**[92] revealed the *trans*-stereochemistry of the hydrogenation product, indicating a predominance of carbomethoxy group directing effect[93] over steric effects.[94]

Reversal of the stereochemistry of hydrogenation was made possible by utilizing the haptophilic properties of the hydroxyl group[95] employing the substrate 10-hydroxymethyl compound **184**. The desired material **184** could be obtained either by reduction of **177** to the diol **185** with lithium aluminum hydride followed by allylic oxidation with activated manganese dioxide or ketalization to the isomeric mixture **186** followed by reduction and deketalization. The former method suffers from the need for large amounts of manganese dioxide, the latter from the extreme sensitivity of the ketals. Neither is suitable for large-scale preparation. Catalytic hydrogenation over 10%-palladium-on-carbon yielded the decalone **176** (68%) and some hydrogenolysis product (10%).

The authors proposed that alkylation at C-1 should proceed from the open β-face of the hinge-like *cis*-fused decalin. The molecule proved particularly resistant to C-1 alkylation, although a product tentatively assigned structure **188** was obtained from enol acetate **187** and allyl bromide.

Studies on the ABC + E ring construction (**189**) of the pentacyclic triterpene skeleton were recently described by ApSimon et al.[96] The route has the advantage of rapid provision of a polycyclic species with preformed stereochemistry, countered by the disadvantage of uncontrolled stereochemistry in the cyclization step.

A further problem in the synthetic plan lies in the alkylation step between racemates **190**[96] and **191**,[97] which would reasonably be expected to lead to a

1:1 mixture of diastereomers (**192**) resulting in an immediate 50% loss of material. However, it was believed[98] that one of the diastereomers would cyclize preferentially with respect to the other and lead to **193** possessing the correct stereochemistry for elaboration to friedelin **126**. This point appeared to outweigh the disadvantage of 50% material loss, so the route was pursued as planned.

The tosylate **191** was synthesized from the malononitrile **194**[97] obtained in modest yield by potassium fluoride catalyzed Michael addition of malononitrile to 3,6,6-trimethyl-2-cyclohexenone. A series of unexceptional steps led from **194** to **191** obtained in overall yield of 26%.

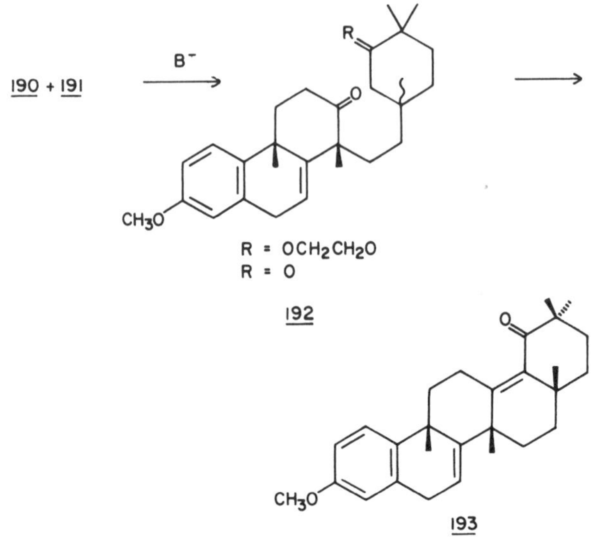

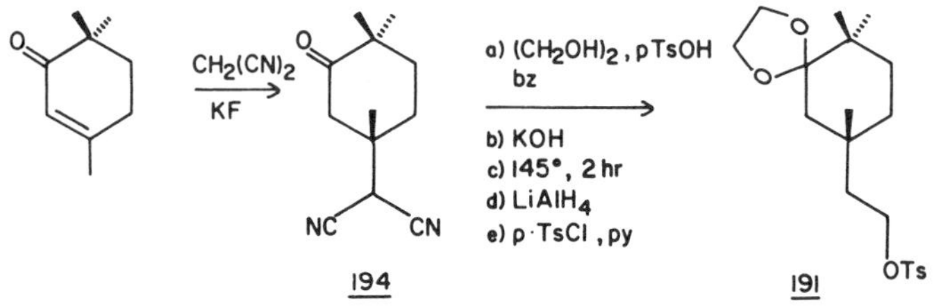

190 + 191 →[B⁻]

R = OCH₂CH₂O
R = O

192

193

CH₂(CN)₂ / KF

a) (CH₂OH)₂, pTsOH
   bz
b) KOH
c) 145°, 2 hr
d) LiAlH₄
e) p·TsCl, py

194

191

The dienolate anion of **190** generated with potassium *t*-amyloxide was alkylated with **191** to yield the mixture **192** (R = OCH₂CH₂O) in 54% yield. Acid-catalyzed hydrolysis gave the crystalline mixture of ketones. The stereochemistry at C-1 was assigned on the basis of previous observation,[97,99,100] and the benzene induced shifts of methyl signals in the nmr spectrum,[97] confirmed by the X-ray structure determination.

The ketone mixture is stable in crystalline form but undergoes an oxidation at the benzylic-allylic position (C-9) to enone **195** in solution.

Attempts to effect intramolecular aldol condensation using a variety of bases under a range of conditions were unsuccessful. Examination of acid-induced reactions led to the isolation of a crystalline product **196**. The physical properties of this material indicated that it was most certainly not the desired pentacycle **193**.[101] Single crystal X-ray structure determination confirmed the material as the dihydroanthracene **196**. The formation of this material was explained by a complex stepwise acid catalyzed transformation.

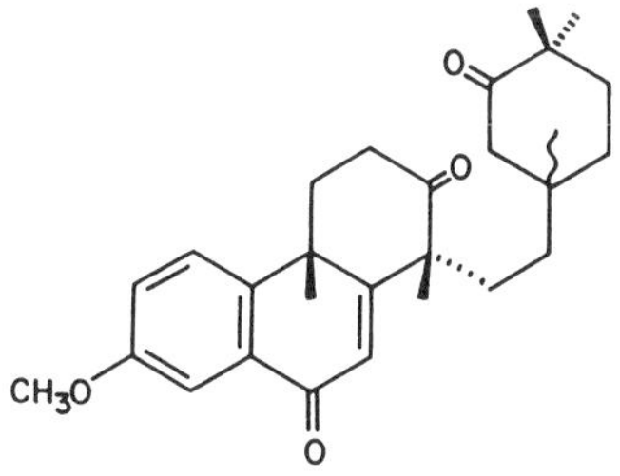

195

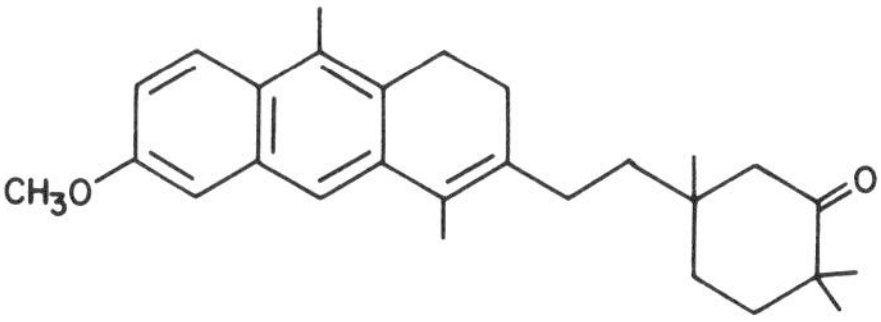

196

Postulating that the failure of cyclization was due to the lability of cation **197** leading to more stable allylic or benzylic species, attention was turned to cyclization in the absence of the double bond between C-1a and C-10. Previous work having shown that hydrogenation of related compounds proceeds stereoselectively,[97,101] reduction in xylene with palladium-on-carbon yielded the mixture of diketones **198**.

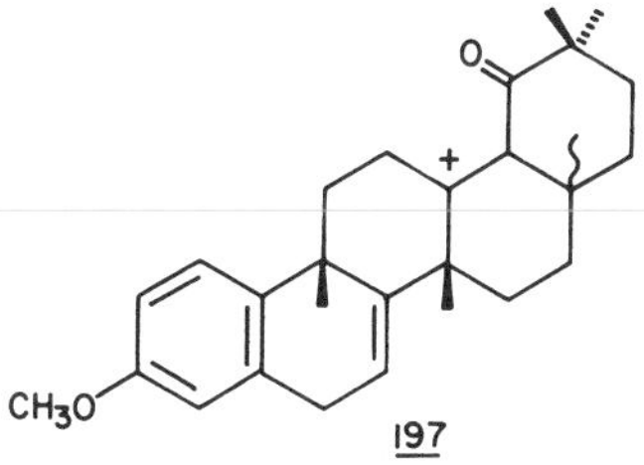

197

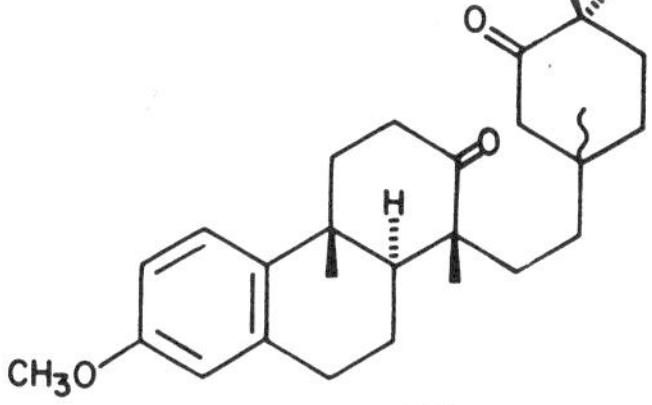

198

Contrary to the prediction that one diastereomer would cyclize more readily than the other, reaction of **198** in xylene-*p*-toluenesulfonic acid yielded two products, one of which crystallized after chromatography. The crystalline material obtained in 90% yield, based on cyclization of one diastereomer, was assigned structure **199** based on spectral data and confirmed by X-ray diffraction. The structure of the other product is still unknown.

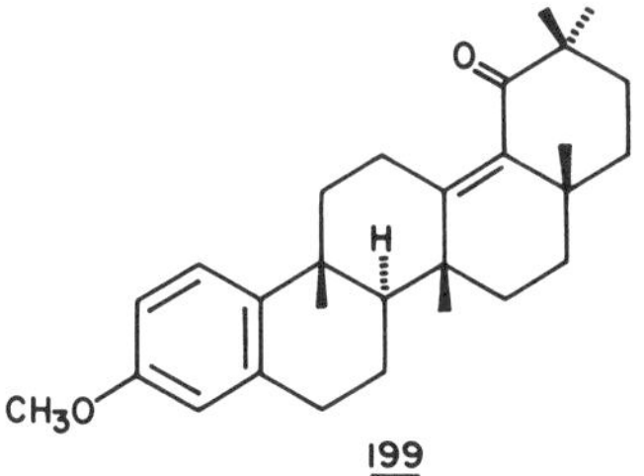

199

In another ABC + E ring approach also under investigation by the same group,[102] a novel rearrangement was observed under the hydrogenation conditions previously described[96,97,101] (e.g. **193** → **198**).

Precursor **201** was synthesized from dienone **200**, and hydrogenation of both double bonds was required. Reduction of **201** under the usual conditions, hy-

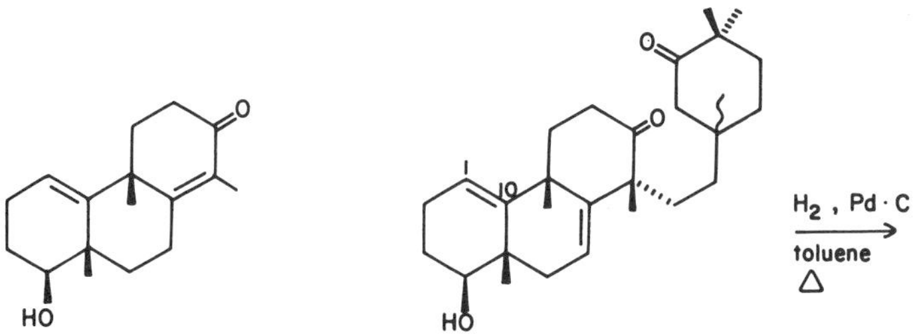

200

201

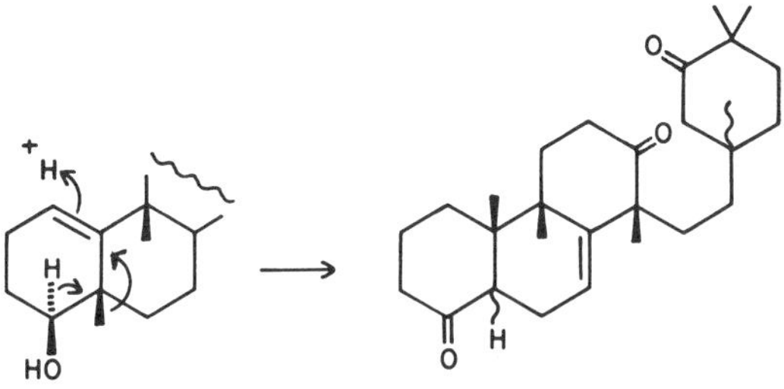

H₂ , Pd·C
toluene
Δ

202

drogen over refluxing toluene with palladium on carbon catalyst, yielded 70% of rearranged material **202**. Compound **200** showed similar behavior under the reaction conditions. It appears that prolonged hydrogenation leads to slow protonation of the 1,10-double bond, followed by methyl migration and hydride shift.

The bond formation shown below (Figure 1) between rings A and C generates the desired *trans*-fused BC ring system when subjected to cycloalkylation.[44] An appropriate CDE ring unit (**203**) with the correct stereochemistry at the DE ring junction and opportunity for further elaboration at the C ring was recently reported by ApSimon and co-workers.[103]

Alkylation of 2-carbomethoxy-5,5-dimethylcyclohexanone[87] with 2-(3-methoxyphenyl) ethyl bromide in the presence of potassium *t*-butoxide produced the

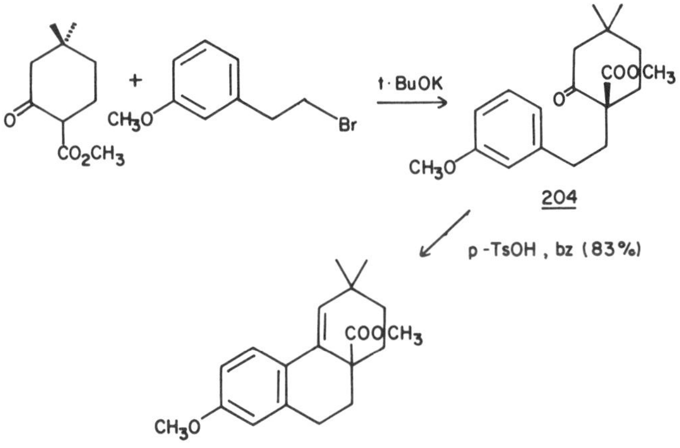

204

203

bicyclic intermediate **204** in modest (31%) yield. Acid-catalyzed cyclization then proceeded readily to give the CDE ring precursor **203**.

Catalytic hydrogenation of the ester **204** yielded only product with the *trans*

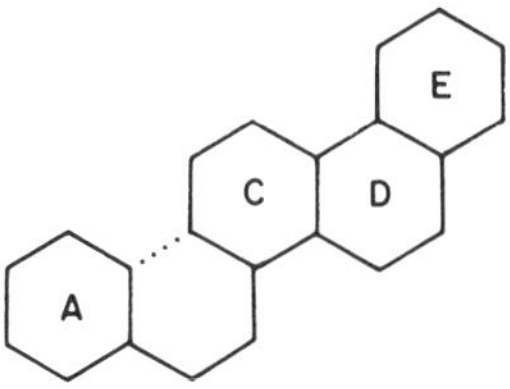

Figure 1

ring junction. Use of the alcohol to direct hydrogenation, a technique used successfully in the synthesis of decalone **176**, yielded a mixture of *cis* and *trans* isomers.

Investigation of metal-ammonia reduction of the carboxylic acid,[104] however, proved quite successful, the *cis*-saturated acid (**205**) being formed in 74% yield. Structural assignments were made on the basis of the comparative ^{13}C chemical shifts of the angular hydroxymethyl groups[105] of the *cis* and *trans* isomers.

205   206

Having obtained the desired *cis* stereochemistry, the alcohol was oxidized to the aldehyde with pyridinium chlorochromate and subjected to Wolff-Kishner reduction to yield the CDE moiety in overall yield of 30% from the unsaturated tricylic ester. Lithium-ammonia reduction of the aromatic nucleus gave the β,γ enone **206**.

(c) Serratenediol

Serratenediol **207**[106] is a member of a recently discovered class of pentacyclic triterpenes,[106,107] which have a seven-membered C ring and an entirely different skeletal structure from the triterpenoids described above. The compounds may be classified as a modified hopane group with a close biogenetic relationship, serratenediol being chemically interrelated with α-onocerin by a simple transformation.[106]

Prestwick and Labovitz's[108] recent total synthesis of serratenediol is based on a biogenetic type polyene cyclization of the tetraenic alochol **208**.

The known tricyclic enone **215**[19] was prepared by a new method (Scheme 24). Conversion of the ketone **215** to keto acid **218** via **216** and **217** produced the potential DE ring, onto which the polyene chain was attached.

The allylic alcohol **209** obtained from 3-(*m*-methoxyphenyl) propanol and isopropenyl magnesium bromide in THF at − 78°C was converted to the ketone **210** by a chloroketal Claisen reaction.[109] Reaction with isopropenyllithium followed by reduction of the resulting allylic epoxide (**211**) resulted in formation of the tetramethyl allylic alcohol **212**. Cyclization with stannous chloride[110] afforded an 85:15 mixture of **213**:**214**; ozonolysis of **213** afforded **215**, the route

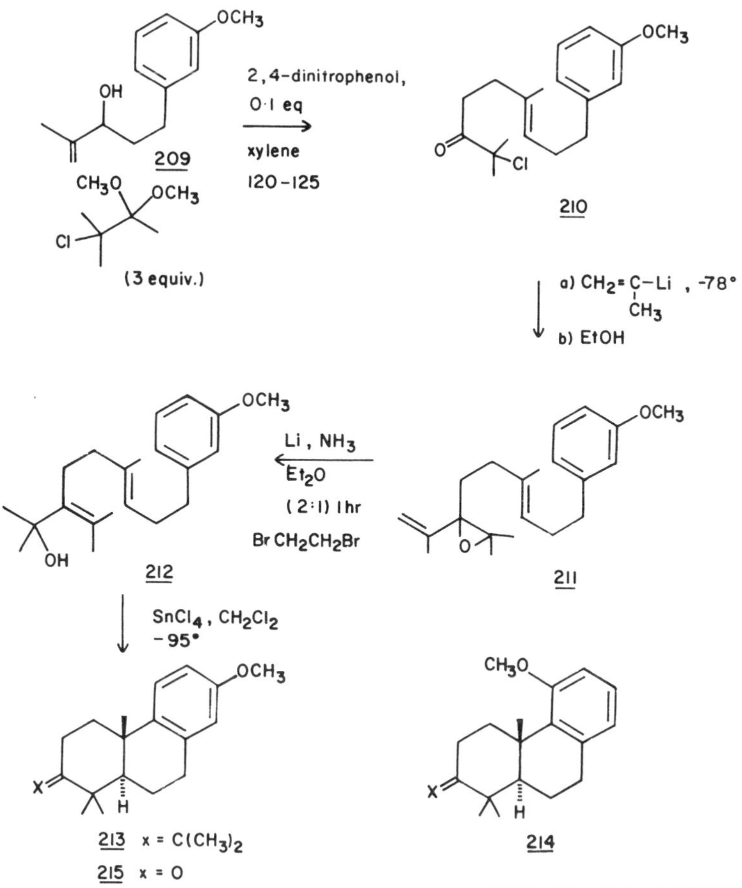

Scheme 24

providing improvements in both yield and stereospecificity over the previous approach.[19]

Tricyclic ketone **215** was converted by previously reported procedures[19] to the ketoacid **218**, which was treated with methylenetriphenylphosphorane to introduce the exocyclic methylene on the D ring (**219**) (Scheme 25). The ester was then converted to the aldehyde (**220**), treated with isopropenylmagnesium bromide to yield the allylic alcohol, subjected to Claisen rearrangement with methyl or thioacetate, and the resulting aldehyde converted to the tetramethyl allylic alcohol **208** by the previously described sequence (**210** → **212**, Scheme 24). The polyolefinic precursor to the serratene skeleton was obtained in overall yield of 3.1% from *m*-methoxycinnamic acid.

Cyclization of tetraene **208** with trifluoroacetic acid at −78° gave the pen-

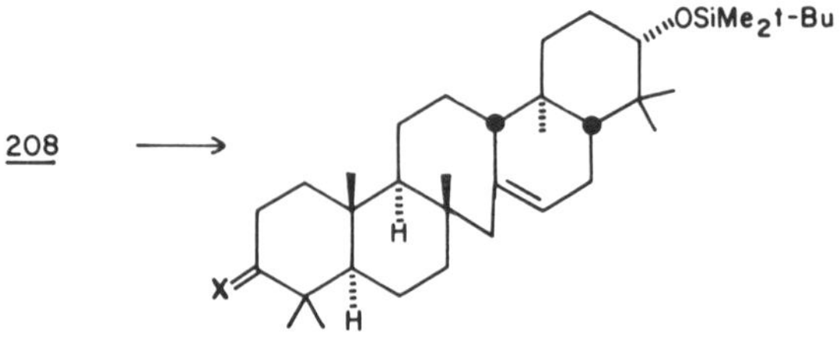

208 $\longrightarrow$

221 X = C(CH$_3$)$_2$
     X = O

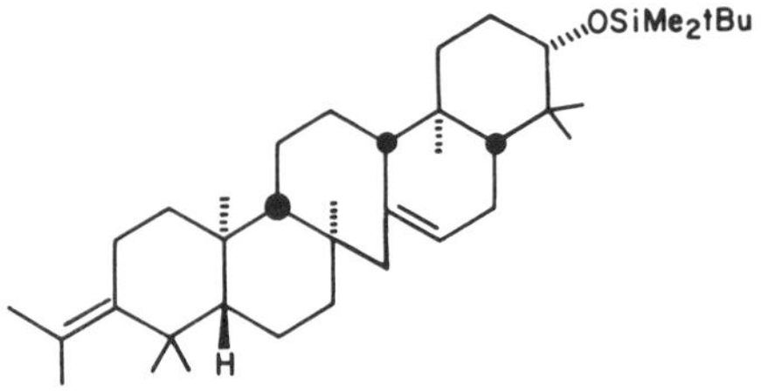

222

tacyclic material **221** (20%) and an isomeric substance (14%) tentatively assigned the structure **222** based on its nearly identical nmr and mass spectra. Ruthenium tetroxide converted **221** to the ketone, which was reduced to the C-3 equatorial alcohol (**223**). The C-21 hydroxyl protecting group was removed to give the first totally synthetic sample of *dl*-serratenediol (**207**).

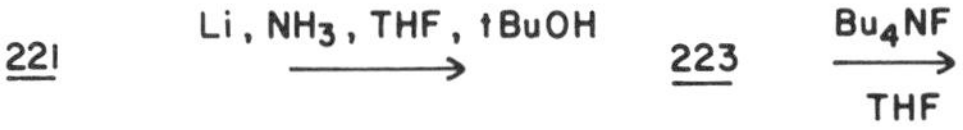

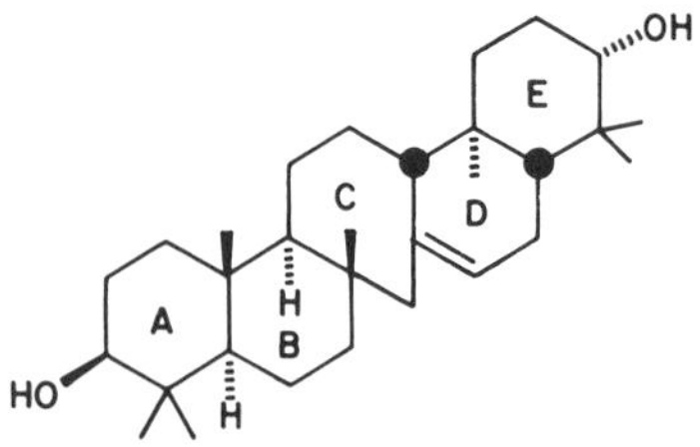

$$ \underline{221} \quad \xrightarrow{\text{Li , NH}_3\text{ , THF , tBuOH}} \quad \underline{223} \quad \xrightarrow[\text{THF}]{\text{Bu}_4\text{NF}} $$

207

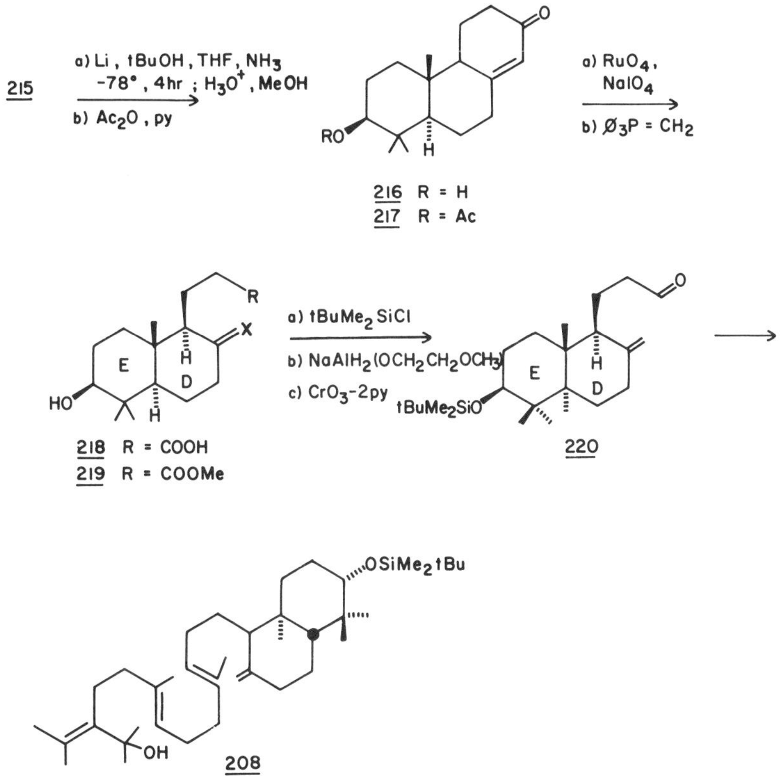

Scheme 25

REFERENCES

1. J. W. ApSimon and J. W. Hooper, in *The Total Synthesis of Natural Products,* Vol. 2, Wiley-Interscience, New York, 1973, p. 559.

2. S. Yamada, N. Oh-hashi, and K. Achiwa, *Tetrahedron Lett.,* 2557 (1976) and references cited therein.

3. S. Yamada, N. Oh-hashi, and K. Achiwa, *Tetrahedron Lett.,* 2561 (1976).

4. E. E. van Tamelen, P. McCurry, and V. Huber, *Proc. Nat. Acad. Sci. USA,* **68,** 1294 (1971).

5. J. F. Biellmann and J. B. Ducep, *Tetrahedron,* **27,** 5861 (1971).

6. R. B. Boar and K. Damps, *Tetrahedron Lett.,* 3731 (1974); D. H. R. Barton, T. R. Jarman, K. E. Watson, D. A. Widdowson, R. B. Boar, and K. Damps, *Chem. Comm.* 861 (1974); R. B. Boar and K. Damps, *J. Chem. Soc., Perkin I,* 709 (1977).

7. J. Staunton and E. J. Eisenbraun, *Org. Synth.* **42,** 4 (1962).

8. D. H. R. Barton, T. R. Jarman, K. C. Watson, D. A. Widdowson, R. B. Boar, and K. Damps, *J. Chem. Soc., Perkin I,* 1134 (1975).

9. T. Shishibori, T. Fukui, and T. Suga, *Chem. Lett.*, 1137 (1973).

10. A. D. Josey, *J. Org. Chem.*, **39**, 139 (1974); K. Takahashi, G. Hata, and A. Miyake, *Bull. Chem. Soc. Japan*, **46**, 600 (1973); J. Berger, Ch. Duscheck, and H. Reichel, *J. Prakt. Chem.* **315**, 1077 (1973); M. Anteunis and A. DeSmet, *Synthesis*, 800 (1974).

11. T. Mandai, H. Yamaguchi, K. Nishikawa, M. Kawada, and J. Otera, *Tetrahedron Lett.*, 763 (1981).

12. T. Mandai, H. Yokoyama, T. Miki, H. Fukuda, H. Kobata, M. Kawada, and J. Otera, *Chem. Lett.*, 1057 (1980).

13. E. Bertele and P. Schudel, *Helv. Chim. Acta*, **50**, 2445 (1967).

14. E. J. Corey, R. L. Danheiser, and S. Chandrasekaran, *J. Org. Chem.*, **41**, 260 (1976).

15. P. R. Ortiz de Montellano, R. Castillo, W. Vinson, and J. S. Wei, *J. Am. Chem. Soc.*, **98**, 2018 (1976).

16. E. J. Corey, P. R. Ortiz de Montellano, and H. Yamamoto, *J. Am. Chem. Soc.*, **90**, 6254 (1968).

17. R. A. Packer and J. S. Whitehurst, *J. Chem. Soc., Perkin I*, 110 (1978); *Chem. Comm.*, 757 (1975).

18. F. H. Howell and D. A. H. Taylor, *J. Chem. Soc.*, 1248 (1958).

19. G. Stork, A. Meisels, and S. E. Davies, *J. Am. Chem. Soc.*, **85**, 3419 (1963).

20. K. E. Pfitzner and J. G. Moffatt, *J. Am. Chem. Soc.*, **87**, 5661, 5670 (1965).

21. H. C. Brown and H. M. Hess, *J. Org. Chem.*, **34**, 2206 (1969).

22. S. Winstein and J. Sonnenberg, *J. Am. Chem. Soc.*, **83**, 3235 (1961).

23. K. G. Orrell, R. A. Packer, V. Sik, and J. S. Whitehurst, *J. Chem. Soc., Perkin I*, 117 (1978).

24. W. Reusch, K. Grimm, J. E. Karoglan, J. Martin, K. P. Subrahamanian, Y-C. Toong, P. S. Venkataramani, J. D. Yordy, and P. Zoutendam, *J. Amer. Chem. Soc.*, **99**, 1953 (1977).

25. W. Reusch, K. Grimm, J. E. Karoglan, J. Martin, K. P. Subrahamanian, P. S. Venkataramani, and J. D. Yordy, *J. Amer. Chem. Soc.*, **99**, 1958 (1977).

26. J. L. Martin, J. S. Ton, and W. Reusch, *J. Org. Chem.*, **44**, 3666 (1979).

27. R. A. Benkeser, *Synthesis*, 347 (1971).

28. E. C. Levy and D. Lavie, *Israel J. Chem.*, **8**, 677 (1970); I. G. Guest and B. A. Marples, *J. Chem. Soc., C*, 1468 (1971); O. E. Edwards and Z. Paryzek, *Can. J. Chem.*, **51**, 3866 (1973).

29. O. E. Edwards and Z. Paryzek, *Can. J. Chem.*, **53**, 3498 (1975).

30. Z. Parysek, *Tet. Lett.* *1976*, 4761, *J. Chem. Soc., Perkin I*, 1222 (1979).

31. W. O. Godtfredsen, W. von Daehne, S. Vangedal, A. Marque, D. Arigoni, and A. Melera, *Tetrahedron*, **21**, 3505 (1965).

32. R. E. Ireland, P. Beslin, R. Giger, U. Hengartner, H. A. Kirst, and H. Maag, *J. Org. Chem.*, **42**, 1267 (1977).

33. R. E. Ireland and U. Hengartner, *J. Am. Chem. Soc.*, **94**, 3652 (1972).

34. R. E. Ireland, S. W. Baldwin, and S. C. Welch, *J. Am. Chem. Soc.*, **94**, 2056 (1972).

35. D. Felix, J. Schreiber, G. Ohloff, and A. Eschenmoser, *Helv. Chim. Acta*, **54**, 2896 (1971).

36. R. K. Muller, R. Joos, D. Felix, J. Schreiber, C. Winter, and A. Eschenmoser, *Org. Synth.*, **55**, 114 (1976).

37. G. Stork and R. Borch, *J. Am. Chem. Soc.*, **86**, 935 (1964).

38. R. E. Ireland, D. C. Muchmore, and U. Hengartner, *J. Am. Chem. Soc.*, **94**, 5098 (1972).

39. S. Iwasaki, M. I. Sair, H. Igarashi, and S. Okuda, *Chem. Commun.*, 1119 (1970).

40. D. N. Kirk and J. M. Miles, *Chem. Commun.*, **518**, 1015 (1970).

41. C. Amendolla, G. Rosenkranz, and F. Sondheimer, *J. Chem. Soc.*, 1226 (1954).

42. K. B. Sharpless, R. F. Laurer, and A. Y. Teranishi, *J. Am. Chem. Soc.*, **95**, 6137 (1973); H. J. Reich, I. L. Reich, and J. M. Renga, *ibid*, **95**, 5813 (1973).

43. R. E. Ireland, R. Giger, and S. Kamata, *J. Org. Chem.*, **42**, 1276 (1977).

44. R. E. Ireland, M. I. Dawson, C. S. Kowalski, C. A. Lipinski, D. R. Marshall, J. W. Tilley, J. Bordner, and B. L. Trus, *J. Org. Chem.*, **40**, 973 (1975).

45. G. F. Hennion and F. R. Kupieki, *J. Org. Chem.*, **18**, 1601 (1953).

46. E. Wenkert and D. A. Berges, *J. Am. Chem. Soc.*, **89**, 2507 (1967).

47. J. A. Marshall and G. L. Bundy, *J. Am. Chem. Soc.*, **88**, 4291 (1966).

48. H. C. Brown and W. C. Dickason, *J. Am. Chem. Soc.*, **92**, 709 (1970).

49. H. E. Simmons and R. D. Smith, *J. Am. Chem. Soc.*, **81**, 4256 (1959).

50. For a review, see W. Nagata, *Org. Reactions,* **25**, (1976).

51. W. Nagata and M. Yoshioka, *Org. Synth.*, **52**, 9, 100 (1972).

52. For recent reviews, see W. S. Johnson, *Bioorg. Chem.*, **5**, 51 (1976); *Angew. Chem. Int. Ed. Engl.*, **15**, 9 (1976).

53. E. J. Corey and H. A. Kirst, *Tetrahedron Lett.*, 5041 (1968).

54. H. M. Schmidt and J. F. Arens, *Rec. Trav. Chim. Pays-Bas*, **86**, 1138 (1967).

55. E. J. Corey, J. A. Katzenellenbogen, and G. H. Posner *J. Am. Chem. Soc.*, **89**, 4245 (1967).

56. R. E. Ireland, C. J. Kowalski, J. W. Tilley, and D. M. Walba, *J. Org. Chem.*, **40**, 990 (1975); R. E. Ireland, C. A. Lipinski, C. J. Kowalski, J. W. Tilley, and D. M. Walba, *J. Am. Chem. Soc.*, **96**, 3333 (1974).

57. A. Eschenmoser, D. Felix, and G. Ohloff, *Helv. Chim. Acta,* **50**, 705 (1967).

58. P. E. Peterson and R. J. Kamat, *J. Am. Chem. Soc.*, **91**, 5421 (1969).

59. a) W. S. Johnson, M. B. Gravestock, R. J. Parry, R. F. Meyers, T. A. Bryson, and D. H. Miles, *J Am. Chem. Soc.*, **93**, 4330 (1971); b) P. T. Lansbury and G. E. DuBois, *Chem. Commun.*, 1107 (1971).

60. G. W. Moersch and W. A. Neuklis, *J. Chem. Soc.*, 788 (1965).

61. R. E. Ireland and L. N. Mander, *J. Org. Chem.*, **32**, 689 (1967); H. O. House and T. M. Bore, *ibid.*, **33**, 943 (1968).

62. W. Nagata and Y. Hayase, *J. Chem. Soc., C,* 460 (1969).

63. I. Ojima, T. Kogure, and Y. Nagai, *Tetrahedron Lett.*, 5035 (1972).

64. B. M. Trost, D. F. Taber, and J. B. Alper, *Tetrahedron Lett.*, 3857 (1976).

65. J. Cologne and G. Poilane, *Bull. Soc. Chim. France,* 1953 (1955).

66. R. E. Ireland, P. Bey, K.-F. Cheng, R.J. Czarny, J.-F. Moser, and R. I. Trust, *J. Org. Chem.*, **40**, 1000 (1975).

67. R. E. Ireland, M. I. Dawson, S. C. Welch, A. Hagenbach, J. Bordner, and B. Trus, *J. Am. Chem. Soc.*, **95**, 7829 (1973).

68. (a) W. S. Johnson, M. B. Gravestock, and B. E. McCarry, *J. Am. Chem. Soc.*, **93**, 4332 (1971). (b) R. L. Carney and W. S. Johnson, *J. Am. Chem. Soc.*, **96**, 2549 (1974).

69. R. E. Ireland, T. C. McKenzie, and R. I. Trust, *J. Org. Chem.*, **40**, 1007 (1975).

70. H. C. Brown and A. W. Moerikofer, *J. Am. Chem. Soc.*, **85**, 2063 (1963).

71. R. E. Ireland and D. M. Walba, *Tetrahedron Lett.*, 1071 (1976).

72. a) E. J. Corey and J. J. Ursprung, *J. Am. Chem. Soc.*, **78**, 5041 (1956); b) G. Brownie, F. S. Spring, R. Stevenson, and W. S. Strachan, *J. Chem. Soc.*, 2419 (1956).

73. T. Kametani, Y. Hirai, F. Satoh, and K. Fukumoto, *J. Chem. Soc., Chem. Commun.*, 16 (1977); T. Kametani, Y. Hirai, K. Fukumoto, and F. Satoh, *J. Am. Chem. Soc.*, **100**, 554 (1978).

74. W. Oppolzer, *J. Am. Chem. Soc.*, **93**, 3833, 3834 (1971); *Tetrahedron Lett.*, 1001 (1974).

75. I. L. Klundt, *Chem. Rev.*, **70**, 471 (1970).

76. T. Kametani, H. Nemoto, H. Ishikawa, K. Shiroyama, and K. Fukumoto, *J. Am. Chem. Soc.*, **98**, 3378 (1976).

77. T. Kametani, Y. Kato, T. Honda, and K. Fukumoto, *J. Am. Chem. Soc.*, **98**, 8185 (1976) and references cited therein.

78. W. E. Parham and E. L. Anderson, *J. Am. Chem. Soc.*, **70**, 4187 (1948).

79. T. Kametani, M. Kajiwara, and K. Fukumoto, *Chem. Ind.* (London), 1165 (1973); *Tetrahedron*, **30**, 1053 (1974); T. Kametani, H. Nemoto, H. Ishikwaka, H. Matsumoto, K. Shiroyama, and K. Fukumoto, *J. Am. Chem. Soc.*, **99**, 3461 (1977).

80. W. Nagata, H. Itazaki, *Chem. Ind.* (London), 1194 (1964).

81. J. S. Dutcher, J. G. MacMillan, and C. Heathcock, *J. Org. Chem.*, **41**, 2663 (1976).

82. F. Sondheimer and D. Elad, *J. Am. Chem. Soc.*, **80**, 1967 (1958).

83. G. Stork, P. Rosen, N. Goldman, R. V. Coombs, and J. Tsuji, *J. Am. Chem. Soc.*, **87**, 275 (1965).

84. (a) C. H. Heathcock, J. E. Ellis, and J. S. Dutcher, *J. Org. Chem.*, **41**, 2670 (1976); C. H. Heathcock and J. E. Ellis, *Chem. Commun.*, 1474 (1971); J. E. Ellis, J. S. Dutcher, and C. H. Heathcock, *Synth. Commun.*, **4**, 71 (1974). (b) E. E. van Tamelen, M. P. Seiler, and W. Wierenga, *J. Am. Chem. Soc.*, **94**, 8229 (1972).

85. T. G. Halsall and D. B. Thomas, *J. Chem. Soc.*, 2431 (1956).

86. I. Sircar and P. C. Mukharji, *J. Org. Chem.*, **45**, 3744 (1977).

87. J. W. ApSimon, S. Badripersaud, T. T. Nguyen, and R. Pike, *Can. J. Chem.*, **56**, 1646 (1978).

88. T. K. Devon and A. I. Scott, *Handbook of Naturally Occurring Compounds*, Vol. 2, Academic Press, New York, 1972.

89. G. Buchi, O. Jeger, and L. Ruzicka, *Helv. Chim. Acta*, **31**, 241 (1948); H. O. House and W. F. Fischer, *J. Org. Chem.*, **33**, 949 (1968).

90. W. G. Dauben and J. B. Rogan, *J. Am. Chem. Soc.*, **79**, 5002 (1957).

91. W. G. Dauben, J. W. McFarland, and J. B. Rogan, *J. Org. Chem.*, **26**, 297 (1961).

92. C. S. Huber and E. J. Gabe, *Acta Crystallogr., B*, **30**, 2519 (1974).

93. L. S. Minckler, A. S. Hussey, and R. H. Baker, *J. Am. Chem. Soc.*, **78**, 1009 (1956).

94. R. L. Augustine, Catalytic Hydrogenation, Mercet.

95. H. W. Thompson and R. Naipawar, *J. Am. Chem. Soc.*, **95**, 6379 (1973); H. W. Thompson and E. McPherson, *ibid.*, **96**, 6232 (1974); H. W. Thompson, E. McPherson, and B. L. Lences, *J. Org. Chem.*, **41**, 2903 (1976).

96. J. W. ApSimon, S. Badripersaud, J. W. Hooper, R. Pike, G. I. Birnbaum, C. Huber, and M. L. Post, *Can. J. Chem.*, **56**, 2139 (1978).

97. J. W. ApSimon, P. Baker, J. Buccini, J. W. Hooper, and S. MacCaulay, *Can. J. Chem.,* **50,** 1944 (1972).

98. H. Burgi, J. D. Dunitz, J. M. Lehn, and G. Wipff, *Tetrahedron,* **30,** 1563 (1974); H. Burgi, *Angew. Chem. Int. Ed. Engl.,* **14,** 460 (1975); W. T. Wipke and P. Gund, *J. Am. Chem. Soc.,* **98,** 8107 (1976).

99. G. Stork and J. W. Schulenberg, *J. Am. Chem. Soc.,* **84,** 284 (1962).

100. E. Wenkert, A. Alfonso, J. Brendenberg, C. Kaneko, and A. Tahara, *J. Am. Chem. Soc.,* **86,** 2038 (1964); R. E. Ireland and R. C. Kierstead, *J. Org. Chem.,* **31,** 2543 (1966); C. L. Graham and F. J. M. McQuillin, *J. Chem. Soc.,* 4521 (1964); V. Permutti and Y. Mazur, *J. Org. Chem.,* **31,** 705 (1966); G. Just and K. St. C. Richardson, *Can. J. Chem.,* **42,** 464 (1964).

101. R. E. Ireland, D. A. Evans, D. Glover, G. M. Rubottom, and H. Young, *J. Org. Chem.,* **34,** 3717 (1969).

102. J. W. ApSimon and I. Toth, *J. Chem. Soc., Chem. Commun.,* 67 (1979).

103. J. W. ApSimon, D. Moir, and K. Yamazaki, *Can. J. Chem.,* **59,** 1010 (1981).

104. H. W. Thompson, E. McPherson, and B. L. Lences, *J. Org. Chem.,* **41,** 2903 (1976).

105. D. Caine and T. L. Smith Jr., *J. Org. Chem.,* **43,** 755 (1978).

106. a) Y. Tsuda, T. Sano, K. Kawaguchi, and Y. Inubushi, *Tetrahedron Lett.,* 1279 (1964); b) Y. Inubushi, T. Sano, and Y. Tsuda, *Tetrahedron Lett.,* 1303 (1964).

107. J. W. Rowe, *Tetrahedron Lett.,* 2347 (1964).

108. G. D. Prestwick and J. N. Labovitz, *J. Am. Chem. Soc.,* **96,** 7103 (1974).

109. L. Werthemann and W. S. Johnson, *Proc. Natl. Acad. Sci. USA,* **67,** 1465 (1970).

110. P. A. Bartlett and W. S. Johnson, *J. Am. Chem. Soc.,* **95,** 7501 (1973).

The Total Synthesis of Carbohydrates 1972–1980

A. ZAMOJSKI and G. GRYNKIEWICZ

Institute of Organic Chemistry,
Polish Academy of Sciences,
Warsaw, Poland

1. INTRODUCTION

During the past nine years the subject of total synthesis of sugars and cyclitols has been vividly developed. The discovery of a large number of unusual sugars in nature—deoxy, amino, branched chain sugars (many of them components of antibiotics)—made their synthesis attractive. On the other hand, improvements in organic reagents and synthetic methods allowed a return to older preparations which could be now better performed. Finally, some synthetic ideas already described in this series, in the first chapter[1] of the volume on the total synthesis of carbohydrates, are further expanded.

Many sugars, especially of the deoxy type, can be obtained nowadays in higher yields and more readily by total synthesis than by the classical routes involving transformations of natural carbohydrates. The problem of optical activity of the sugars prepared remains incompletely solved, although—as shown below—satisfactory results have been achieved in many preparations.

The formulas of racemic products in this chapter are shown in the D form for reasons of simplicity. When optically active sugars are obtained, their formulas reflect their real configuration.

2. BASE CATALYZED CONDENSATIONS WITH CARBON–CARBON BOND FORMATION

The formose reaction has remained a subject of current interest in connection with possibility of its industrial application[2] as well as with the speculations on the prebiotic synthesis of carbohydrates.[3–6] Although the complex mixture of sugars and alditols produced in the base catalyzed condensation of formaldehyde, so-called "formose," was shown to be toxic when fed to mammals,[7] there are several reports on the utilization of formose by microorganisms[8–11] and some insects.[12] Alternatively, processes for the production of various polyhydric alcohols have been developed on the basis of the formose reaction.[13]

The subject of the chemical synthesis of formose sugars, including the historical background, the results of kinetic studies and a résumé of biological tests, have been summarized by Mizuno and Weiss.[14] Some chemical aspects of the formose reaction have also been surveyed in Japanese papers.[15,16] The review[14] covers the literature up to 1971, which roughly coincides with the period covered in the previous report in this series. In the subsequent years, efforts have been continued to evaluate the influence of various factors affecting the yield and composition of the formose sugars.

A series of reports have appeared from the Weiss laboratory, devoted to studies on the formose condensation performed in a continuous stirred tank reactor (CSTR). Two reaction paths, both first order with calcium hydroxide were recognized: (i) first order in formaldehyde reaction, leading to Cannizzaro products; (ii) formose condensation reaction, which is autocatalytic at low conversion levels and independent of the concentration of organic reactants and products at intermediate conversion levels.[17] The composition of terminal products determined by gas-liquid chromatography (glc) showed 50% C_6, 30% C_5, 10% C_4 and 5% of the higher sugar content.[17] At very low formaldehyde conversion, the main process is the Cannizzaro reaction, whereas near 50% conversion only 5% of formaldehyde is consumed in this way. Rate instabilities occur both at low and high catalyst concentration levels at a fixed formaldehyde addition rate.[18] The nature of the temperature and concentration rate instabilities under CSTR isothermal reaction conditions have been discussed. It has been postulated that both the formose and the Cannizzaro reaction proceed through a common intermediate.[19]

Considerable attention has been given to the catalytic species active in the formose condensation reaction. Alkaline earth hydroxide catalysts were applied with a glycolaldehyde or a glucose co-catalyst.[20,21] It was found that a complex of calcium hydroxide with glucose brought about formaldehyde condensation to sugars, but was inactive as a catalyst of the Cannizzaro reaction. Isolated solid complexes of calcium hydroxide with glucose and dihydroxyacetone were found to be amorphous by X-ray examination. On the basis of electron spectroscopy for chemical analysis (ESCA) and ir data it was concluded that a complex, dynamic mixture was formed of 1:1 $Ca(OH)_2$-carbohydrate adducts having a weakly polarized, coordinated character.[22] The results were not contradictory with ene-diol structure of the catalytic complex (1) postulated by Fujino and coworkers.[23] It was shown that sugar formation occurs only within the formaldehyde or carbohydrate complex, since the reaction products obtained in D_2O medium did not contain C-D bonds.[22]

$$
\begin{array}{c}
\text{H} \\
| \\
\text{C} - \text{O}^{-}\cdots \\
|| \qquad\qquad\;\; \text{Ca}\big\langle{}^{\text{OH}}_{\text{OH}} \\
\text{C} - \text{O}^{-}\cdots \\
| \\
\text{H}
\end{array}
$$

I

A remarkably selective conversion of formaldehyde into glucose was reported by Weiss and co-workers.[24] The reaction, carried out at 98°, was initiated by adding aqueous sodium hydroxide to a calcium chloride-formaldehyde solution. Under these conditions the induction period was 15 seconds, and at 18% formaldehyde conversion level hexoses were formed with 84.3% selectivity. Analysis by glc of the products [in the form of trimethyl silyl (TMS) derivatives] revealed that no branched sugars were formed and glucose constituted about 90% of the reaction mixture. It should be noted that analysis—gas chromatography-mass spectrometry (gc-ms) of the corresponding alditol acetates—of the mixtures obtained in the formose reaction carried out under normal conditions revealed the presence of 33-37 components.[14]

The catalytic activity of different hydroxides in the presence of a co-catalyst (glycolaldehyde or glucose) was examined. In the presence of glycolaldehyde, the activity of strontium hydroxide was found comparable to that of calcium hydroxide. Catalysis with barium or strontium hydroxide, however, led to a significant increase in the yield of heptoses. Calcium hydroxide showed the best selectivity as a catalyst for normal chain (nonbranched) sugar synthesis.[25] The rate of the autocatalytic condensation of formaldehyde in the range of 30 to 90% conversion was found to be $3.5 \cdot 10^{-3}$ mol min^{-1}, independent of the glucose co-catalyst concentration. The energy of activation was estimated as being close to that for glycolaldehyde condensation (21 kcal mol^{-1}). Increase in the concentration of formaldehyde from 0.8 to 6.5 M, while maintaining a constant concentration of calcium hydroxide (0.19 M) and glucose (4.16 $\cdot$ 10^{-2} M), led to a drop in formaldehyde condensation with a simultaneous increase in the rate of the Cannizzaro reaction. Experiments with ^{14}C labeled glucose indicated that a co-catalyst takes part in both the synthesis and breakdown of the carbohydrates. Both processes can proceed in absence of formaldehyde.[26] The catalytic activity of lead oxide in the formose reaction was investigated. A long induction period was observed but this could be overcome by addition of a monosaccharide. The solubility of lead oxide in the reaction medium could be increased significantly by the addition of glucose.[27] The reaction of a 1.53 M solution of formaldehyde in the presence of 0.187 M calcium hydroxide at 40° leads mainly to glyceraldehyde (89% of total sugar content). Addition of 0.0416 M glycolaldehyde to the reaction mixture results in the formation of C_4-C_7 sugars in 83% yield. The role of the co-catalyst possessing an α-hydrogen atom, which can be eliminated under basic catalysis, was discussed.[28] Specific base catalysis of the formose condensation, carried out in the pH range 10-12.5, was suggested on the basis of kinetic measurements.[29] The "crossed-formose" reaction between formaldehyde and acetaldehyde was examined and found to produce, in addition to formose sugars, 2-deoxy-DL-ribose, 2-deoxy-DL-xylose and 2-deoxy-DL-glucose.[30] Recent study on the kinetics and mechanism of the sugar-forming step of the formose condensation reaction, carried out with the aid of high performance liquid chromatography (hplc) and GC-MS methods, led to the conclusion that formaldehyde adds mainly to the intermediates and that formation of *trans*-addition products (e.g. xylose) is favored. In the presence of excess formaldehyde only a minor amount of branched monosaccharides was formed. The Cannizzaro reaction was found to produce polyols as well as aldonic acids, and saccharinic acids were also found in the reaction mixture. Almost no cleavage of the intermediate and secondary reaction end products was observed.[31]

Shigematsa and co-workers have shown that changes in the oxidation-reduction potential (ORP) correspond to the characteristic phases of the formose reaction. In a particular induction period, the saccharide forming and saccharide decomposing stages (as recognized by chemical determination of total sugar content) were found to correlate well with characteristic changes of the ORP curve.[32,33] By considering the minimum of the ORP curve as the beginning of the formose-forming step and the maximum as the end of the reaction (the so-called yellowing point), the influence of the various factors (formaldehyde concentration, calcium hydroxide quantity and initial pH on the induction period, sugar-forming step, and sugar yield) was evaluated.[34] The kinetics of the formose condensation under homogeneous conditions, catalyzed by calcium formate-potassium hydroxide at pH 10.5-12.0, was examined by ORP measurement. A linear correlation of the concentration of the dissolved calcium species with induction period (T min) and sugar forming period (T_{max}-T_{min}) at a given pH was observed. It was concluded that $CaOH^+$ constitutes the principal catalytic species during the induction period.[35]

The removal of calcium ions from the reaction mixture at the point corresponding to the ORP maximum, by precipitation with oxalic acid, resulted in selective formation of branched sugar alcohols: 2-hydroxymethylglycerol (**2**) and 2,4-bis(hydroxymethyl)1,2,3,4,5-pentanepentaol (**3**). Another major component of the reaction mixture consisted of diastereoisomeric mixture of 3-hydroxy-methyl-1,2,3,4,5-pentanepentaol (**4**).[36]

```
                            CH2OH                    CH2OH
                             |                        |
                        HO-C-CH2OH                   CHOH
         CH2OH              |                         |
          |            H-C-OH                    HO-C-CH2OH
    HO-C-CH2OH              |                         |
          |           HO-C-CH2OH                   CHOH
        CH2OH               |                        |
                           CH2OH                   CH2OH

         2                   3                       4
```

It is well known that addition of methanol to the formose reaction suppresses the Cannizzaro and the crossed-Cannizzaro reactions. In the course of a systematic study by Shigemasa's group on the effects of organic solvents on the formose condensation, the reaction performed in methanol was compared with that performed in aqueous solution. In aqueous methanol, the reaction can occur at a lower ratio of catalyst to formaldehyde ($\sim$0.05) than in water. Contrary to the standard conditions, in methanol the sugar yield becomes higher with increased formaldehyde concentration. The rates of formaldehyde consumption and sugar decomposition increase with pH. The distribution of the products obtained in water and methanol was different, but both reactions were found to be nonselective.[37] An attempt was made to correlate solvent polarity and its ability to form adducts with the catalyst, using the results of the formose reaction. It was found that the induction period (T_i) of reactions performed in alcoholic solvents increased with decreased water content. The T_i value depended on the kind of alcohol used, in the following order: t-BuOH > i-PrOH > EtOH > MeOH. The sugar yield obtained in aqueous alcohols increases in the reverse order. With the exception of methanol, the reaction did not occur in nonaqueous hydroxylic solvents. The reaction, when carried out in pure methanol, shows a remarkable decrease in the induction period compared with that in 90% aqueous methanol, suggesting that the formation of the formaldehyde-hemiacetal may play an important role in sugar formation. The reactions performed in aqueous ethylene glycol, ethylene chlorohydrin, and methyl cellosolve afforded sugars in 46%, 37%, and 13% yield respectively, as compared with 43-44% obtained in pure methanol or water.[38] It was found that the condensation of methanolic formaldehyde, catalyzed by barium chloride-potassium hydroxide at pH 12, leads to the selective formation of the branched ketone: 2,4-di-C-(hydroxymethyl)-2-pentulose (5). The yield of 5, determined by glc, was 33% (total yield of sugars 50%). The compound was isolated by chromatography on cellulose and was compared, after reduction with sodium borohydride, with an authentic sample of 2,4-di-C-(hydroxymethyl) pentitol.[39]

The irradiation of aqueous formaldehyde with the UV light in the presence of an inorganic base also produced a mixture of sugars and sugar alcohols, but the distribution of products was entirely different from that obtained in thermal

$$
\begin{array}{l}
\text{CH}_2\text{OH} \\
| \\
\text{HO}-\text{C}-\text{CH}_2\text{OH} \\
| \\
\text{C}=\text{O} \\
| \\
\text{HO}-\text{C}-\text{CH}_2\text{OH} \\
| \\
\text{CH}_2\text{OH}
\end{array}
$$

5

reaction. The reaction carried out in 0.62 M sodium carbonate solution resulted in the formation of 8.7% sugars (calculated as glucose). The major products were pentaerytritol and 2-hydroxymethylglycerol.[40]

Some solid catalysts have been tried for the selective condensation of formaldehyde to polyols. CSH gel (tobermorite) calcinated at 600° was found to be very active as a formose catalyst. Calcium oxalate precipitated at the minimum ORP in the formose reaction and heated at 40-300° for 4 hours promoted the selective formation of the branched sugar alcohols 2-4. An aqueous 30% solution of formaldehyde was converted into glycolaldehyde on contact with NaX, 5Å, molecular sieves, or Na Mordenite at 94° with the pH maintained at about 11 by the addition of a sodium hydroxide solution. A crossed-Cannizzaro side reaction converted a fraction of the product into ethylene glycol.[41] It was shown that among the arbitrarily chosen collection of 34 minerals about 50% were catalytically active in the condensation of formaldehyde leading to formose sugars.[5] Triose aldol condensation, a possible secondary process in the formose reaction, has been reinvestigated using gc-ms of isopropylidene derivatives for determination of the hexuloses formed. The reaction between glyceraldehyde and 1,3-dihydroxy-2-propanone catalyzed by sodium, calcium, or barium hydroxide was found to give racemic uloses: dendroketose (10-18%), psicose (8-12%), tagatose (2-5%), fructose (40-45%), and sorbose (25-32%). Glyceraldehyde alone, taken as a substrate, gives a similar distribution of the products due to its rapid isomerization to 1,3-dihydroxy-2-propanone. The use of strongly basic anion exchange resins favors the formation of fructose, which amounts to 67-78% of the reaction mixture.[42] The formation of sugars in the aldol condensations of lower aldehydes or oxy-oxo compounds, promoted by diluted calcium or barium hydroxide, as well as in the reaction of acrolein dibromide with barium hydroxide was again discussed.[43]

3.   SYNTHESES FROM VINYLENE CARBONATE

The dienophilic properties of vinylene carbonate (1,3-dioxol-2-one, 6) have been repeatedly exploited for the synthesis of cyclic polyols.[1] Recently a number of ribofuranose derivatives have been synthesized starting from 6. The known adduct 7, obtained from 6 and furan, was *cis*-hydroxylated, and after acetonation was hydrolyzed with base and cleaved with permanganate to give the dicarboxylic acid 8. The corresponding anhydride 9 gave, on treatment with trimethylsilyl azide, the isocyanate 10, which in turn was converted into the carbonate 11 on addition of methanol. The monoester 12, obtained in the reaction of anhydride 9 with isopropanol, was effectively resolved into enantiomers by the use of brucine or (R)-1-(2-naphthyl)ethylamine.[44]

The simple synthesis of DL-apiose, completed by Ishido, also makes use of **6** as the substrate. Photochemical cycloaddition of 1,3-diacetoxy-2-propanone to **6** gave an oxetane derivative **13,** which on alkaline hydrolysis afforded apiose in 23% overall yield.[45]

Another idea using **6** as an equivalent of 1,2-di-hydroxyethane unit in the construction of a carbohydrate chain has been put forward by Tamura, Kunieda,

$$6 \;+\; \text{AcOH}_2\text{C} \overset{\text{O}}{\overset{\|}{\text{C}}} \text{CH}_2\text{OAc} \;\longrightarrow\; \text{AcOH}_2\text{C} \overset{\text{CH}_2\text{OAc}}{\underset{}{}} $$

13

$$13 \;\longrightarrow\; \begin{array}{c} \text{CHO} \\ | \\ \text{H--C--OH} \\ | \\ \text{C--OH} \\ \text{HOH}_2\text{C} \quad \text{CH}_2\text{OH} \end{array}$$

and Takizawa.[46,47] These authors have found that **6** reacts with polyhalogeno-methanes in the presence of free radical initiators to give a mixture of telomers of the general formula **14**.

The products containing up to three dioxolanone units can be separated from higher telomers by extraction with hot dichloromethane. The lower telomers thus obtained are readily separable by column chromatography on silica gel. It has been found that the distribution of products in the telomerization reaction depends strongly on the kind of the telogen (halogenomethane), as well as on the ratio of the reagents used. Thus, the reaction of **6** with four equivalents of bromo-trichloromethane gives **21** in 92% yield, whereas the reaction of the same sub-strate with five equivalents of carbon tetrachloride affords a mixture comprising comparable amounts of the telomers containing one and two dioxolanone units (**15**, $n = 1,2$). From the same reaction 6% of the corresponding compound having a seven-carbon chain (**15**, $n = 3$) was isolated.[48] It has been found that chloroform reacts with **6** by hydrogen transfer to give **16**, whereas dibromo-methane or tribromomethane gives two series of telomers, each (**17-19**) resulting from abstraction of a bromine or hydrogen atom. In addition, the reaction of tribromomethane and tetrabromomethane with **6** gave "twofold addition" product (**20**).[49] These compounds arise from a 1,3-rearrangement of the intermediate

$$6 \;+\; \text{R X} \;\longrightarrow\; R\!-\!\left[\begin{array}{c} \\ \\ \text{O} \quad \text{O} \\ \text{O} \end{array}\right]_n \!\!-\!X$$

14

$R = CCl_3, CBr_3, CHBr_2, CHBr$

$X = H, Cl, Br$

$n = 1, 2, 3$

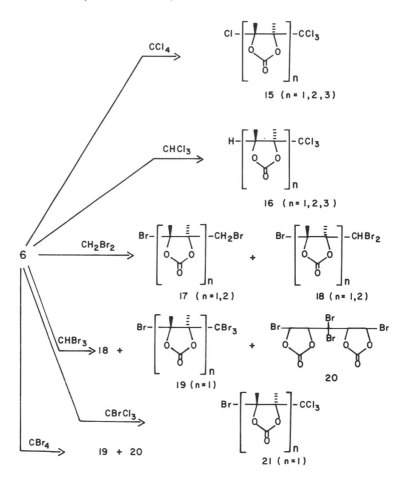

radical. They can also be obtained in high yield by the reaction of **6** with **19**. The variety of the lower telomers obtained from different telogens is presented (**15-21**).

The formation of the aforementioned compounds proceeds with remarkable selectivity. Only one stereoisomer has been identified for each constitutionally different "mono" telomer (**14**, $n = 1$), whereas for compounds with $n = 2$ and $n = 3$, two and four stereoisomeric compounds have been isolated, respectively.[50] It has been assumed that *trans* addition to **6** is strongly favored under the conditions used. This assumption was substantiated by the small coupling constant (Jvic $\approx$ 2.0 Hz) that was consistently observed for all telomers.[48] Rigorous proof of the *trans* stereochemistry comes from the chemical transformations of the telomers. Thus, compounds of the general formula **14** contain protected α-hydroxyaldehyde function, and in principle they should be readily transformable into racemic monosaccharides or their derivatives. In fact, telomers **15** and **22** are quantitatively hydrolyzed, on dissolving in water, to afford derivatives of 3-deoxy-glyceraldehyde, which were isolated in their dimeric form. The trichloromethyl compound **15** selectively forms a five-membered ring dimer **23**, whereas the dichloromethyl telomer **22** yields a six-membered cyclic dimer **24**. The direct reduction of the telomers (**15, 22**) with sodium borohydride gives alditols possessing a terminal halogenomethyl group, in high yield. The hydride reduction of **22** followed by hydrolysis with aqueous silver nitrate gave DL-glyceraldehyde, albeit in low yield. The treatment of the telomers with methanol, on the other hand, produces the corresponding dimethylacetals with high efficiency.[51] These transformations, which are characteristic also for $n = 2$ and $n = 3$ telomers, are summarized in Scheme 1.

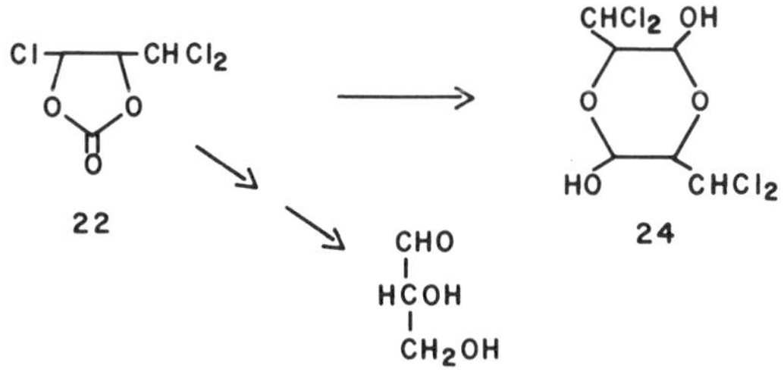

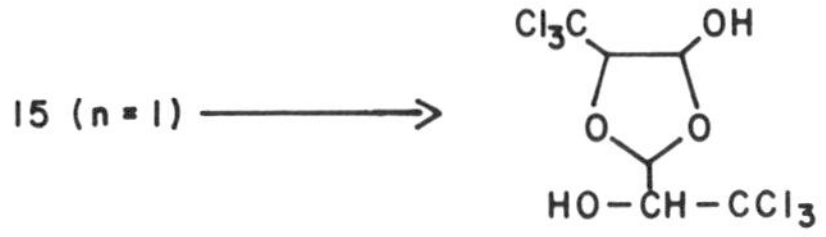

The isomeric "twofold addition" products **20** also gave cyclic acetals on hydrolysis. The ^{1}H nmr analysis of the products (**25, 26**) obtained after hydrolysis and methylation allowed for the assignment of their configuration.[52]

It is evident from Scheme 1, formulas **15-24**, and formulas **20a & b** that both ends of telomers **14** can be converted into aldose C-1, although the efficiency

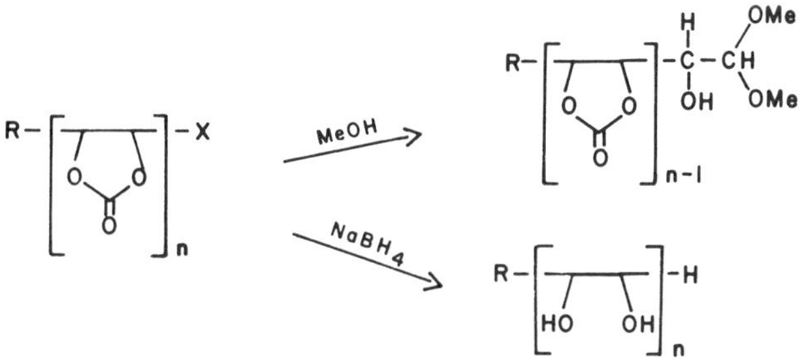

Scheme 1

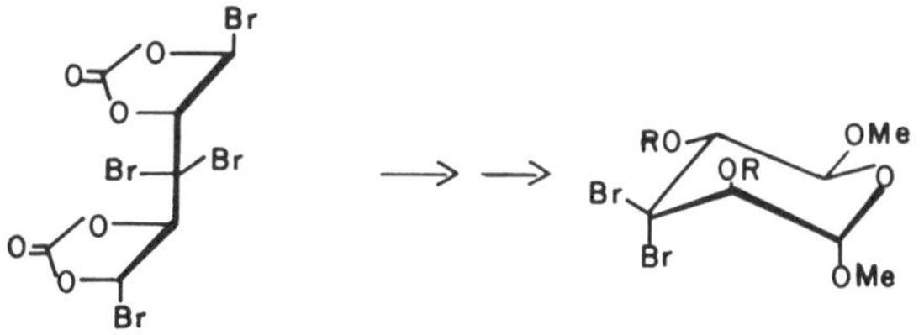

20a 25

R= H, COOCH3

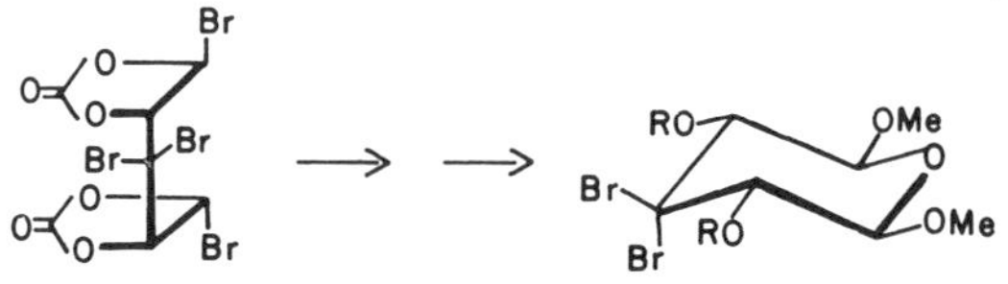

20b 26

of the route involving hydrolysis of the terminal dichloromethyl group seems higher. Two ways have been devised for converting readily available telomers containing trichloromethyl group into the corresponding dichloromethyl compounds, directly hydrolyzable to aldoses. The irradiation of a -CCl_3 containing telomer with Hg-lamp in oxolane brings about the selective photoreduction to the dichloromethyl derivative.[53] The same goal can be achieved by treatment with nickel carbonyl.[54,55] Application of the latter reagent also allows for the selective conversion of dibromomethyl compounds into bromomethyl ones. The last reaction has been used for the correlation of stereochemistry of the telomers obtained from **6** and bromomethanes (**17, 18**, $n = 2$). The isomeric monobromoderivatives **27** and **28** were hydrolyzed to afford 5-bromo-5-deoxy-DL-lyxose

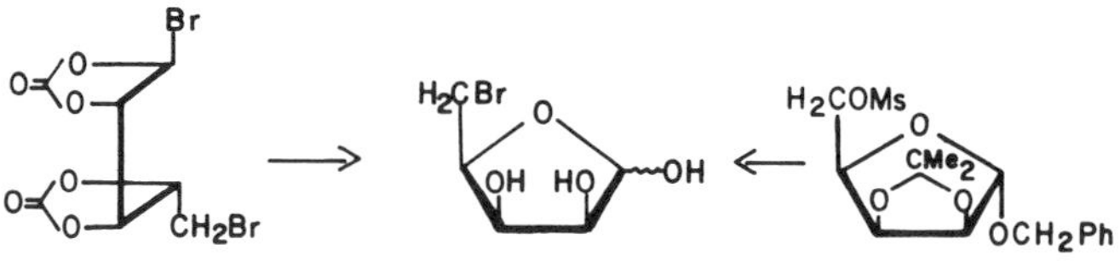

27 29 31

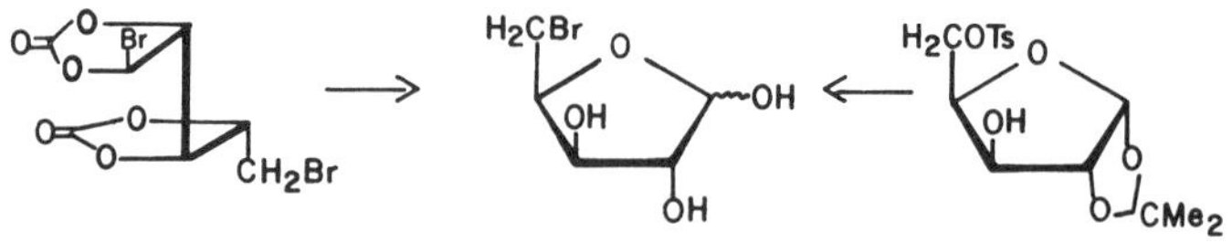

28 30 32

29 and 5-bromo-5-deoxy-DL-xylose **30**. Authentic samples of these compounds were prepared from D-mannose and D-xylose by way of derivatives **31** and **32**.[48]

The reaction route involving the dihalogenomethyl intermediate has been used for preparation of racemic aldopentoses. Thus, the telomers obtained from **6** and chloroform or carbon tetrachloride (**15,16**, $n = 2$) were converted into DL-arabinose and DL-xylose by a two-step procedure comprising photoreduction (~80% yield) and hydrolysis with aqueous silver nitrate (~55% yield).[56] In an analogous way, selected seven-carbon telomers gave DL-glycero-DL-*ido*-heptose (30%), characterized as the corresponding hexaacetate. All four individual telomers **15** ($n = 3$) were reduced photochemically to corresponding dichloromethyl compounds and were hydrolyzed to heptoses, which were then converted to alditols and identified as: DL-glycero-LD galacto, DL-glycero-DL-ido, DL-glycero-LD-ido and DL-glycero-DL-galacto heptoses.[57,58]

It has been demonstrated that a halogen at the secondary carbon atom of the telomers **15** and **17-21** can be readily substituted by a variety of nucleophiles. Particular attention has been paid to the reaction of telomers containing terminal trichloromethyl groups with cyanide anion, because it opens access to the "odd carbon" sugars. The formation of nitriles by nucleophilic substitution of a halogen atom in the telomers, induced by phase transfer catalyst, is a nonselective process leading to equal amounts of epimeric cyanides. Separation of the isomers **33** and **34** obtained from **22** followed by esterification with methanol-hydrogen chloride, reduction with sodium borohydride, and hydrolysis in the presence of silver nitrate afforded both tetroses. Following the same scheme, isomeric nitriles obtained from telomers **15** ($n = 2$) were converted into racemic hexoses. The *trans-syn* (**35**) and *cis-syn* (**36**) nitriles gave rise to galactose and altrose, whereas the *anti-trans* (**37**) and *anti-cis* (**38**) isomers afforded idose and glucose respectively.[56]

The most abundant telomer **15,** containing three dioxolanone units, was converted into a pair of epimeric nitriles, which could not be separated, but the differences in their reactivity toward methanolic hydrogen chloride allowed the isolation of a single compound: DL-threo-DL-*ido*-octose, characterized as the heptaacetate and identified by direct comparison as the corresponding octitol.[58] Among the other reactions of the telomers, which are of potential value for the synthesis of carbohydrate analogs, those with tervalent phosphorus compounds and with amines should be mentioned. The telomers **14** ($n = 2,3$) readily reacted with phosphites and phosphonites to afford the *trans* enolphosphates in a high yield. This reaction has been used for correlation of configuration of stereoisomeric telomers.[59,60] 4-Halogeno-1,3-dioxolan-2-ones react with ammonia and primary aliphatic amines in methanol with the formation of *trans* 4-hydroxy-2-oxazolidones (for example, **39**). The latter react smoothly with benzene in the presence of sulfuric acid, affording **40**. On the other hand, the reaction of **39** with trifluoroacetic acid results in the dehydration product **41**.[61] Dehydrohalogenation of the starting telomers and following transformations are also of synthetic interest.[50] The ionic addition of methanol or ethanethiol to **6** in the presence of unspecified metallic compounds has been reported. Products of the general formula **42** were formed. It was shown that **42** ($n = 2$) products consist mainly of *"cis-syn"* telomers leading to erythrose on hydrolysis.[62] The possibility of the synthesis of aminodeoxy sugars by telomerization of carbon tetrachloride

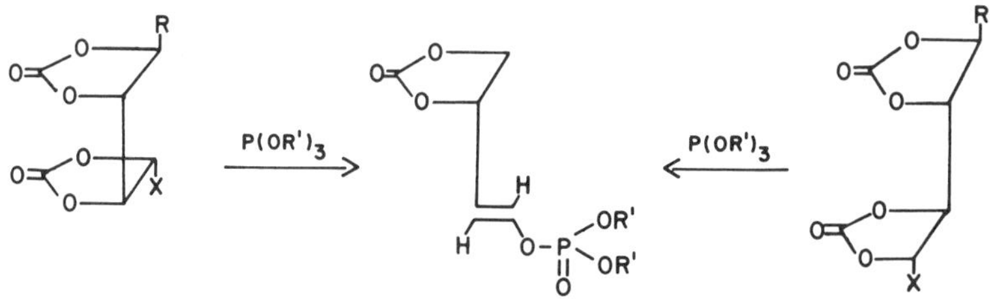

Scheme 2

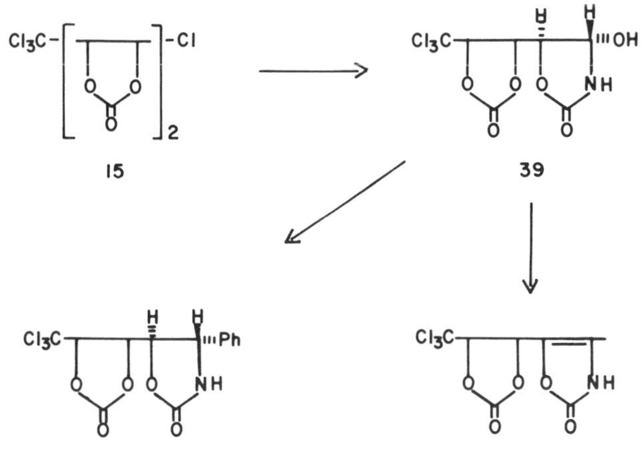

15

39

40

41

and 3-acetyl-2-oxazolone, a close structural analog of vinylene carbonate, has also been considered.[63]

4. SYNTHESES FROM ACETYLENIC AND OLEFINIC PRECURSORS

An olefin of defined stereochemistry about the double bond can readily be converted into a *threo* or *erythro* diol by direct *cis*-hydroxylation or by epoxidation followed by hydrolytic opening of the epoxide obtained ("*trans*-hydroxylation"). *cis*-Hydroxylation of a *cis* olefin leads to an *erythro* diol and formation of an epoxide and its hydrolysis gives a *threo* stereoisomer. Conceivably, from a *trans* olefin both stereoisomeric diols may be obtained by *cis*- or "*trans*"-hydroxylation applied in the reversed way.

The stereochemistry of many open chain olefins is known. From acetylenic substrates *cis* or *trans* olefins can be conveniently prepared by partial hydrogenation over the well-known Lindlar catalyst or by lithium aluminum hydride reduction (in cases when a hydroxyl group is present next to the triple bond).

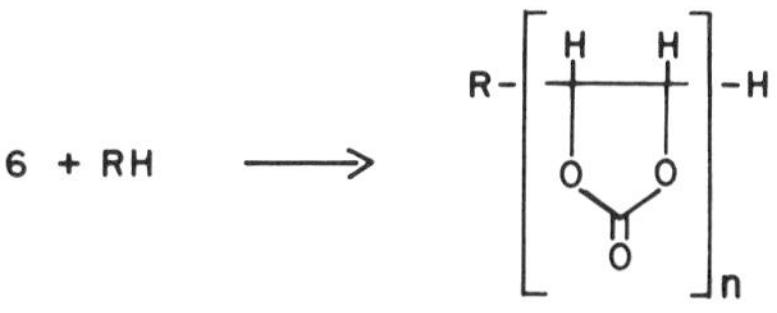

6 + RH

42   R = OMe, SEt
        n = 1, 2, 3

These diol syntheses have found wide application in synthetic carbohydrate chemistry since the early 1930s and were employed many times in the past decade.

A. Acetylenes

Nagakawa and co-workers investigated the use of simple acetylenic alcohols for the synthesis of trioses, tetroses, and 2-deoxy-*erythro*-pentose.

Dihydroxyacetone (**44**) was conveniently prepared[64] from halogenopropargyl alcohol (**43**) by hydration of the triple bond followed by substitution of a halogen atom for the OH group in a reaction with potassium formate. DL-Erythrulose was obtained from 2-butyne-1,4-diol (**45**) in an improved way involving conversion to 1-acetoxy-3-buten-2-one (**46**), formation of the bromohydrin (**47**), substitution of the bromine atom for an OAc group in a reaction with silver acetate, and final deacetylation.

The two racemic tetroses have been prepared[65] from **43** by a sequence of reactions involving the extension of the carbon atom chain by a Grignard reaction with ethyl orthoformate to four atoms, acetylation, and half-reduction of the triple bond to a *cis* olefin (**48**). *cis*-Hyroxylation of **48** with potassium permanganate, peracetylation (for the purpose of isolation), and final hydrolytic deprotection gave DL-erythrose. "*trans*-Hydroxylation" of **48** with peroxyacids failed. Therefore acetal **48** was hydrolyzed with 70% aq. formic acid to 4-acetoxycrotonaldehyde (**49**). Reacetalization of **49** followed by *cis*-hydroxylation led to DL-threose in about 14% yield.

Barton's method of the conversion of alcohols into aldehydes, consisting of the reaction of the chloroformate ester with dimethylsulfoxide, was successfully applied[66] to the preparation of 2,3-isopropylidene-DL-glyceraldehyde (**51**) from 1,2-*O*-isopropylidene-glycerol (**50**). The same oxidation method was used for the preparation of 2-deoxy-DL-*erythro*-pentose from 5-(2'-tetrahydropyran-yloxy)-3-pentyn-1-ol (**52**). The synthesis was not satisfactory because the yields of some steps were low. An alternative approach to the same sugar involved[66] the direct hydroboration of the triple bond in *erythro* 4-pentyn-1,2,3-triol (**53**).

B.   Olefins

The olefinic substrates employed in carbohydrate synthesis can be roughly divided into three basic groups comprising alcohols, aldehydes or hydroxyaldehydes, and hydroxy acids or lactones.

(a)   Unsaturated Alcohols

Substrates of this type were utilized for the synthesis of alditols. Schneider and co-workers[67] prepared from *trans* and *cis* 2-buten-1,4-diols 2,3-anhydro-DL-erythritol (54) and -threitol (55) by direct epoxidation with peroxybenzoic acid. This method of epoxide formation failed in the case of 2E,4E-hexadiene-1,6-diol; from the 2Z,4Z isomer the corresponding diepoxide (57) was obtained in a 2% yield only. However, from 1,6-dibromo-and 1,6-dibenzyloxycarbonyloxy-2E,4E-hexadienes the corresponding derivatives of 2,3:4,5-dianhydro-DL-galactitol (56) could be obtained in moderate yields. Compounds of this type are active against some forms of sarcoma.

It has been found that aldehydes of the cinnamaldehyde type undergo an interesting biotransformation[68] in the medium of growing baker's yeast yielding optically active methyl diols (e.g., 58, 59).

The diol 59 was used for an ingenious synthesis of L-mycarose (62, 2,6-dideoxy-3-C-methyl-L-*ribo*-hexose) and olivomycose (63, 2,6-dideoxy-3-C-methyl-L-*arabino*-hexose). The sequence of reactions consisted in: 1° epoxidation of the double bond and separation of both stereoisomeric epoxides formed, 2° reduction of the oxirane ring to form two triols 60 and 61, and 3° ozonolytic degradation of the aromatic ring to a carboxyl group. Each trihydroxyacid was converted into a lactone and the lactones were eventually reduced to the required branched-chain sugars 62 and 63.

Hydrogenation of the double bond in 58 afforded the diol 64, which was ozonized to 4,5-dihydroxy-hexanoic acid, isolated in the form of a γ-lactone. Reduction of the lactone grouping to the lactol stage afforded 2,3,6-trideoxy-L-*erythro*-hexopyranose (65, R = H, L-amicetose).

Olivomycose (63) also was synthesized[69] from another methyldiol (66), which was biosynthesized by baker's yeast from 2-methyl-5-phenyl-2E,4E-pentadienol.

$$CH_2OH$$
$$\begin{array}{c} H-C \\ H-C \end{array}\!\!\!>\!\!O$$
$$CH_2OH$$

54

$$CH_2OH$$
$$\begin{array}{c} C-H \\ H-C \end{array}\!\!\!>\!\!O$$
$$CH_2OH$$

55

$$CH_2OX$$
$$H-C>O$$
$$C-H$$
$$C-H$$
$$H-C>O$$
$$CH_2OX$$

56

$$CH_2OH$$
$$H-C>O$$
$$H-C$$
$$H-C$$
$$H-C>O$$
$$CH_2OH$$

57

X = Br , OCO_2CH_2Ph

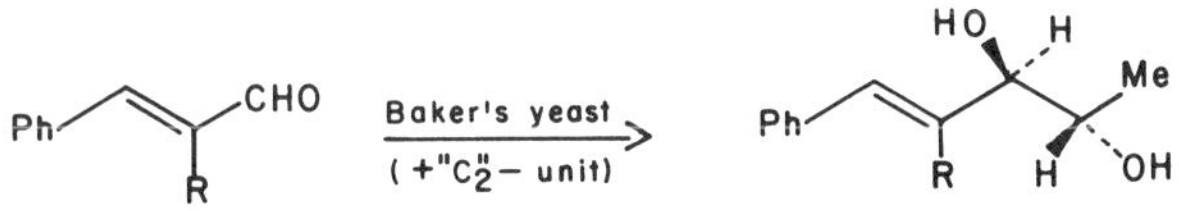

58 R = H
59 R = Me

In this case, diol **66** was converted into the monobenzoate **67** and was epoxidized at the trisubstituted double bond. The epoxide was reduced and the triol
obtained was ozonolytically degraded to the lactone of olivomycosonic acid (**68**).
The half-reduction of the lactone **68** with diisobutylaluminum hydride (DIBAH)
furnished **63**.

(b) Unsaturated Aldehydes and Hydroxyaldehydes

Dyong and Wiemann[70] elaborated a general approach to stereoisomeric alkyl 3-
amino-2,3,6-trideoxyhexopyranosides. The most prominent member of this class
is daunosamine having L-lyxo configuration. The substrate for this synthesis,
4E-hexenal (**70**), was prepared by the Claisen rearrangement of 3-buten-2-yl
vinyl ether (**69**). Acetalization with DL-2,3-butanediol gave dimethyldioxolane
(**71**). Allylic amination of **71** with Sharpless' selenium-chloramine T reagent led

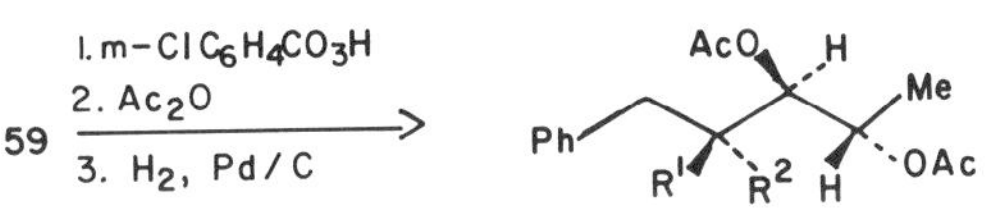

60 R^1 = Me , R^2 = OH
61 R^1 = OH , R^2 = Me

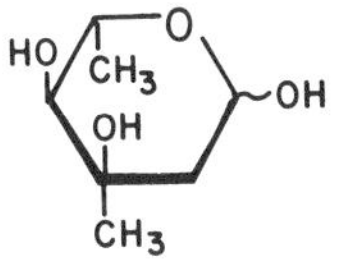

62

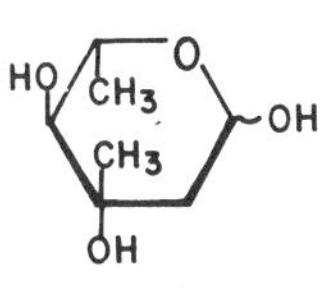

63

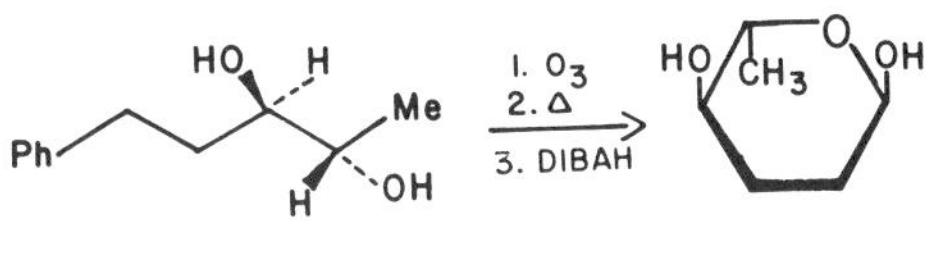

64

65

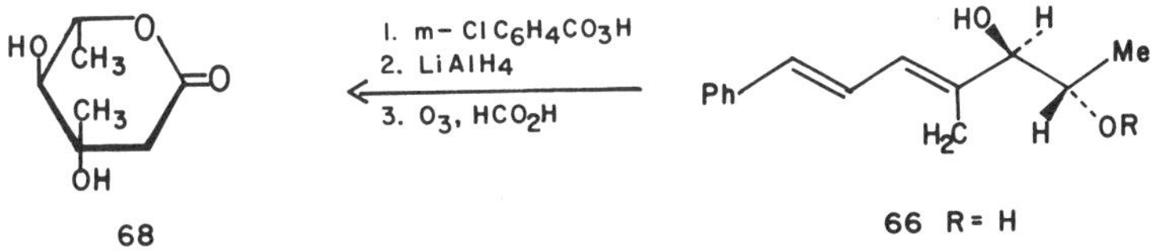

68

66 R = H
67 R = PhCO

CH₃ / O
69

CH₃ / CHO
70

2,3-Butane-diol / TsOH

CH₃ / CH / O-CH₃ / O-CH₃
71

Se, Chloramine T →

CH₃ / CH / O-CH₃ / O-CH₃ / NHR
72 R = Ts
73 R = Ac

CH₃ / N / O / OsO₄ →

CH₃ / OH / HO / CH / O-CH₃ / O-CH₃ / NHAc

1. 2N HCl
2. Me₂CHOH
3. Ac₂O

CH₃ / CHO / CH₃
75

→

CH₃ / AcO / O / OAc / NHAc / CH₃
76

CH₃ / R'O / O / OR² / NHAc
74 R¹ = Ac
      R² = i - Pr

mainly to the 3-*p*-toluene-sulfonamide derivative **72**. Detosylation of **72** with sodium in liquid ammonia followed by acetylation gave the acetamide **73**. *cis*-Hydroxylation of **73** afforded a mixture of stereoisomeric acetamidodiols. Hydrolytic removal of the dioxolane grouping followed by formation of isopropyl glycosides and their acetylation gave crystalline isopropyl 3,4-di-N,O-acetyl-DL-daunosaminide (**74**, R^1 = Ac, R^2 = i − Pr).

In a similar way, 3-acetamido-1,4-di-O-acetyl-3-C-methyl-2,3,6-trideoxy-β-DL-*lyxo*-hexopyranose (**76**, N-acetyl-1,4-di-O-acetyl-β-DL-vancosamine) was synthesized[71] starting from 3-methyl-4E-hexenal (**75**).

A simple synthesis of a glycoside of DL-daunosamine has been devised by Matsumoto and his co-workers.[72] In this approach 1-chloro-1,4-hexadien-3-one (**77**, obtained from crotonyl chloride and vinyl chloride) was converted into 1,1-ethylenedioxy-4E-hexen-3-one (**78**). *cis*-Hydroxylation of **78** afforded *threo*-diolon (**79**). Oximation of **79** and reduction of the oxime gave a single stereoisomeric aminodiol (**80**) which, after treatment with methanolic hydrogen chloride, yielded methyl α-DL-daunosaminide (**74**, R^1 = H, R^2 = Me) in 84% yield. Makin and co-workers[73] described two methods leading to both stereoisomeric 2-deoxy-DL-pentoses in the form of their diethyl acetals. In the first method, the readily available 1,1-diethoxy-3-penten-5-ol (**81**) was directly *cis*-hydroxylated furnishing the *threo* stereoisomer **82** in a 61% yield. The *erythro* compound **83** was obtained by epoxidation of the substrate followed by alkaline hydrolysis of the epoxide.

In the second synthesis, 1,1,3-triethoxy-4-pentene (**84**) served as substrate. Epoxidation of the double bond followed by hydrolysis afforded mixture of both stereoisomeric 2-deoxy-DL-pentoses in the form of their 1,1,3-tri-0-ethyl derivatives (**85** and **86**).

P. Chautemps[74] obtained diepoxides from 1,4-pentadien-3-one (**87**) as a mixture of DL and *meso* compounds (**88:89** as 13:7). Reduction of the carbonyl group in DL-diepoxide (**88**) followed by alkaline hydrolysis gave DL-arabinitol. Analogously, from *meso* diepoxide **89** a mixture of DL-ribitol and DL-xylitol was obtained.

(c)   Unsaturated Hydroxyacids and Lactones

Many unsaturated hydroxyacids, with the stereochemistry of the double bond known, are readily available. Direct hydroxylation converts them into trihydroxy acids (aldonic acids). A number of effective reagents have been developed within the past 15 years for the half-reduction of the carboxyl group to the aldehyde function stage. These are the reasons for particularly wide application of these substrates in sugar synthesis.

Nakagawa and co-workers[75] performed a synthesis of racemic and both enantiomeric forms of 2-deoxy-*erythro*-pentose from 3-hydroxy-4-pentenoic acid (**90**). The synthesis consisted of the reaction of **90** with N-bromosuccinimide (NBS) to give the bromolactone **91**, substitution of the bromine atom for a hydroxyl group to form **92** and reduction of the lactone grouping in **92** with bis[2-(3-methylbutyl)] borane to give the desired product. Optically active forms of the acid **90,** which served for the synthesis of enantiomeric 2-deoxy-*erythro*-pentoses, were readily obtained by resolution of the racemate with quinine.

Dyong and Glittenberg[76] synthesized evermicose (**96**, 2,6-dideoxy-3-C-methyl-D-*arabino*-hexose) along the pattern elaborated earlier by Grisebach and co-workers[77] for the synthesis of DL-mycarose and its 3-epimer. The synthesis started[76] from ethyl 3-hydroxy-3-methyl-4E-hexenoate (**93**), which was used in racemic as well as in enantiomeric (3R) form. The ester **93** was epoxidized with peroxyacetic acid to form, almost exclusively, a *xylo* epoxide **94**. Treatment of **94** with diluted sulfuric acid yielded the γ-lactone (**95**). Reduction of the lactone with diisobutylaluminum hydride afforded smoothly 3-epi-DL-mycarose, whose D-form (**96**) bears the name evermicose. The reasons for the remarkable stereospecificity of the epoxidation reaction (**93** → **94**) are, according to Dyong,[76] the steric factors connected with the alkoxy group present in the ester grouping.

In an attempt to obtain garosamine (**99**, 3-deoxy-4-C-methyl-3-methylamino-L-*arabino*-pentopyranose) by a totally synthetic method, Dyong and Jersch[78] reacted 4-methyl-3-penten-5-olide (**97**) with chloramine T-osmium tetraoxide reagent ("oxyamination" reagent) and obtained the expected N-tosylaminoalcohol **98** in 50% yield. However, the next steps of the synthesis, N-methylation and acetoxylation at C-2, proceeded with very low yields; therefore, this synthetic route was not completed.

Dyong's group elaborated a highly successful approach to several antibiotic sugars starting from sorbic acid (**100**). The essential step of this approach consists[79] in the monoepoxidation of the substrate with peroxyacetic acid, which leads exclusively to the 4,5-epoxide (**101**). This epoxide can be opened with a number of nucleophiles. Acidic hydrolysis gives[79] *erythro* 4,5-dihydroxy-(hex-2-enol) acid (**102**). Hydrogenation of the double bond in **102** followed by lactonization to a γ-lactone (**103**), and reduction of the lactone carbonyl group with DIBAH, gave DL-amicetose (2,3,6-trideoxy-DL-*erythro*-hexopyranose) in 67% yield.

Opening of the epoxide **101** with dimethylamine leads[80,81] to two regioisomeric dimethylamino alcohols, **104** and **105**, in a ratio of 4:1. Catalytic hydrogenation of **104**, lactonization by means of acetic anhydride, and DIBAH reduction of the lactone **106** completed a synthesis of racemic form of another antibiotic sugar, forosamine (**107**, 2,3,4,6-tetradeoxy-4-dimethylamino-DL-*erythro*-hexopyranose). When the starting epoxide **101** was resolved by means of its (+)-phenylethylammonium salt into enantiomers, the same reaction sequence furnished the pure D-form of the sugar.

The racemic and L forms of the epoxide **101** were next employed[82] for the preparation of DL- and L-acosamines (**114**, 3-amino-2,3,6-trideoxy-DL- and L-*arabino*-hexoses), stereoisomers of daunosamine.

The reaction of the methyl ester of the racemic epoxide **101** with acetone and aluminum chloride gave a 4,5-O-isopropylidene derivative **108**. Michael addition of ammonia to the conjugated double bond of **108** with concomitant ammonolysis of the ester group afforded a mixture of 3-amino-2,3,6-trideoxy-4,5-O-isopropylidene-DL-*arabino*- and -*ribo*-hexonamides (**109**). For the purpose of isolation the mixture was N-acetylated. Acidic hydrolysis of the amide groupings gave a mixture of γ- and δ-aminolactone hydrochlorides (**110** and **111**). Careful N-acetylation and reduction of the lactone carbonyl groups in **112** and **113** afforded N-acetyl-DL-acosamine (**115**) in 36% yield. The stereoisomeric potential partner of **115**, N-acetyl-DL-ristosamine (**116**), was not isolated in this procedure.

In the synthesis of N-acetyl-L-acosamine (**115**), the first step, consisting in formation of **108**, could not be used because of the potential danger of racemization. Therefore, the methyl ester of the optically active epoxide **101** was regiospecifically opened with *t*-butanol in the presence of boron trifluoride to methyl *trans* 4-O-*t*-butyl-2,3,6-trideoxy-L-*erythro*-hex-2-enoate (**117**). The free OH group of **117** was protected with 2'-tetrahydropyranyl and the remaining steps of the synthesis were performed as in the case of the racemic compound **108**.

The methyl ester of the racemic epoxide **101** was used[83] also for a simple synthesis of DL-megosamine (3-dimethylamino-2,3,6-trideoxy-DL-*ribo*-hexose) which could be isolated in the form of the 1,4-di-O-acetate (**120**). In this case, a low-temperature (− 10°) Michael addition of dimethylamine succeeded without opening the epoxide ring and yielded a mixture of methyl 4,5-anhydro-3-dimethylamino-2,3,6-trideoxy-DL-*lyxo* and -*xylo* hexonoates (**118**). The products, after hydrolysis, were converted into γ-lactones. Fractional crystallization gave only the *ribo* stereoisomer (**119**). The carefully controlled reduction of the carbonyl group in **119**, followed immediately by acetylation, furnished **120**.

Optically active parasorbic acid (**121**, 2-hexen-5*S*-olide), of natural origin, has been employed as substrate in sugar synthesis since the early 1960s by Lukeš and Jarý.[84] In continuation of earlier works Jarý and his co-workers[85] investigated the reaction of the readily available epoxide of parasorbic acid (**122**, 2,3-anhydro-4,6-dideoxy-L-*ribo*-hexonolactone) with ammonia and dimethylamine. With liquid ammonia at elevated temperature 3-amino-3,4,6-trideoxy-L-*xylo*-hexonamide (**123**, R = H) was the single reaction product. Mild hydrolysis of **123** gave the free acid, which, after neutralization of the amino group with hydrochloric acid, readily formed the δ-lactone (**124**). The reaction of **122** with dimethylamine and further transformation proceeded essentially in the same way. Lactone **124** (R = H) was used[86] for the synthesis of L-desosamine (**125**), an enantiomer of the natural sugar. Half-reduction of the carbonyl group followed by dimethylation of the amino group yielded the desired monosaccharide.

Methyl 2Z,5-hexadienoate (**126**) undergoes cyclization to *RS*-parasorbic acid when treated with polyphosphoric acid. Torssell and his co-workers employed the substrate thus obtained for a new series of sugar synthesis. DL-Desosamine was synthesized[87] in an essentially similar way to that described above. Analogously, DL-chalcose (**127**, 4,6-dideoxy-3-O-methyl-DL-*xylo*-hexopyranose), was obtained,[87] that is, by oxirane ring opening in racemic epoxy-lactone **122** with methanol in the presence of *p*-toluene-sulfonic acid.

Some basic reactions of RS-parasorbic acid were investigated,[88] such as the partial reduction of the lactone grouping to form 2-hydroxy-6-methyl-5,6-dihydro-2H-pyran (128), glycosidation of 128, and addition reaction of some nucleophiles to the double bond to yield 4-substituted lactones (129).

The base-catalyzed isomerization of RS-parasorbic acid involves[88,89] opening of the six-membered ring and leads to 2Z,4E-hexadienoic acid (130), an isomer of sorbic acid, in good yield. Hydroxylation of this acid with m-chloroperoxybenzoic acid gives in a smooth reaction DL-osmunda δ- and γ-lactones (131 and 132) in high yield.

Dimethylamine readily reacted[89] with 131 and 132 to form two products: a keto-amide (133) and an addition product 134. The first was derived from aminolysis of the substrate followed by isomerization of the double bond. The second had the correct stereochemistry of a precursor of megosamine (**136**). Although cyclization to a megosamine hexonolactone (**135**) could be achieved, its reduction to **136** did not succeed.

Butyl 2-methoxy-5,6-dihydro-2H-pyran-6-carboxylate (**279**, R^1 = Me, R^2 = CO_2Bu), a substrate for a total synthesis of monosaccharides (cf. Section 6), readily hydrolyzes under weakly acidic conditions to butyl *trans*-2-hydroxy-6-oxo-4-hexenoate (**137**). M. Chmielewski[90] has demonstrated a highly successful application of this compound as a substrate for the synthesis of a variety of deoxy sugars. All stereoisomeric 3-deoxy and 3,6-dideoxy-DL-hexoses were prepared[91] by a sequence of reactions involving reduction of the aldehyde group of the substrate to hydroxymethyl, or further reduction to the methyl group to form **138**. *Cis*-hydroxylation of the double bond furnished polyols **139** (X = OH or H) and epoxidation gave anhydro compounds **140** (X = OH or H). The treatment of products **139** or **140** with 50% acetic acid followed by acetylation furnished γ-lactones of all stereoisomeric 3-deoxy- and 3,6-dideoxy-DL-hexonic acids (**141-144**). Partial reduction of the lactone carbonyl group gave free sugars, which were isolated by chromatography.

2,6-Dideoxy-DL-hexoses were prepared[92] from **137** in the following way: the carbon atom chain of the substrate was extended to a seven-carbon atom chain by a low temperature Grignard reaction with methylmagnesium bromide. The product obtained, **145**, was separately cis-hydroxylated (to stereoisomeric tetraols **146**) or epoxidized (to epoxides **147**). Products **146** and **147** were hydrolyzed and subsequently subjected to Ruff degradation. Pairs of stereoisomeric 2,6-dideoxy-DL-hexoses could be separated by column chromatography. Essentially the same approach served[93] for a synthesis of 2-deoxy-DL-erythro-pentose. In this case compound **138** (X = OH) was used as a substrate for epoxidation.

The Michael addition of hydrazoic acid to αβ-unsaturated aldehyde system of **137** gave[94] all four stereoisomeric butyl 4-azido-2-hydroxy-tetrahydropyran-6-carboxylates (**148**). Glycosidation followed by aminolysis and simultaneous reduction of the amide and azide groupings gave, after acetylation, a mixture of stereoisomeric methyl 3,6-diacetamido-2,3,4,6-tetradeoxy-DL-hexopyranosides (**149**). The α-*threo* stereoisomer (**150**) could be obtained in the crystalline state directly from the mixture: it is a precursor for the synthesis of negamycin.[95]

5. SYNTHESES FROM FURAN DERIVATIVES

In the period of 1972-1980 many syntheses of monosaccharides, starting from simple furan derivatives, have been completed. Only a few, however, consisted of a furan-into-furanoside transformation, in which the heterocyclic ring was maintained throughout. The majority of the syntheses described in this section made use of a furan derivative only as a substrate for construction of a principal intermediate: unsaturated bicyclic, pyranoid, or acyclic compound, which was eventually converted into the desired monosaccharide or its simple derivative.

2,5-Dihydro-2,5-dimethoxy derivatives of furan (**151,152**) are easily generated[96] by the action of bromine in methanol on 2-substituted furans, or by their electrolytic methoxylation, and have been exploited as a source of furanosides via 3,4-double bond functionalization.

The isomeric acetals **151** and **152** are readily separable by chromatography. Assignment of their configuration can be safely based on ¹H nmr data.[97] *cis*-Hydroxylation of isomeric 2-benzyloxy-2,5-dihydro-2,5-dihydro-2,5-dimethoxyfurans (**151, 152,** R = CH₂OCOPh) with potassium permanganate gave[98] a modest yield (22-26%) of the corresponding diols (in each case one chromatographically homogenous compound of unspecified relative configuration was obtained). Saponification of the ester group in the diol obtained from *trans*-acetal (**152,** R = CH₂OCOPh), followed by hydrolysis in the presence of Dowex W-50, afforded[98] a high yield of *erythro*-pentopyranos-4-ulose in the form of a hydrate. Oxidation of 2,5-dihydro-2,5-dimethoxy-2,5-dimethylfuran (**153,** *cis-trans* mixture) with potassium chlorate in the presence of catalytic amounts of osmium tetraoxide resulted in acetal ring opening and formation of *erythro*-3,4-dihydroxyhexane-2,5-dione **154**, which was converted into the natural flavor compound, furaneol (**155**), by treatment with a base.[99]

The direct epoxidation of **151** and **152** by the action of hydrogen peroxide or

151   152

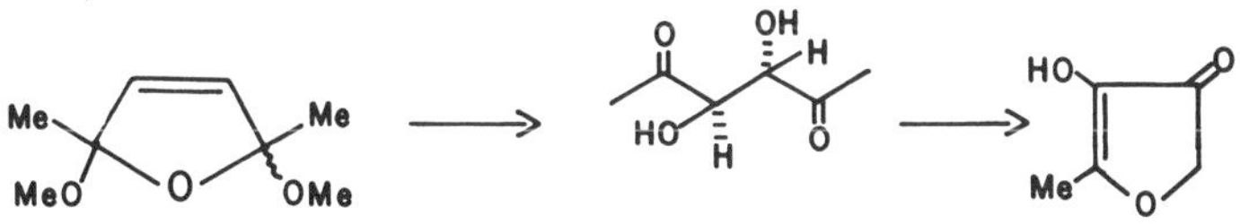

153   154   155

organic peroxyacids failed. 2,5-Dimethoxy-3,4-epoxytetrahydrofurans (**156-160**) were, however, obtained[100] by successive addition of hypochlorous acid to **151-152** and treatment of the unstable chlorohydrines thus formed, with powdered potassium hydroxide in ether. Only for the parent compounds in the series (**156**) was the configuration assigned and the reaction carried out on the epoxide ring with amines, which leads to the aminodeoxyfuranoid compounds **161-165**.[101]

A number of 3-substituted furans were subjected to electrolytic methoxylation followed by *cis*-hydroxylation of the 3,4-double bond. Crystalline **164** (configuration assignment based on the Overhauser effect) was obtained from 3-acetylfuran.[102]

3-Formyl-2-methylfuran, via its dimethylacetal and corresponding 2,5-dihydro-2,5-dimethoxy and 3,4-dihydroxy-2,5-dimethoxy derivatives **165** and **166**, was converted into DL-streptose tetramethylacetal **167**, accompanied by corresponding *ribo*-compound **168**.[103] Analysis by gc-ms method of racemic products related to streptose as well as methods for their preparative separation were described in some detail.[104]

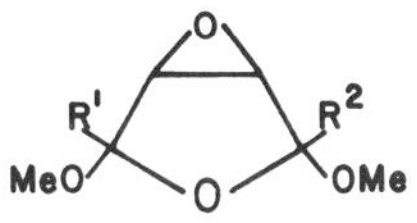

156 $R^1 = R^2 = H$
157 $R^1 = H$, $R^2 = CO_2Me$
158 $R^1 = Me$, $R^2 = CO_2Me$
159 $R^1 = Me$, $R^2 = H$
160 $R^1 = CH_2OH$, $R^2 = H$

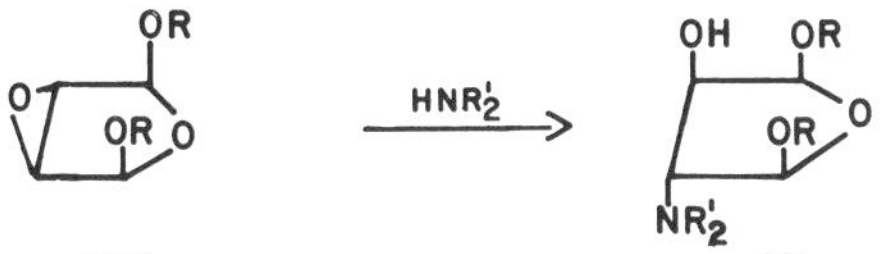

156a   161

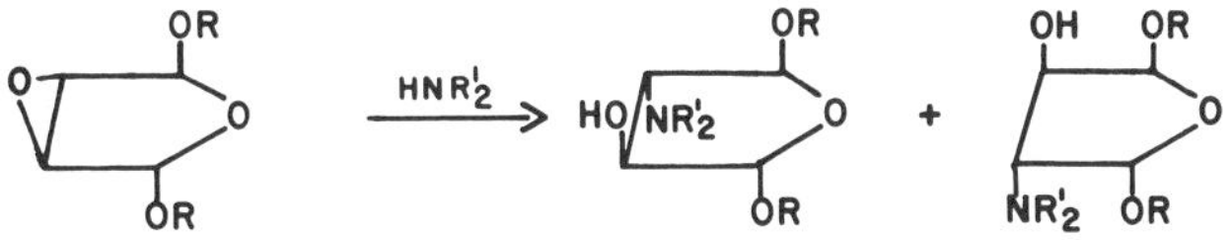

156b   162   +   163

R, R' = Me, Et

HO   OH

CH–CH₃
OH

MeO   O   OMe

164

Kinoshita and Miwa have found[105] that 3-furoic acid easily undergoes Birch reduction, affording the 2,3-dihydroderivative, which was isolable as the corresponding methyl ester **169**. The reduction, when performed in the presence of alcohols, leads to addition products **170**. These compounds can be readily converted into the 2,5-dihydrofuran derivatives by the bromination-dehydrobromination procedure. Compound **171**, obtained in this way, was cis-hydroxylated and the corresponding isopropylidene derivative reduced to give methyl 3-C-hydroxy methyl-2,3-O-isopropylidene-β-DL-erythrofuranoside **172**; it was identified by comparison with the compound obtained from natural apiose.[106] An analogous sequence of reactions was applied to the synthesis of dihydrostreptose from 2-methyl-3-furoic acid.[107]

Dehydrobromination of compound **173** gave isomeric 2,5-dihydrofurans **174** and **175**. Upon oxidation with osmium tetraoxide, the cis compound (**174**) gave exclusively **176**, whereas the trans isomer (**175**) afforded approximately equal amounts of **177** and **178**. The methyl esters **176-178** were reduced to the corresponding alcohols **179-181**. A mixture of **179** and **180** was directly correlated with compound **182** obtained from D-xylose, whereas hydrolysis of **181** in concentrated hydrochloric acid gave α,β-DL-dihydrostreptose.

Tetrahydrofurfuryl alcohol was transformed directly into a mixture of diastereoisomeric 3-deoxypentofuranoses (**183-186**) by the action of iodine tris(trifluoroacetate). The main component: the trifluoroacetate of 3-deoxy-α-DL-threo-pentofuranose (**183**) crystallized out of the reaction mixture.[108]

Among the monosaccharide syntheses in which a furan precursor is amenable to ring opening, the one based on transformation of 2-furylmethanols into methyl 2,3-dideoxy-DL-glyc-2-enopyranosid-4-uloses seems to be most prolific. The idea of this method, elaborated by Achmatowicz, consisted of the stepwise functionalization of the cyclic enone; it was outlined in a previous report in this series.[1] Recently it was shown that furfuryl alcohols can be converted directly

H   CH(OMe)₂
CH₃
MeO   O   OMe

165

HO   OH
CH(OMe)₂
CH₃
MeO   O   OMe

166

CH(OMe)₂
H–C–OH
(MeO)₂CH–C–OH
R–C–R'
CH₃

167 R = OH, R' = H
168 R = H, R' = OH

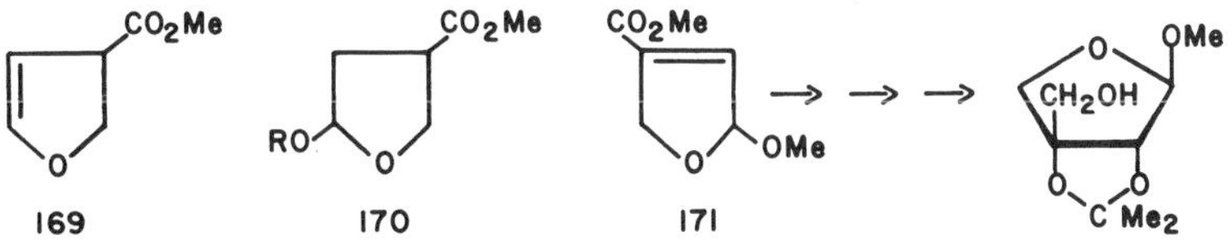

169   170   171   172

R = Me, Et, iPr

into glyc-2-enopyranos-4-uloses by the action of *m*-chloroperoxybenzoic acid[109,110] or pyridinium chlorochromate.[111] Another improvement of the three-step procedure[112] leading to the unsaturated methyl glyculosides (e.g., **187**, which are principal intermediates in the synthesis), allows one to obtain them directly from 2,5-dihydro-2,5-dimethoxy derivatives by treatment with methanol in the presence of trifluoroacetic or formic acid.[113] A selection of enulosides, including unsaturated disaccharides, was obtained by the exchange of glycosyl ester group in the presence of stannic chloride[114] or by coupling of the pentenulose with an alcohol in the presence of diethyl azodicarboxylate-triphenylphosphine-mercuric bromide reagent.[115]

It was found that the reduction of methyl 2,3-dideoxy-DL-pent-2-enopyranosid-4-ulose (**187**, R = H) with complex metal hydrides proceeds with high selectivity (affording about 90% of *erythro* compound **188**) as a result of the predominately axial approach of the hydride anion.[116] The same tendency was observed for conformationally stable[117] α-anomers bearing substituents at C-5. The minor processes accompanying the formation of unsaturated alcohols included 1,4-addition of hydride anion to the enone, leading to saturated pyranoside **190,** and reductive rearrangement of the main reduction product (**188**), affording glycal **191**.[118]

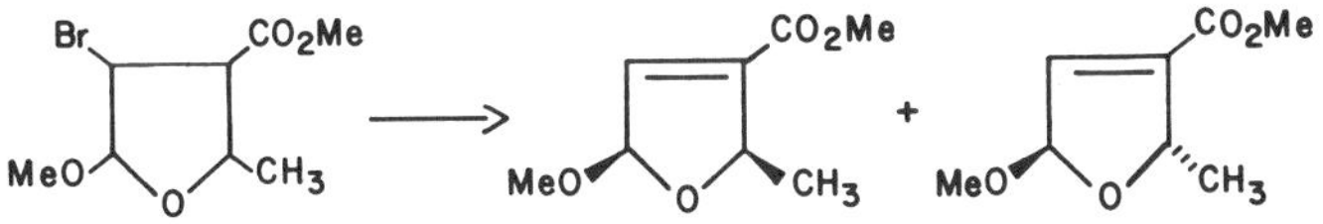

173   174   175

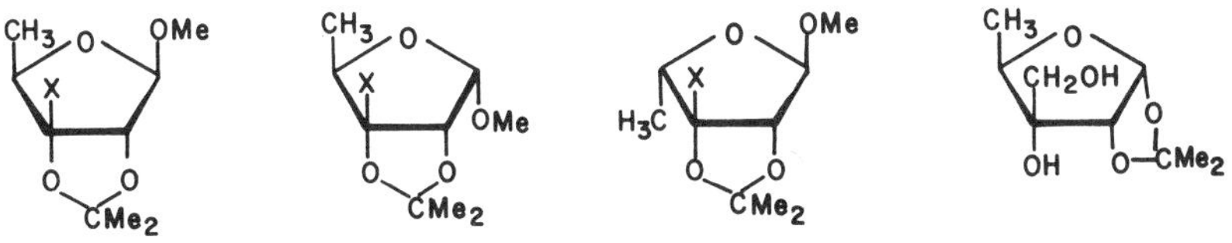

176   X = CO$_2$Me   177   X = CO$_2$Me   178   X = CO$_2$Me   182
179   X = CH$_2$OH   180   X = CH$_2$OH   181   X = CH$_2$OH

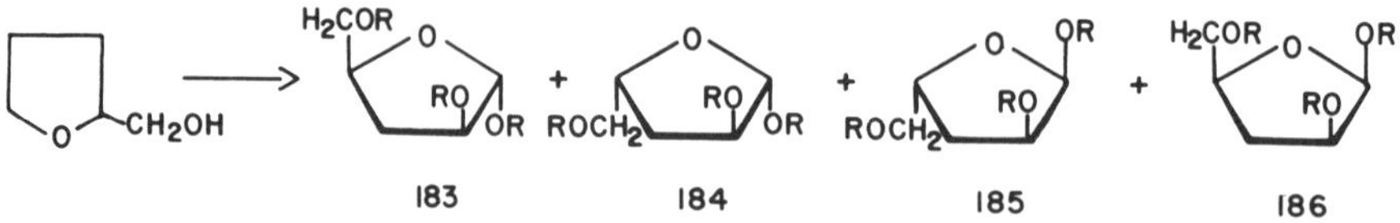

183   184   185   186

R = CF$_3$CO

Methyl pentenopyranoside **188** (R = H) gave on hydroxylation with Milas reagent methyl α-DL-lyxopyranoside in 42% yield. Epoxidation of the acetate of **188** with *m*-chloroperoxybenzoic acid gave methyl 2,3-anhydro α-DL-*ribo*-pyranoside and the corresponding *lyxo*-isomer in 9:1 ratio. The epoxide ring opening by alkaline hydrolysis of the *ribo*- compound yielded 73% of two methyl pentopyranosides in a 4:1 ratio. The main component was identified as α-DL-xyloside, whereas the minor product was assigned as the α-DL-arabino-configuration by direct comparison with authentic samples.[119] Methyl β-*lyxo* and β-*ribo-pento*-pyranosides were obtained by hydroxylation of **189** (R = H). Stereoselective synthesis of racemic ribose derivatives required introduction of *cis*-diol grouping to **187** (R = H) prior to the carbonyl group reduction. Starting from methyl pentenuloside **187** (R = H), or the corresponding glycosyl benzoate, methyl 2,3-*O*-isopropylidene-β-DL-ribopyranoside and 2,3-*O*-isopropylidene-DL-ribofuranose were prepared.[120] Methyl glycosides of all racemic 6-deoxyhexoses were obtained from (2-furyl)ethanol by way of the unsaturated pyranosides **188** and **189** (R = Me), and their β-*erythro* analog. *Cis*-Hydroxylation of methyl 2,3,6-trideoxy-α-DL-*erythro*-hex-2-enopyranoside (**188**, R = Me) with Milas reagent gave exclusively methyl α-*manno*-glycoside (methyl α-DL-rhamnoside) in 39% yield, whereas the substrate with β-DL-*erythro* configuration afforded on treatment with the same reagent methyl 6-deoxy-β-DL-allopyranoside (36%). Unsaturated pyranoside **189** (R = Me) yielded under the same conditions a mixture of methyl 6-deoxy hexopyranosides of the α-DL-*talo* and α-DL-*gulo* configurations.[121]

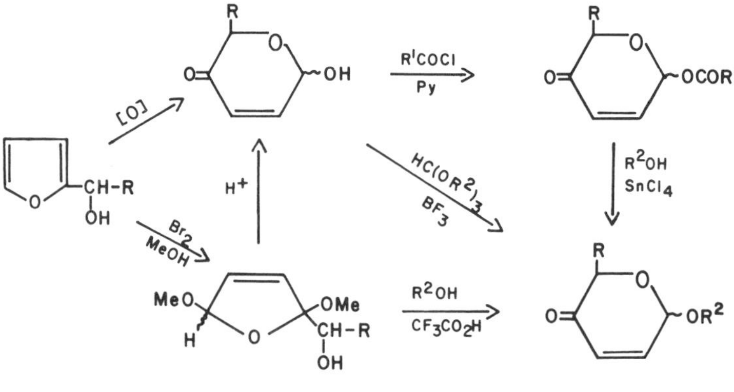

Scheme 3

All the stereoisomeric methyl 2,3-anhydro-6-deoxy-DL-hexopyranosides were obtained from the unsaturated pyranosides by epoxidation with hydrogen peroxide-benzonitrile. Their conformation was discussed on the basis of ^{1}H nmr spectra of their 4-O-acetyl derivatives.[122] Alkaline hydrolysis of methyl 2,3-anhydro-6-deoxy-α-DL-*manno*hexopyranoside proceeded with high regioselectivity affording methyl 6-deoxy-α-DL-altropyranoside as the sole product. The α-*allo* epoxide underwent nonselective ring opening at C-2 and C-3 in acidic media, giving rise to methyl 6-deoxy-α-DL-*altro* and 6-deoxy-α-DL-*gluco*hexopyranosides in approximately equal amounts. The α-*gulo* epoxide rearranged to the 3,4-anhydro sugar in basic medium, but was cleaved selectively to 6-deoxy galactoside (α-fucoside) on acidic hydrolysis. Methyl 6-deoxyhexopyranoside of α-*ido* configuration was obtained selectively, by alkaline hydrolysis of methyl 2,3-anhydro-6-deoxy-α-DL-taloside. The yields of oxirane ring opening reactions were in the range of 53-66%. Methyl 6-deoxypyranosides obtained were characterized as tri-O-acetyl derivatives.[123]

Synthesis of racemic ketohexosides, namely methyl α-sorboside, α-tagatoside, β-fructoside, and α-psicoside was achieved in an analogous way, starting from the monobenzyl ether of 2,5-di(hydroxymethyl) furan.[124] The inversion of configuration at C-5, necessary for completion of some of these syntheses, was performed at the stage of the unsaturated pyranoside by means of the benzoic acid-triphenylphosphine-ethyl azodicarboxylate reagent.[125]

1-(2-Furyl)-1,2-dihydroxyethane, obtained in two steps from furan and butyl glyoxylate[126] served as the substrate in the total syntheses of methyl hexopyranosides. The diacetate of the unsaturated α-*erythro* pyranoside (**188**, 4-O-Ac, R = CH₂OAc), obtained in the usual way, gave on *cis*-hydroxylation a single product: methyl α-DL-mannoside in 80% yield. The isomeric β-*erythro* compound afforded on treatment with Milas reagent methyl β-DL-allopyranoside in 70% yield. The epoxidation of **188** (4-O-Ac, R = CH₂OAc) resulted in the formation of two 2,3-anhydro glycosides: methyl α-*manno*- and α-*allo*- in comparable amounts. The latter compound gave, on treatment with aqueous sodium hydroxide, methyl α-DL-altropyranoside, which was characterized as the tetra-O-acetyl derivative.[127] The analogous synthesis of methyl hexopyranosides, starting from enantiomerically pure methyl (R)- and (S)-(2-furyl)glycolates,[128] was completed. Methyl 6-O-benzyl-2,3-dideoxy-α,β-L-*erythro*-hex-2-enopyranosides (**192, 193**) obtained in eight steps from methyl (R)-2-furyl-glycolate were reduced without separation. *cis*-Hydroxylation, carried out on the mixture of unsaturated pyranosides containing mainly **194** and **195** afforded methyl 6-O-benzyl-α-L-mannopyranoside, which was, after hydrogenative debenzylation, directly compared with methyl α-D-mannopyranoside. Epoxidation of the same mixture of unsaturated alcohols gave two 2,3-anhydro glycosides, to which α-*allo* (**196**) and β-*allo* (**197**) configuration was assigned, on the basis of the ^{1}H nmr spectra. The reaction of methyl 2,3-anhydro-6-O-benzyl-α-L-allopyranoside with aqueous ammonia at 125° gave predominantly one compound identified, after removal of the benzyl group and acetylation, as methyl 3-acetamido-2,4,6-tri-O-acetyl-3-deoxy-α-L-glucopyranoside (kanosaminide).[129]

In a similar sequence of reactions, methyl α-L- and α-D-glucopyranosides were synthesized from their respective 2-furylglycolates without loss of enantiomeric purity.[130]

A number of modified sugars have been obtained by using the aforementioned approach. For example, methyl 4-O-methyl-5,5-di-C-methyl-β-DL-lyxoside, the glycoside of the sugar component of the antibiotic novobiocin, was synthesized with high selectivity from 2-(2-furyl)propan-2-ol.[131] Methyl 6-acetamido-6-deoxy and 6-deoxy-6-nitro-α-DL-mannopyranosides were prepared from suitably substituted 2-furylethanols.[132] 1,4-Addition of an active methylene compounds to

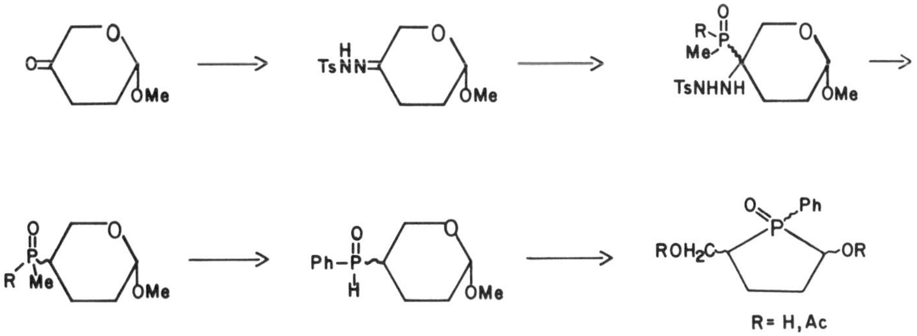

Scheme 4

enuloside **187** (R = H) leading to 1,2-trans branched pyranosiduloses was described.[133] Direct epoxidation of enones **187** (R = H, Me, Et) was reported in connection with maltol synthesis.[134] The reaction of pentenuloside **187** (R = H) with iodine azide was described.[135] The saturated analog of the same substrate served as the starting material in the synthesis of sugar analogs, containing a phosphorus atom in the place of the hemiacetal ring oxygen.[136]

Disaccharides containing D- or L-pent-2-enopyranoside units at the nonreducing end were obtained by combining racemic pentenulose with appropriately protected natural monosaccharide derivatives, followed by separation of the diastereoisomeric disaccharide precursors and reduction of the keto group.[137] It was found that the glycosyl ester **198** can alkylate aromatic substrates in the

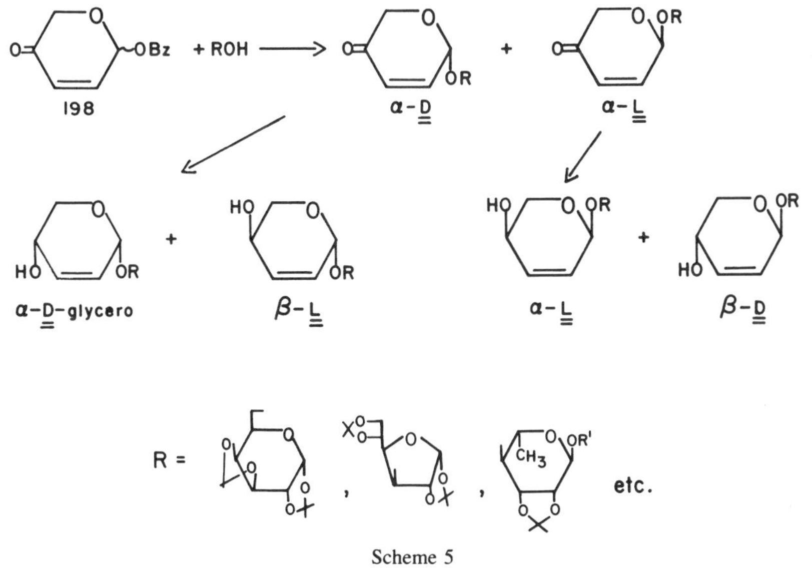

Scheme 5

198 + PhOMe $\longrightarrow$

199

198 + [furan structure]—Me $\longrightarrow$

200   Ar = 2(5methylfuryl)

presence of Lewis acid catalysts, with formation of C-glycosylic compounds **199**. However, reaction with 2-methylfuran led, under the same conditions, to the C-2-substituted derivative **200**.[138]

An idea of building up the ribofuranoside portion on a frame of a bicyclic (for example Diels Alder adduct) precursor, obtained from furan, have found wide applications in the syntheses of C-glycosyl compounds. The transformation of Diels–Alder adducts of furan with 2-nitroacrylate and acetylenedicarboxylate described by Just and Martel[139] constitutes a good example of this approach. Regioisomers **201** and **202** were converted by way of cis-hydroxylation and acetonide formation, followed by elimination of nitrous acid molecule, into unsaturated compounds **203**. The action of ozone on **203** followed by reduction of the ozonide with sodium borohydride afforded a mixture of epimeric triols **204**. Periodate oxidation of **204** gave 3,4-isopropylidene-2,5-anhydro-DL-allose **205** (overall yield ~ 15%). The same compound was obtained by a series of analogous transformations from adduct **206**, prepared by addition of acetylenedicarboxylate to furan. The stable ozonide **207**, obtained in the course of this synthesis was found to produce triol **204** in addition to the expected tetrol **208**, when treated with lithium aluminum hydride. Thermal decomposition of **207** was studied and an explanation of the loss of skeletal carbon atom was offered.[140] Oxalate **209** obtained during decomposition of **207** gave[141] after hydrogenation and methanolysis a mixture of methyl 2,3-O-isopropylidene-β-DL-talofuronate **210** and the isomeric allo compound **211**.

Various ways of functionalization of the lactol **205** leading to C-nucleoside analogs were described.[142] Compound **205** readily reacted with carboethoxymethylenetriphenylphosphorane with formation of trans ester **212**, whereas the Wittig reagent prepared from pyruvate ester gave[143] a product of an intramolecular Michael reaction (**221**). Acrylate ester **212** was converted on addition of diazomethane into pyrazoline **213**, and further, by successive bromination-dehydrobromination followed by acid hydrolysis, was transformed into DL-3-(carboxamido)-4-β-ribofuranosylpyrazole **214**. Another precursor of heterocyclic C-ribofuranosides (**215**) was obtained in the reaction of lactol **205** with

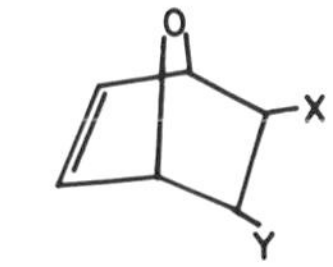

201 X= CO₂Me, Y= NO₂
202 X= NO₂, X= CO₂Me

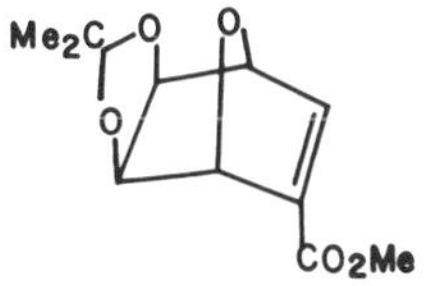

203

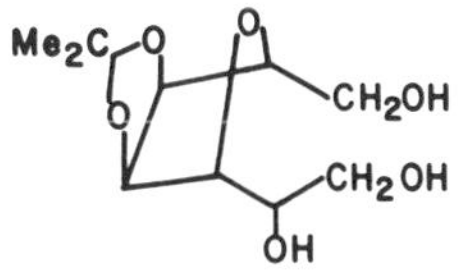

204

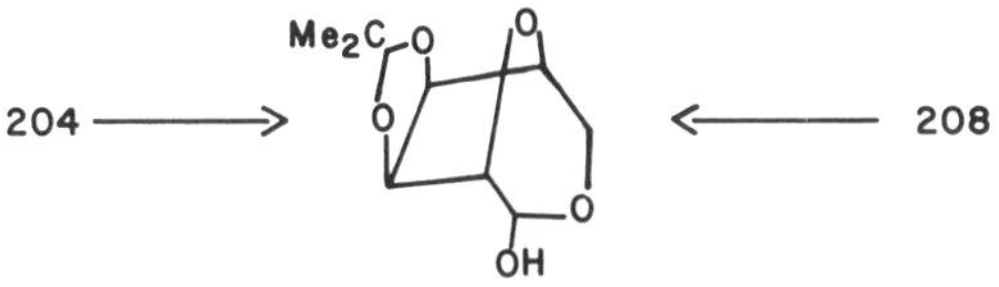

204 ——→   ←—— 208

205

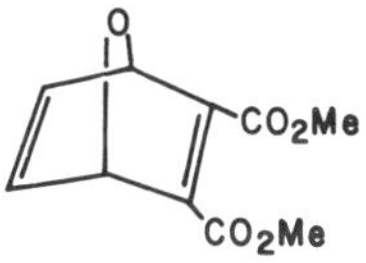

206 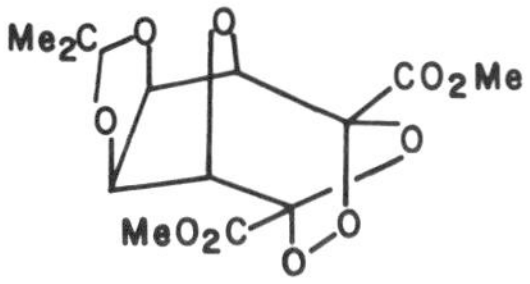   207 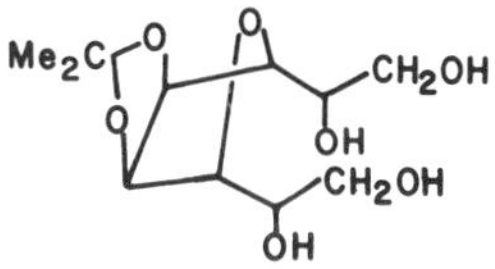   208

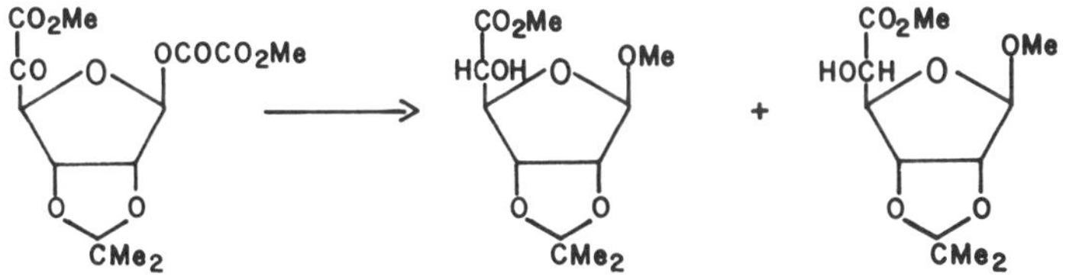

209   210   211

semicarbazide. Treatment of **215** with lead tetraacetate resulted in formation of oxadiazole **216**.[144] *Allo*lactone **217**, obtained by oxidation of **205** with Fetizon reagent, gave on reaction with aminoguanidine, triazole **218**. Since lactol **205** did not react directly with any unstabilized Wittig reagents it was converted into the unstable aldehyde **220** by way of oxazolidine **219**. A number of derivatives of **221** denoted as bis-homo anhydro-*C*-nucleosides (for example **222-223**) were obtained.[143]

In order to prepare *C*-glycosidic compounds of *arabino* configuration the

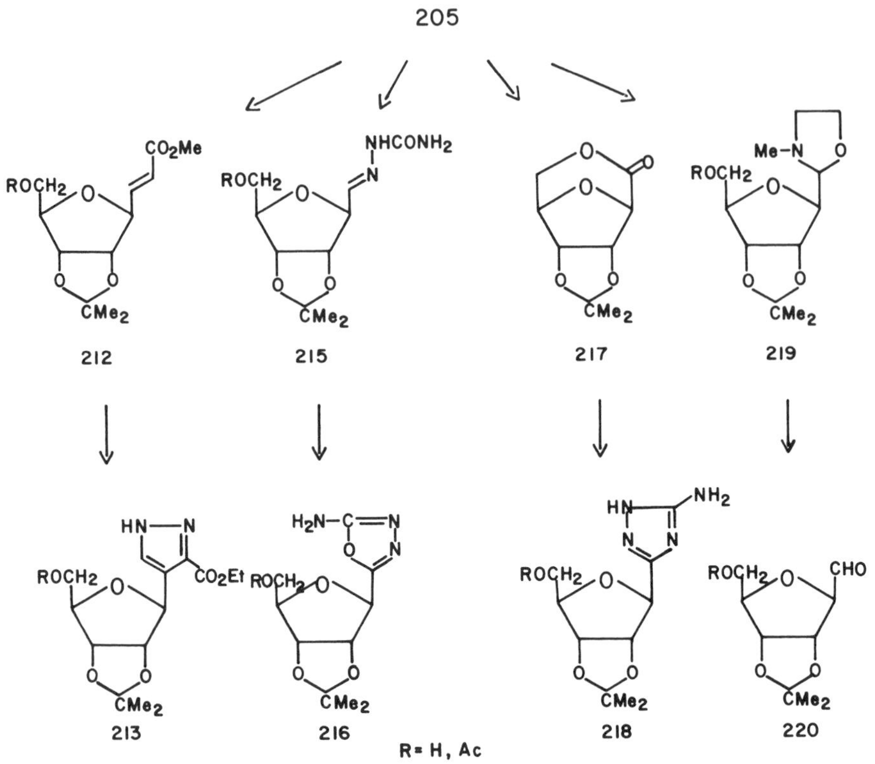

205

212   215   217   219

213   216   218   220

R = H, Ac

213 ⟶

214

221 X = CH₂COCO₂Et

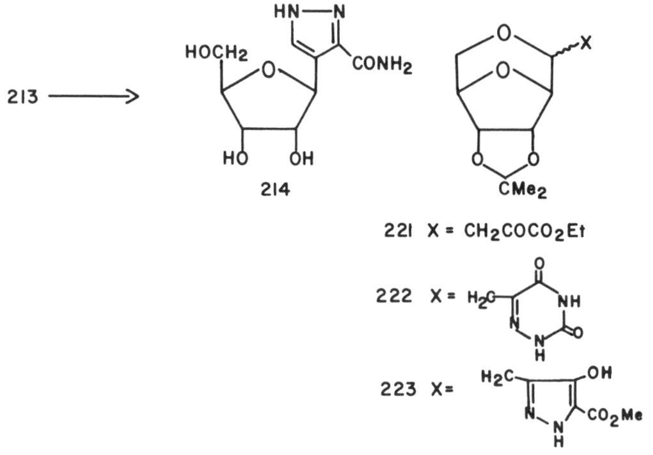

222 X =

223 X =

adducts **201** and **202** were epoxidized with *m*-chloroperoxybenzoic acid (*m*CPBA), followed by treatment with diazabicycloundecane (DBU), to give the olefinic *exo*-epoxide **224**. Further transformation of **224** involving oxirane ring splitting and (after protection as the *t*-butyldimethylsilyl ether), ozonization, reduction and oxidation, gave keto ester **225** (in 6.5% overall yield from nitroacrylate adduct), which was reacted with carbamoylmethylenetriphenyl-phosphorane, thus completing the synthesis of DL-2′-epi-showdomycin (**226**).[145] The same inter-mediate **225** was converted into 2′-*epi*-pyrazofurin A derivative **227**. By a similar sequence of reactions DL-2′-deoxyshowdomycin **230** was obtained.[146] The *exo*-nitro adduct **202** was hydroborated, and after elimination of nitrous acid mole-cule, the regioisomeric alcohols (**228, 229**) were separated as the corresponding acetates. The required isomer **228** was converted into **227** by way of the cor-responding ozonide.

It was found that 1,3-diethoxycarbonylallene **231** readily enters [4 + 2] cy-cloaddition with heterocyclic dienes.[147] The adduct (**232**) obtained from **231** and furan was easily converted into ketoester **233**. Treatment of the latter compound with lithium diisopropylamide and benzenediazonium fluoroborate resulted in the formation of the α-phenylazo compound **234**, which in turn underwent sodium borohydride promoted scission of carbon–carbon bond affording the C-glycosylic compound **235**[148]

A new general approach to the synthesis of C-glycosidic compounds and C-nucleosides was elaborated by Noyori.[149] In his approach the starting material, unsaturated bicyclic ketone **236**, was prepared in a cyclocoupling reaction be-tween 1,1,3,3-tetrabromoacetone and furan, followed by a zinc-copper couple

224

225 R = Me$_3$C Si (Me)$_2^-$

226 R = OH
230 R = H

227

228   R = H, Ac

229

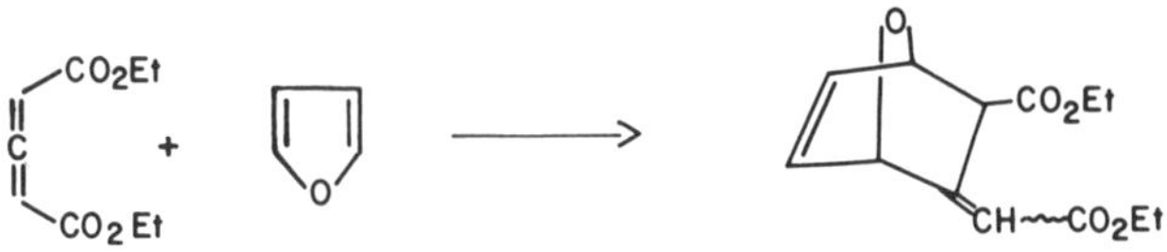

231   232

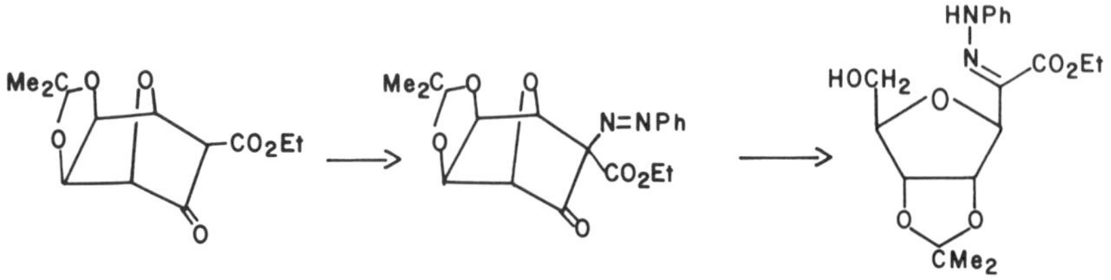

233   234   235

reduction.[150] Compound **236** was converted into the key intermediate **238** by highly stereoselective *cis*-hydroxylation, followed by acetonide formation and Bayer–Villiger oxidation. For numerous syntheses of C-nucleoside analogs, enantiomerically pure dextrorotatory lactone **238** was obtained by resolution of the corresponding hydroxy acid through cynchonidine salt and recyclization.[151] Condensation of lactone **238** with bis(dimethylamino)-tert-butoxymethane produced dimethylaminomethylene lactone **237** (*E* and *Z* mixture), which was converted into a uracil derivative by treatment with urea in the presence of sodium ethoxide. Reactions of **237** with thiourea and guanidine, leading to another C-nucleoside analog, was also described.[151] For the purpose of showdomycin synthesis, lactone **238** was condensed with furfural in the presence of lithium cyclohexylisopropylamide, and the aldol condensation product was dehydrated to afford **239** (R = CH-2-furyl). Treatment of **239** with methanolic sodium methoxide gave the methyl ester **240**, which was subjected to ozonolysis after silylation. Reductive work up of the ozonide gave the unstable keto ester **241**, which, after

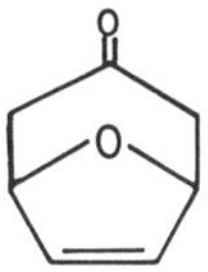

236

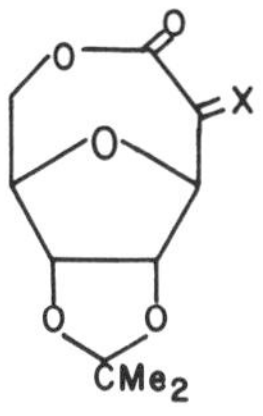

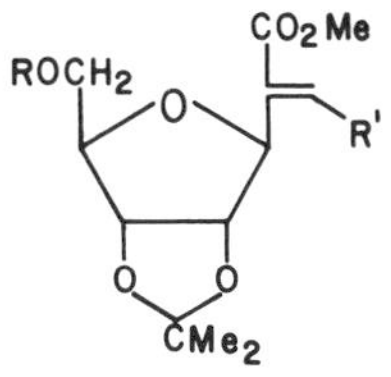

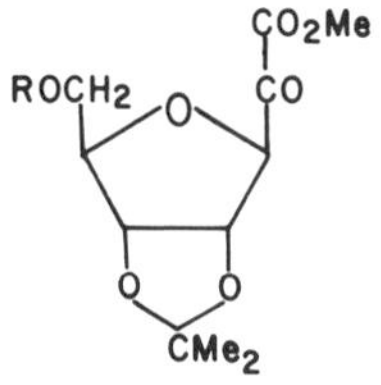

237 X = CHNMe₂
238 X = H₂
239 X = CH-(2-furyl)
242 X = O

240   241

R' = 2-furyl, R = Me₃CSi(Me)₂⁻

243

R = Me$_3$CSi(Me)$_2^-$

244 R' = OH
245 R' = OSO$_2$C$_6$H$_4$Me(p)
246 R' = CN
247 R' = CO$_2$Me

248

Tr = (C$_6$H$_5$)$_3$C−

249

250

251

reaction with triphenylmethylene-carbamoylphosphorane and deprotection, afforded showdomycin (in 29% yield, based on silylated ester **241**). 6-Azapseudouridines were also obtained from the keto ester **241** by way of the corresponding semicarbazone.[152] Showdomycin was also prepared from **237** by way of the unstable keto lactone **242**, which was reacted with the aforementioned Wittig reagent.

The general scheme of synthesis presented above was applied to the preparation of modified C-nucleosides. Thus, instead of the derivative of acetone, the use of 1,1,3-tribromo-3-methyl-butan-2-one in reaction with furan led to pseudouridines modified at the C-5 position.[153] A route to C-4 alkylated pyrimidine C-nucleosides was explored by applying tetrabromoacetone–2-methylfuran adduct as the starting material.[154] The chiral lactone **238** was converted into pyrimidine homo-C-nucleosides in two ways: treatment of **238** with sodium methanolate in methanol gave ester **243**, homologated via reduction to the corresponding primary alcohol **244**, tosylation (**245**) cyanide displacement (**246**), hydrolysis, and methylation. The ester **247** was further functionalized by carbanion generation formylation, and methylation to give the enol ether **248**, which when heated with an excess of urea afforded the uracil derivative **249**.[155] Alternatively, ester **243** (R = t-C$_4$H$_9$(CH$_3$)$_2$Si) was reduced with diisobutylaluminum hydride to the corresponding aldehyde, which gave,[156] upon reaction with methoxycarbonylmethylenetriphenylphosphorane, the unsaturated ester **250**, readily convertible into 3-carbamoyl-4-[(β-D-ribofuranosyl)methyl] pyrazole **251**.

It has been demonstrated that products of the photochemical addition of carbonyl compounds to furan[157] can serve as substrates in syntheses of branched sugars. In particular, adduct **252** was converted into 3-deoxystreptose derivative **255**. Both *cis*-hydroxylation and epoxidation of **252** favored *exo*-attack of the oxidizing species leading to products with the required configuration of the nonhemiacetal hydroxyl group (**253,254**). Hydrolytic cleavage of the bicyclic precursors **253** or **254** under conditions of isopropylidene derivative formation led to **255**, apparently by epimerization at C-3, and **256**. Configuration of **255** was established by correlation with 1,2-O-isopropylidene-3,5-dideoxy-3-C-methyl–D-arabinopentofuranose.[158]

An acyclic precursor devised by Bognár and Herczegh,[159] prepared via oxidative cleavage of the furane nucleus, has been applied in the synthesis of racemic pentoses. 2-(2-Furyl)-4,4,5,5-tetramethyl-1,3-dioxolane **257** obtained in reaction of furfural with 2,3-dimethylbutane-2,3-diol was chosen as a substrate, owing to its stability in acidic media (at pH > 3).[159] Oxidation of **257** with bromine water gave the unstable endialone derivative **258**, which was reduced with sodium borohydride to **259** (isolated as crystalline dibenzoate). Epoxidation of **259** gave 3,4-anhydro-DL-arabinose (**260**) exclusively, whereas *cis*-hydroxylation of its dibenzoate afforded a mixture of DL-ribose (**261**) and DL-arabinose (**262**) derivatives. Free sugars were obtained by successive alkaline and acidic hydrolysis. Compound **260** hydrolyzed with diluted sulfuric acid gave DL-xylose, identified as the tetraacetate and toluene-*p*-sulfonylhydrazone.[160] Analogous transformation of the tetramethyldioxolane obtained from 5-methyl-2-furaldehyde led to a mixture of unsaturated diols **263**, which was, after chromatographic separation, converted to methyl 3,4,6-trideoxy-α-DL-*threo*-hex-3-enopyranoside (**264**) and methyl 3,4,6-trideoxy-β-DL-*erythro*-hex-3-enopyranoside (**265**) by boiling with methanolic hydrogen chloride.[161]

6. SYNTHESES FROM PYRAN DERIVATIVES

3,4-Dihydro- and 5,6-dihydro-2H-pyrans (**266** and **267**) continued to be substrates for sugar synthesis. These ring systems are readily available and can be functionalized in a number of ways. In consequence, as in the case of furfuryl alcohols, general methods for carbohydrate synthesis have been developed.

A. 3,4-Dihydro-2H-pyran and Its Derivatives

Hydroxylation of the double bond in 3,4-dihydro-2H-pyran (**266**, X = Y = H), providing a ready access to 3,4-dideoxy-DL-pentoses (**268**), has been studied for

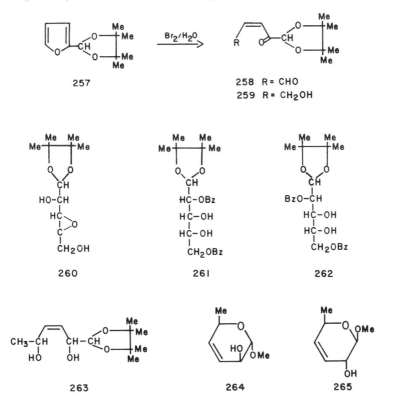

257

Br$_2$/H$_2$O

258 R = CHO
259 R = CH$_2$OH

260

261

262

263

264

265

some time.[162] In the past decade the preparation of the 2-ulose (269) has been described.[163–165] Compound 269 served[166] as substrate for the synthesis of a 3,4-unsaturated methyl pentuloside (270) via formation of a 2,2-di-O-methyl derivative, bromination in position 3, elimination of elements of hydrogen bromide, and final hydrolysis of the acetal grouping at C-2.

Full experimental details have been published of the syntheses of kasugamycin (272, R = C(NH)CO$_2$H) from 3,4-dihydro-6-methyl-2H-pyran-2-one (271)[167] and of methyl α-DL-mycaminoside (274), α-DL-oleandroside (275), α-DL-cymaroside (276), α-DL-tyveloside (277), and α-DL-chromoside C(278) from 3,4-dihydro-2-ethoxy-6-methyl-2H-pyran (273).[168] The schemes of these syntheses have been presented.[169]

B.   Derivatives of 5,6-Dihydro-2H-pyran

The potency of derivatives of 5,6-dihydro-2H-pyran (267) as convenient substrates for the synthesis of a variety of racemic sugars was demonstrated in the first volume[170] of this series. During the past decade these compounds continued to be used for many sugar syntheses, including syntheses of di- and trisaccharides. In fact, a general scheme of monosaccharide synthesis based on derivatives of 2-alkoxy-5,6-dihydro-2H-pyran (279) has been developed in recent years. This scheme comprises the following steps: (i) epoxidation of the double bond to form stereoisomeric alkyl 2,3-anhydro-4-deoxy-DL-glycopyranosides (280); (ii) conversion of the epoxide into an allylic alcohol: alkyl 3,4-dideoxy-DL-glyc-3-enopyranoside (281); (iii) introduction of suitable substituents at C-3 and C-4 via direct addition of groups (e.g., cis-hydroxylation or oxyamination reactions) or via formation of a 3,4-anhydro compound and subsequent opening reactions with nucleophilic reagents that leads eventually to the full sugar structure.

C.   Syntheses and Reactions of 2-Alkoxy-5,6-dihydro-2H-pyrans

The Diels–Alder condensation of 1E,4E-diacetoxy-1,3-butadiene with butyl glyoxylate[171] leads to a 1:1 mixture of butyl 1,4-di-O-acetyl-2,3-dideoxy-α-DL-threo- and -erythro-hex-2-enuronates (282 and 283) in 75% total yield. These compounds appear to be excellent substrates for further conversion to the full sugars by the direct addition of appropriate reagents to the double bond.

1-Methoxycarbamido-1,3-butadienes (284) undergo cycloaddition[172] with compounds containing activated carbonyl group furnishing 6-substituted derivatives of 2-methoxycarbamido-5,6-dihydro-2H-pyran (285).

David and co-workers[173,174] elaborated two syntheses of 1,3-butadienyl ethers

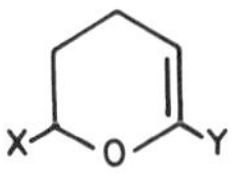

266: X = OR, H
     Y = H, Me

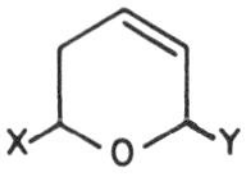

267: X = H, CO₂R, CH₂OH
     Y = H, OR

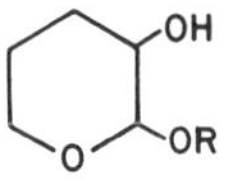

268: R = H, alkyl

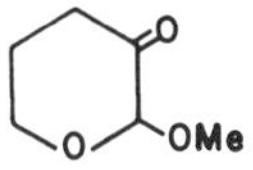

269

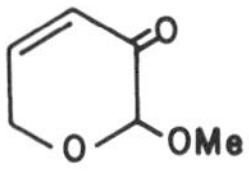

270

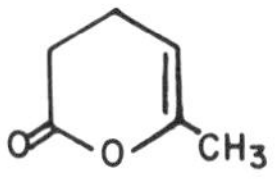

271

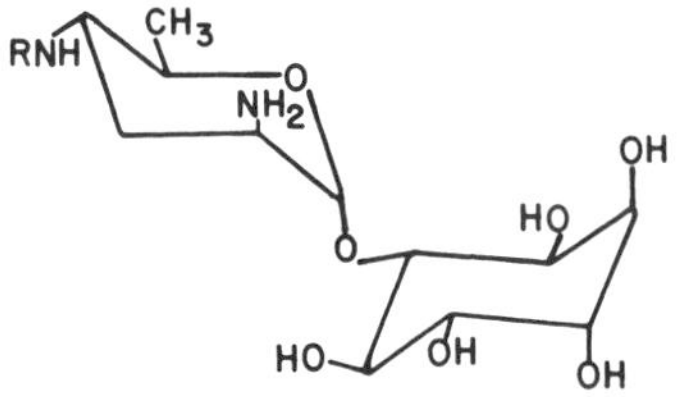

272

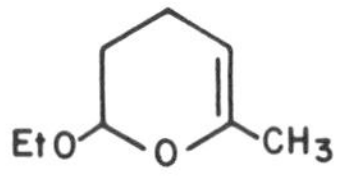

273

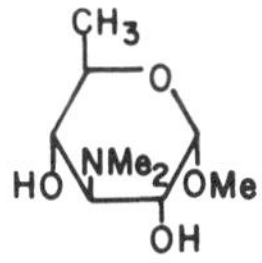

274

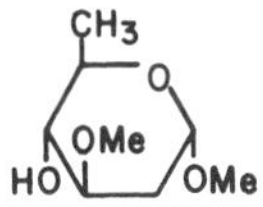

275

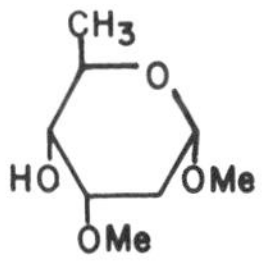

276

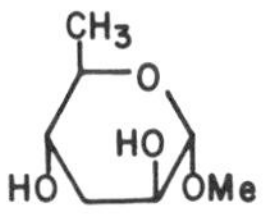

277

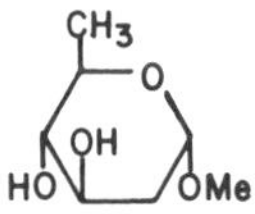

278

attached to sugar residues (e.g., **289** and **290**). These dienes were condensed with butyl glyoxylate furnishing mixtures of all four stereoisomeric butyl 2-alkoxy-5,6-dihydro-2H-pyran-6-carboxylates (**279**, R^1 = sugar residue, R^2 = CO$_2$Bu). These products were further employed for the synthesis of a variety of disaccharides.

The first synthesis of **289** and **290** started from the base-catalyzed addition of a sugar containing an "isolated" hydroxyl group to 2,7-dimethyl-3,5-octa-dyine-2,7-diol (**286**). Two enynyl ethers (**287** and **288**) were formed and separated, the partial hydrogenation over a Lindlar catalyst affording dienyl ethers **289** and **290**. Alternatively, both ethers were prepared[174] by a Wittig reaction; from chloromethyl ether (**291**) a phosphonium salt (**292**) was prepared and converted into a phosphorane. Reaction with acrylaldehyde gave a mixture of **289** and **290**.

R² structures with reaction arrows:

279: R¹= Me,Et,etc.   280   281
R²= H,Me,CH₂OH,etc.

Alkyl
pentopyranosides,
hexopyranosides,etc.

A study of the stereochemical results of cycloaddition reactions between butyl glyoxylate and 1E,3-butadienyl ethers containing different sugar units revealed[175] that the proportion of stereoisomeric products formed depends more on the preferential approach of the carbonyl reagent to one of the faces of the butadienyl system than on the *endo*-addition rule.

Butyl glyoxylate reacts normally only with 1E,3-butadienyl ether **289**. However, David and co-workers[176] have found that with the more active dienophile, diethyl mesoxalate, 1Z,3-butadienyl ether (**290**) could also enter into cycloaddition, although the reaction was appreciably slower. The resulting cycloadducts **293** and **294** were decarboethoxylated according to the Krapcho method, furnishing the α-D (**295**) and α-L (**296**) as the main products.

Further extension of David's syntheses[177] involved preparation of unsymmetrical 1,4-dialkoxy-1,3-butadienes by means of the Wittig reaction. For the condensation of the phosphorane prepared from **292**, 3E-benzyloxy-acrylaldehyde (**297**) was used instead of acrolein. The resulting mixture of 1E,4E- and 1Z,4E-dialkoxy-1,3-butadienes (**298** and **299**) could not be resolved. However, equilibration of the transient betaine during the Wittig reaction raised the proportion of 1E,4E-diene (**298**) from 40 to 78%. Cycloaddition of **298** with L-menthyl glyoxylate gave three adducts, two of them (**300** and **301**) being precursors of disaccharides and the third (**302**) being a precursor of a disaccharide ether.

Almost pure 1E,3-butadienyl ether **289** was obtained[178] by way of the Wittig reaction in which 3-O-substituted acrylaldehyde (**303**) (a sugar residue being the 3-O substituent) was condensed with methylenetriphenylphosphorane. Condensation of **303** with chloromethylenetriphenylphosphorane yielded a 1:1 mixture of 1E-alkoxy-4E-chloro- and 1E-alkoxy-4Z-chloro-1,3-butadienes (**304** and **305**). Although this mixture again could not be separated, it was found that only the 1E,4E-diene entered into cycloaddition with diethyl mesoxalate, giving two adducts, **306** and **307**, in 69% and 13% yield respectively. These compounds are also precursors of disaccharides. It is interesting that the formation of re-

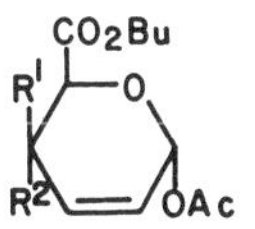

282: R^1 = OAc, R^2 = H.
283: R^1 = H, R^2 = OAc.

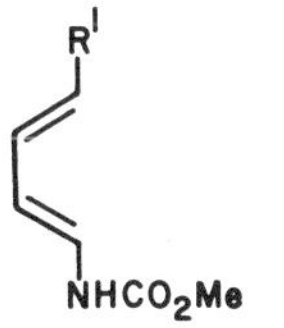

284: R^1 = CH_3, C_6H_5

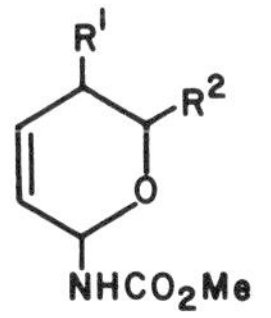

285: R^1 = CH_3, C_6H_5
R^2 = CO_2Et, CCl_3

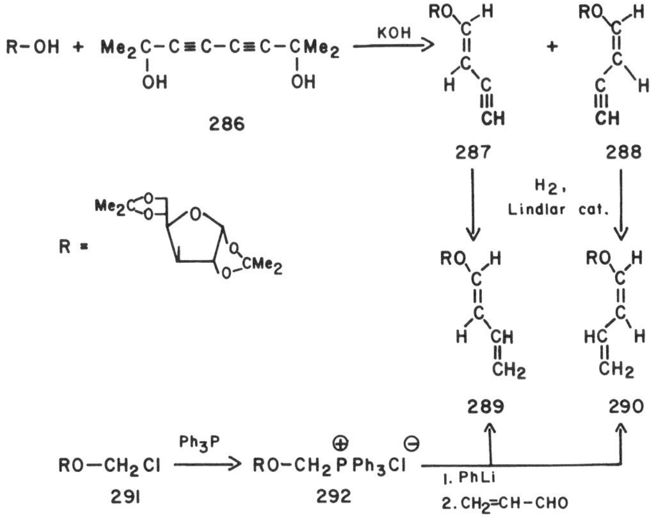

$$R-OH + Me_2C-C\equiv C-C\equiv C-CMe_2 \xrightarrow{KOH}$$
$$\quad\quad\quad\quad | \quad\quad\quad\quad\quad | $$
$$\quad\quad\quad\quad OH \quad\quad\quad\quad OH$$

286

287   +   288

$$\downarrow H_2,\ \text{Lindlar cat.} \downarrow$$

289   290

$$RO-CH_2Cl \xrightarrow{Ph_3P} RO-CH_2\overset{\oplus}{P}Ph_3\overset{\ominus}{Cl} \xrightarrow[\substack{1.\ PhLi \\ 2.\ CH_2=CH-CHO}]{}$$

291   292

gioisomeric compounds (e.g., **308**) was not observed in the cycloaddition reaction. The availability of compounds **306** and **307** opens new possibilities in sugar synthesis by substitution of the chlorine atom and by functionalization of the double bond.

Some general reactions of 2-alkoxy-5,6-dihydro-2H-pyrans have been investigated.

Addition of bromine to 2-ethoxy-5,6-dihydro-2H-pyran (**309**) leads to two stereoisomeric 3,4-dibromo derivatives (**310** and **311**). Dehydrohalogenation of **310** and **311** with sodium ethoxide[179] was found to afford four products, identified as *trans*-5,6-, *trans*-2,5-, *cis*-2,5-diethoxy-, and 3-bromo-2-ethoxy-5,6-dihydro-2H-pyrans (**312-315**), thus correcting an older study.[180]

Olefins **316** and **317** were extensively employed by R.K. Brown and his coworkers[181] as substrates for the syntheses of DL-hexoses. Allylic bromination of **316** and **317** led[182] to the same single compound, 1,6-anhydro-2-bromo-2,3,4-trideoxy-β-DL-*erythro*-hexopyranose (**318**). This result indicated that the allylic

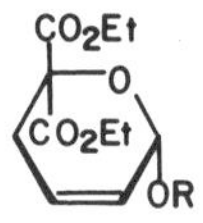

293 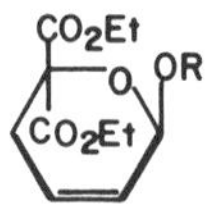   294 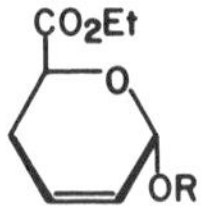   295 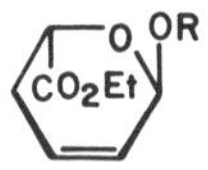   296

R = 3-(1,2:5,6-di-0-isopropylidene-α-<u>D</u>-glucofuranosyl)

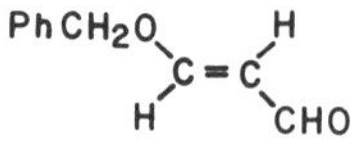

297 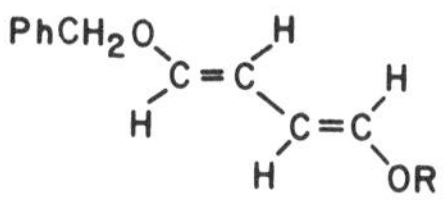   298

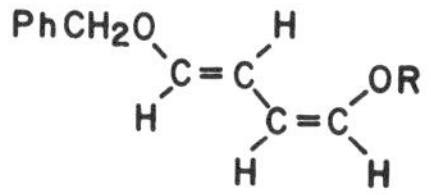

299

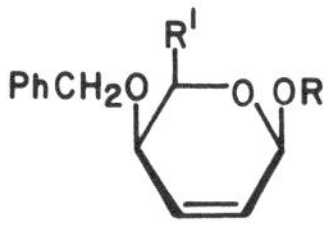

300 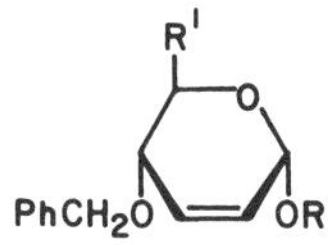   301 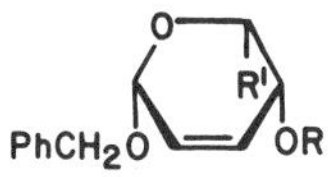   302

R^1 = CO$_2$-L-menthyl

radical (319) was an intermediate in the reaction. Heating of 318 with sodium methoxide led to two methyl ethers, 320 and 321, in a ratio of 2:1.

Reaction of 2-alkoxy-5,6-dihydro-2H-pyrans (e.g., 309) with methyldiphenylsilane leads[183] to the dealkoxylated products 322 in good yields.

The alkoxy group in 2-alkoxy-5,6-dihydro-2H-pyrans (279) can be exchanged[184] for another OR or NR^{1}R^2 group when the substrate is heated at 150-200° with suitable alcohols or amines; this opens the way to the nucleoside type of compound (e.g. 323). The alkoxy group in 279 can be also be exchanged[185] for a thioalkyl group in a substitution reaction with mercaptans. The reactions are catalyzed by stannic chloride or diluted hydrochloric acid. The 2-alkylthio derivatives (324) are often accompanied by allylic rearrangement products (325).

It was shown earlier[186,187] that 2-methoxy-5,6-dihydro-2H-pyrans (279, R^1 = Me;R^2 = H, CO$_2$Bu) readily added methanol across the double bond, forming mixtures of stereoisomeric 2,4-dimethoxytetrahydropyrans. This addition reaction was extended to acetic acid,[188] which was added in the presence of sulfuric acid to 6-acetoxy-methyl-2-methoxy-5,6-dihydro-2H-pyran (326,R-

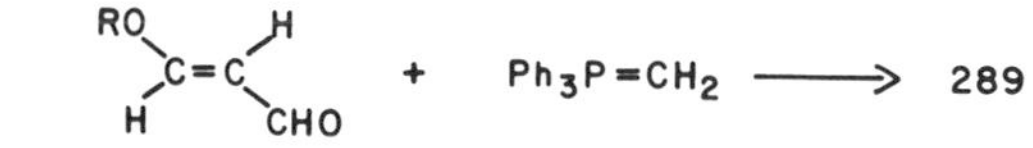

303

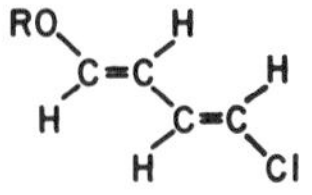

304

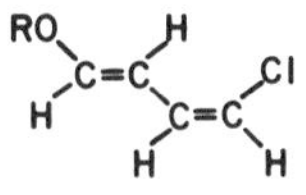

305

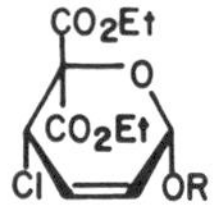

306

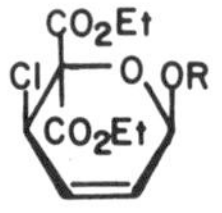

307

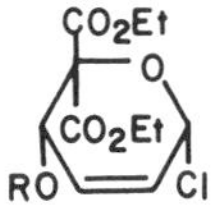

308

Ac) and gave a mixture of 1,3,6-tri-O-acetyl-2,4-dideoxy-α-DL-*threo*, α- and β-DL-*erythro*-hexopyranoses (**327-329**) in the proportions 1:1.5:1. Basic reactions of racemic 2,4-dideoxyhexoses like glycosidation, substitution reactions at C-3 and C-6, and so on, were investigated.[189] BF$_3$-Catalyzed addition of benzyl carbamate to the double bond of 2-ethoxy-6-methanesulfonyloxymethyl-5,6-dihydro-2H-pyran (**279**,R^1 = Et,R^2 = CH$_2$OSO$_2$Me) furnished[190] a product which was directly converted with sodium azide into ethyl 6-azido-3-benzylcarbamido-2,3,4,6-tetradeoxy-α-DL-erythro-hexopyranoside (**330**), an intermediate for the synthesis of the antibiotic negamycin. Alternative substrates for the synthesis of this antibiotic were prepared by the Michael addition of benzylcarbamate or O-benzylhydroxylamine to 5,6-dihydro-6-methanesulfonyloxymethyl-2H-pyran-2-

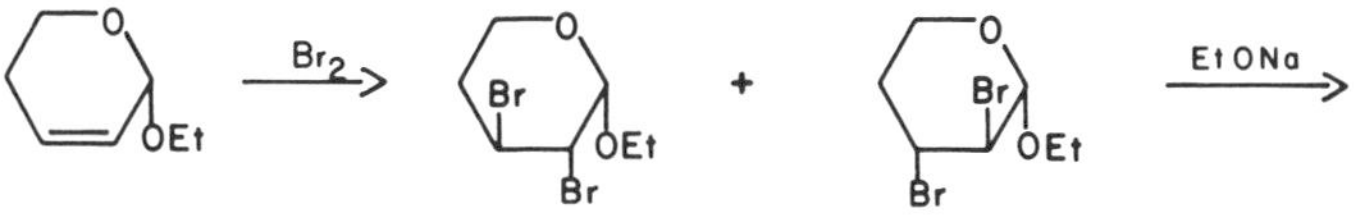

309

310

311

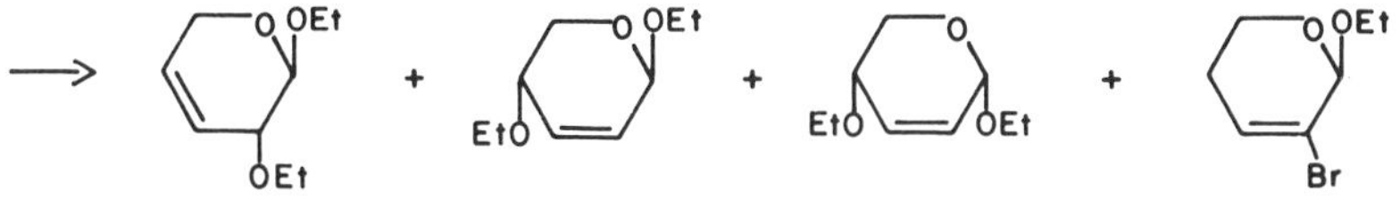

312

313

314

315

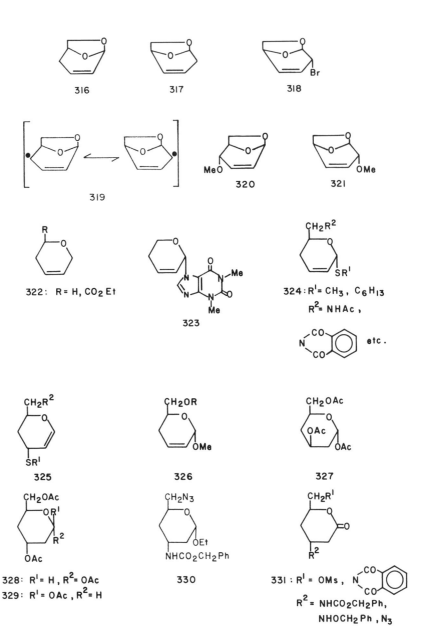

316

317

318

319

320

321

322: R = H, CO₂Et

322 : R = H, CO$_2$Et

323

324 : R¹ = CH₃, C₆H₁₃

324 : R^1 = CH$_3$, C$_6$H$_{13}$
    R^2 = NHAc ,

etc.

325

326

327

328: R¹ = H, R² = OAc
328 : R^1 = H, R^2 = OAc
329 : R^1 = OAc, R^2 = H

330

331 : R^1 = OMs ,
    R^2 = NHCO$_2$CH$_2$Ph,
    NHOCH$_2$Ph , N$_3$

one[190] or by the addition of hydrazoic acid to 5,6-dihydro-6-N-phthalimido-methyl-2H-pyran-2-one.[191] Single-addition products (**331**) were obtained of the *erythro* configuration (cf. Section 4, on addition reactions to the double bond of parasorbic acid).

The substrate for the synthesis of hexoses, *trans*-5,6-dihydro-6-hydroxy-methyl-2-methoxy-2H-pyran (**326**,R = H), was resolved[192] by means of its ω-camphanyl ester into enantiomers. The *levo*-rotating enantiomer was determined to have the 2S:6S configuration.

Extension of the side chain at position 6 in 2-alkoxy-5,6-dihydro-2H-pyrans was achieved[193] by base-catalyzed condensation of fatty acid esters with alkyl 5,6-dihydro-2-methoxy-2H-pyran-6-carboxylates followed by ketone cleavage of the β-ketoesters (**332**) formed. 6-Acyl-5,6-dihydro-2-methoxy-2H-pyrans (**333**) were obtained in 60-80% yield. Ketones **333** could also be obtained,[194] although in lower yields, by a Grignard reaction between alkyl magnesium bromides and 6-cyano-5,6-dihydro-2-methoxy-2H-pyran (**334**).

A full paper was published[195] on the structure of the Diels–Alder adducts (**335** and **336**) obtained by condensation of cyanodithioformate with 1,3-buta-diene or 1-methoxy-1,3-butadiene. Cyanodithioformate was successfully con-densed with 1E,4E-diacetoxy-1,3-butadiene and cyclopentadiene, furnishing ad-ducts **337** and **338**. *cis*-Hydroxylation of **336** with osmium tetroxide gave[196] a dithio sugar, 2,3-di-S,O-methyl-2,6-dithio-α-DL-*arabino*-hex-2-ulopyrano-ni-trile (**339**). The analogous *cis*-hydroxylation of **338** furnished a bridged thiosugar **340**.

The steric effects involved in the *cis*-hydroxylation of 5,6-dihydro-2H-pyrans substituted at positions 6 and/or 2 with aqueous potassium permanganate were studied by Mochalin and co-workers.[197] It was found that the reaction course is dependent on the steric hindrance exerted by the bulky substituent at C-2: both OH groups enter preferentially *trans* with respect to that substituent. In conse-quence, products of α-*lyxo* or β-*erythro* configuration are obtained in the case of di- or mono-substituted substrates. However, in the case of 5,6-dihydro-2-methoxy-2H-pyran, both possible stereoisomeric products, methyl 4-deoxy-α-and β-DL-*erythro*-pentopyranosides, were formed[197] in approximately equal amounts.

332 : R¹ = H, Me, Et, Pr
R² = Me, Et, Bu.

333 : R = Me, Et Pr, Bu.

334

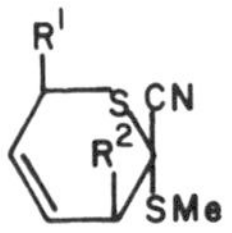

335: R^1 = R^2 = H
336: R^1 = H, R^2 = OMe
337: R^1 = R^2 = OAc

338

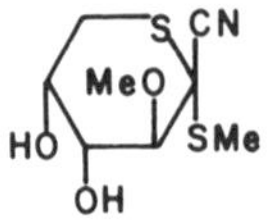

339

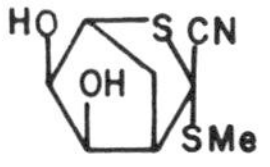

340

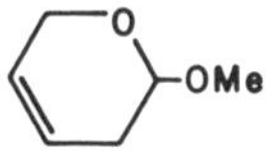

341

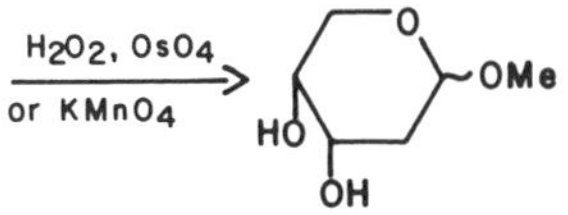

342

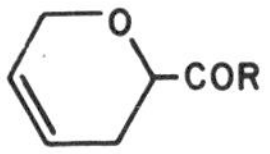

343: R = OEt
344: R = NH$_2$

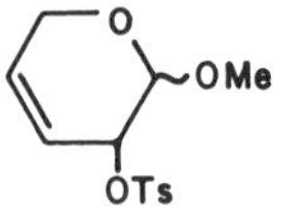

345

Cis-hydroxylation of 3,6-dihydro-2-methoxy-2H-pyran (**341**) afforded[198–200] a mixture of methyl 2-deoxy-α- and β-DL-*erythro*-pentopyranosides (**342**). Hydrolysis of this mixture gave free 2-deoxy-DL-*erythro*pentose (2-deoxyribose). Compound **341** was obtained[198] by the reductive removal of the p-toluenesulfonyloxy group in methyl 3,4-dideoxy-2-O-*p*-toluenesulfonyl-α,β-DL-pent-3-eno-pyranoside (**345**). Mochalin[199,200] obtained **341** in three steps starting from the Diels–Alder adduct **343** via the amide **344** and its Hoffman degradation in methanol solution.

Sharpless' oxyamination reagent, composed of chloramine-T and a catalytic quantity of osmium tetraoxide, was reacted[201] with a series of 6-substituted 5,6-dihydro-2-methoxy-2H-pyrans. Both regioisomeric methyl 2,4-dideoxy-2-*p*-toluenesulfonylamido- and 3,4-dideoxy-3-*p*-toluenesulfonylamido DL-*lyxo*-hexopyranosides (**346** and **347**) were obtained in yields of 40–60%, with the latter distinctly prevailing in mixtures.

The epoxidation of 2-alkoxy-5,6-dihydro-2H-pyrans (**279**) can be effected with all typical epexidizing reagents. Usually, mixtures of stereoisomeric alkyl 2,3-anhydro-4-deoxy-DL-glycopyranosides (**280**) are readily separable by column chromatography.

Descours and co-workers[202,203] obtained several alkyl 3-amino-3,4-dideoxy-γαβ-DL-*threo*-pentopyranosides **348** (cf. Mochalin and co-workers,[204]) by treatment of epoxides **280** (R^1 = Me,Ph;R^2 = H) with ammonia or dimethylamine.

346   347

R = H, OH, NHAc, etc.

348 : R^1 = H, Me
R^2 = Me, Ph

349

Inversion of configuration of the 2-OH group in **348** gave a series of analogous amino sugars of the *erythro* configuration.

The reaction of *t*-butyl (methyl 2,3-anhydro-4-deoxy-α- and β-DL-*ribo*-hexopyranosid) uronates with ammonia[205] gave, after *trans*-esterification, methyl (methyl 3-amino-3,4-dideoxy-α- and -β-DL-*xylo*hexopyranosid) uronates (**349**). 3-Amino-3,4-dideoxy-D-*xylo*-hexuronic acid (ezaminuroic acid) is a component of antibiotic ezomycin A$_2$. The methyl glycoside of another antibiotic sugar belonging to the 4-deoxyhexose family, 4-deoxy-DL-neosamine C (in form of its peracetylated derivative **350**) was synthesized[206] from the butyl ester **279** (R^1 = Me, R^2 = CO$_2$Bu) according to the route shown.

D. Alkyl 3,4-Dideoxy-DL-glyc-3-enopyranosides

For conversion of epoxides **280** into 3,4-unsaturated pyranosides **281**, Brown[207] employed the well-known isomerization of oxiranes into allylic alcohols, which is induced by strong bases, for example, *n*-butyl lithium. It was found[208] that for the isomerization of **351** and **352** to the corresponding allylic alcohols **353** and **354**, lithium diethyl amide was more effective; with *n*-butyl lithium other reactions took place.

For the conversion of **280**→**281**, two alternative methods were developed. The first consists of the opening of the oxirane ring with dimethylamine that leads to alkyl 3,4-dideoxy-3-dimethylamino-DL-glycopyranoside (**355**). Oxidation of **355** to N-oxide (**356**) and its thermal degradation gives **281** in 50–70% yields.

By this method all stereoisomeric methyl 3,4-dideoxy-DL-pent-3-enopyranosides[209] and -hex-3-enopyranosides,[210] 3,4,6-trideoxy-DL-hex-3-enopyranosides,[210] and *t*-butyl (methyl 3,4-dideoxy-DL-hex-3-enopyranosid)uronates[211] were obtained.

The second approach[212] makes use of the Sharpless method, consisting in opening of the oxirane ring with selenophenol to **357**. Oxidation of **357** to the unstable selenooxide (**358**) followed by refluxing in ethanol gives **281** in good yields. This method was used by David for the preparation of 3,4-unsaturated precursors of disaccharides.

E.   Mono-, Di-, and Trisaccharides

(a)   Cis-Hydroxylation

Cis-hydroxylation of alkyl 3,4-dideoxy-glyc-3-enopyranosides (**281**) with typical reagents (osmium tetroxide, Milas reagent) is subjected to steric control, that is, both hydroxyl groups enter preferentially from the less hindered side of the ring. If substituents at C-1 and C-2 are on the same side of the six-membered ring (for example, **359**) their steric effects act in the same direction, and sugars of the β-*arabino* (**360**) or α-*galacto* (**361**) configuration distinctly predominate over their α-*ribo* (**362**) or α-*allo* (**363**) partners.

If the substituents are on opposite sides (for example **364**), their steric influences are divided and both stereoisomeric products are formed in varying amounts, depending on the bulkiness of the C-2 substituent in the first place. Thus, products of the α-*arabino* (**365**) or α-*altro* (**366**) and β-*ribo* (**367**) or α-*talo* (**368**) configuration are obtained.

By this method both anomers of methyl DL-arabinopyranoside and -DL-ribopyranoside were obtained.[213] In a similar manner, normal[214] and 6-deoxy[215] methyl DL-hexopyranosides of the *allo, altro, galacto,* and *talo* configuration, and esters of DL-hexuronic acids[211] of *altro, galacto,* and *talo* configuration were prepared.

Application of *cis*-hydroxylation to 3,4-unsaturated disaccharide precursors led to a series of disaccharides in which both sugar units were optically active. Thus, the first semisynthetic disaccharides prepared by David and co-workers[216] were α-D- and α- -L-altropyranosyl-(1→3)-D-glucoses. Continuation of this work[212] extended the number of hexopyranosyl-(1→3)-D-glucoses synthesized with units of the α-D-*allo* β-L-*altro*, and α-D-*galacto* configuration.

Modification of the unsaturated moiety in 3,4-dideoxy-α-D-hex-3-enopyranosyl-(1→3)-1,2:5,6-di-O-isopropylidene-α-D-glucofuranose (**369**) consisting of the oxidation of the 2-OH group to the corresponding ketone, oximation of the ketone, and *cis*-hydroxylation of the double bond gave rise[217] —after reduction of the oxime grouping, N-acetylation, and removal of other blocking groups— to two disaccharides, 2-acetamido-2-deoxy-α-D-*galacto*- and α-D-*talo*pyranosy (1→3)-D-glucoses. In a similar way, with some modifications, further di- and trisaccharides were prepared: 4-O-[(2-O-α-L-fucopyranosyl)- -β-D-altropyrano-syl]-D-glucose,[218] 4-O-[(2-O-α-L-fucopyranosyl)-β-L-altropyranosyl]-D-glu-cose,[218] 4-O-(β-L-allopyranosyl)-D-glucose[218], 4-O-(β-L-galactopyranosyl)-D-glucose,[218] maltose,[218] 2-O-(α-L-fucopyranosyl)-3-O-(2-acetamido-2-deoxy-α-D-galactopyranosyl)-D-galactose,[219] and 3-O-(4-0-benzyl-β-D- and -L-gulopyr-anosyl)-1,2:5,6-di-O-isopropylidene-α-D -glucofuranoses.[177]

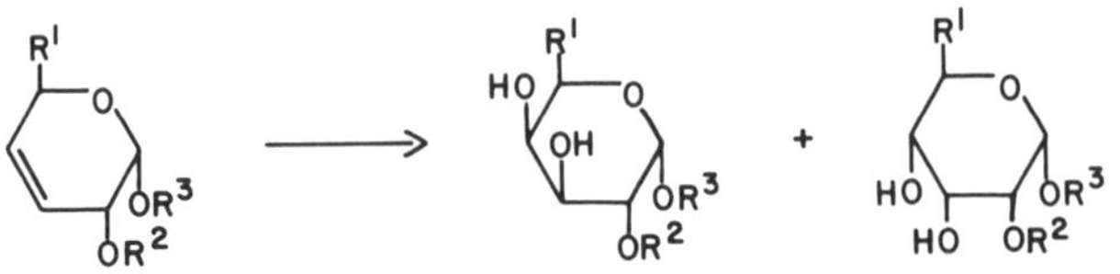

359

360,361

362 , 363

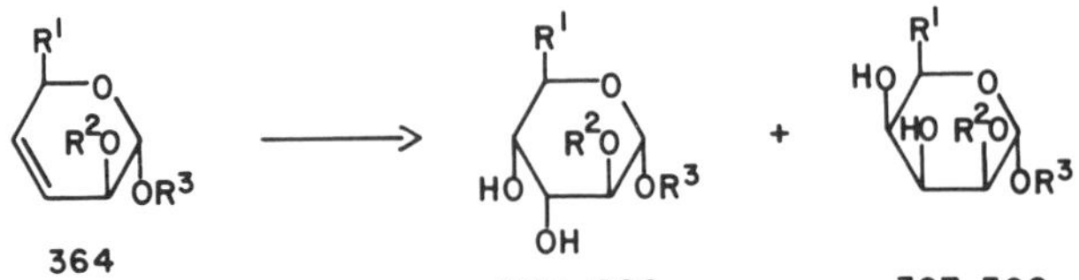

364

365,366

367,368

359,364: R^1 = H, CH$_2$OH, CH$_3$
R^2 = H, Ac
R^3 = Me or a
sugar moiety

360,362
365,367: R^1 = H
R^2= H, Ac
R^3= Me

361,363
366,368:
R^1 = CH$_2$OH, CH$_3$
R^2= H,Ac
R^3= Me or a
sugar moiety

Following the general route shown, N-acetylated methyl 7-deoxy-α-DL-lincosaminide (370), an eight-carbon atom sugar related to the antibiotic lincomycin, was synthesized[220] starting from 5,6-dihydro-2-methoxy-6-propionyl-2H-pyran (333,R = Et); the last step involved *cis*-hydroxylation and also gave—besides the desired "*galacto*" product—some of the "*allo*" stereoisomer (371).

(b) Epoxidation

The steric course of the epoxidation of alkyl 3,4-dideoxy-glyc-3-enopyranosides (281) with peroxyacids is dependent on the configuration and bulkiness of substituents at C-1 and C-2, in rough analogy with the course of *cis*-hydroxylation.

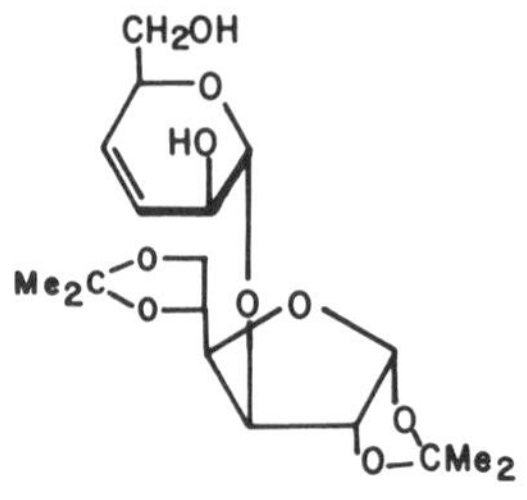

369

An important difference is the course of epoxidation when the 2-OH group is free; in that case the epoxide is formed preferentially from the same side as that occupied by the hydroxyl group. This result is ascribed to the transient hydrogen bond between the OH group and peroxyacid molecule.

Starting from **281** (R^1 = Me, R^2 = H, CH_2OH), all stereoisomeric methyl 3,4-anhydro-α and β-DL-pentopyranosides[221] and α-DL-hexopyranosides[222] were prepared. Also some *t*-butyl (methyl 3,4-anhydro-α-DL-hexopyranosid)uronates were obtained[211] by reaction of **281** (R^1 = Me, R^2 = CO_2Bu) with hydrogen peroxide-acetonitrile reagent.

(c) Hydrolytic Opening of the 3,4 -Anhydro Ring

The regiochemistry of hydrolytic oxirane ring opening in alkyl 3,4-anhydro-DL-glyco-pyranosides is dependent on several factors, three of them playing a particularly important role: preference for the axial attack of the nucleophile, neighboring group participation, and conformational control.[223]

From methyl 3,4-anhydro-DL-pentopyranosides, methyl α- and β-DL-*xylo*-and *lyxo*pyranosides were obtained.[213] Analogously, from methyl 3,4-anhydro-DL-hexopyranosides, normal methyl DL-hexopyranosides were prepared[224] of the *gulo, gluco, manno,* and *ido* configuration. In the same manner, esters of DL-hexuronic acids were obtained[211] having the *manno, gulo,* and *gluco* configuration.

Both components of the nonreducing disaccharide everninose: 2-O-methyl-lyxose (**372**) and 2,6-di-O-methyl-mannose (**373**) were synthesized via the suitable 3,4-anhydro intermediates.[225,226]

$$372$$

$$373$$

(d) Other Syntheses

Hydration of the double bond in 3,4-unsaturated sugars by the oxymercuration-demercuration procedure should lead either to 3- or to 4-deoxy sugars. It was found that this reaction applied to methyl 3,4-dideoxy- and 3,4,6-trideoxy-α-DL-hex-3-enopyranosides[227] proceeded regio- and stereospecifically and furnished methyl 3-deoxy- and 3,6-dideoxy-DL-hexopyranosides. From substrates of the α-*erythro* configuration only α-*xylo* products (374) were obtained. α-*threo* substrates gave products (375) of the *lyxo* configuration.

David and Lubineau[228] presented an efficient synthesis of the LL-enantiomer of kasuganobiosamine (272,R = H). Cycloaddition of a protected L-chiro-inositol butadienyl ether with L-menthyl glyoxylate gave a mixture of four diastereoisomeric dihydropyrans, from which α-L component (376) was isolated and used for further steps. Conversion of the ester group in 376 to a methyl group, and formation of the 3,4-unsaturated system, led to two allylic alcohols (377). Oxidation of 377 to a ketone, addition of hydrazoic acid to the conjugated double bond (an equatorial azide 378 was formed), oximation of the keto group, and reduction of both nitrogen functions gave a mixture of two diamino sugars from which the desired α-L-*arabino* compound (379) was isolated by chromatography. Catalytic hydrogenation followed by mild acidic hydrolysis gave the desired product in 1.3% overall yield.

Purpurosamine B (382), a seven-carbon atom sugar component of the antibiotic gentamicin C_2, was synthesized[229] from 6-acetyl-5,6-dihydro-2-methoxy-2H-pyran (333, R = Me). Oximation of the keto group, reduction of the oxime to the corresponding amine, and N-acetylation gave two stereoisomeric N-acetamides. For further synthesis the *erythro* stereoisomer (380) was taken. Further steps involved the epoxidation of the double bond, isolation of the *manno* epoxide, and its reduction to methyl 6-acetamido-3,4,6,7-tetradeoxy-α-DL-*arabino*-heptopyranoside (381). Oxidation of the 2-OH group, oximation of the ketone, and reduction of the oxime followed by acetylation gave methyl N,N′-diacetyl-α-DL-purpurosaminide B, which was hydrolyzed to N,N′-diacetyl-DL-purpurosamine B (383).

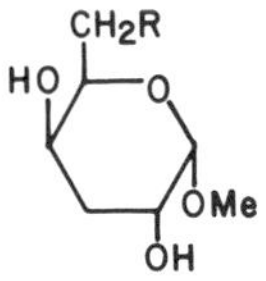

374

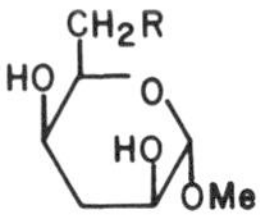

375

R = H, OH

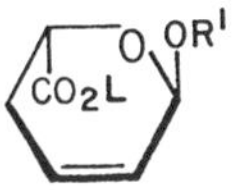

376: L = L-menthyl

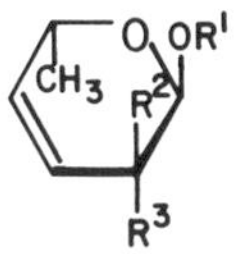

377: R^2 = OH, R^3 = H
and R^2 = H, R^3 = OH

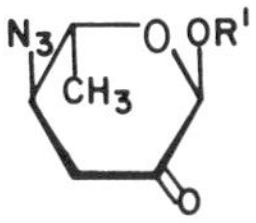

378

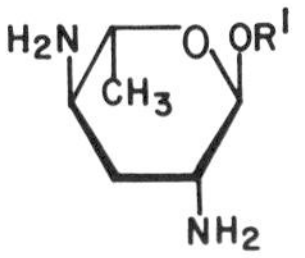

379

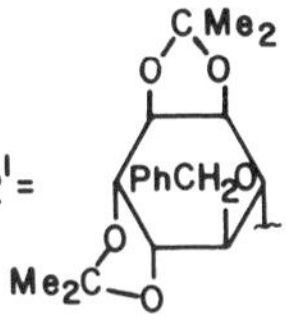

7. MISCELLANEOUS SYNTHESES

It has been demonstrated in previous sections of this chapter that some substrates (e.g., furfuryl alcohols, vinylene carbonate telomers, etc.) are capable of being transformed, in a limited number of synthetic steps, into the full variety of stereoisomeric sugars. These syntheses are fairly general. In this section, carbohydrate preparations are described which start from some peculiar substrates and by exploitation of specific reactions lead to a single monosaccharide or, at most, to a limited number. Natural products have often been used for the purpose. Their structural features usually predetermine the type of sugar that can be obtained. These substrates, being optically active, afford eventually enantiomeric products. The usually troublesome resolution of racemic intermediates of final

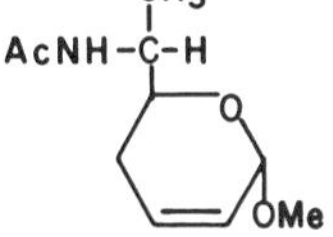

380

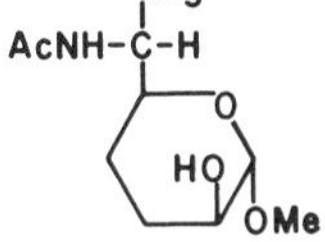

381

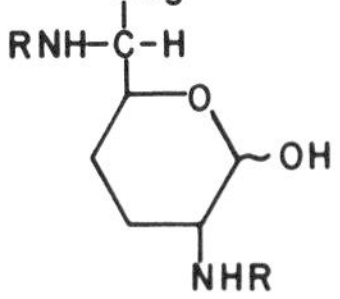

382: R = H
383: R = Ac

products can thus be avoided. This is quite important in carbohydrate chemistry, where practically all compounds occur in enantiomeric form.

A. Tartaric Acid, Glyceraldehyde, and Amino Acids

$2R$:$3R$-Tartaric acid (dextro-tartaric acid) has been employed[230] for a long time as a substrate in carbohydrate synthesis, especially in the field of tetroses. In the past decade Nakagawa and co-workers[231] improved the preparation of L-threose. Methyl hydrogen di-O-acetyl-*dextro*-tartrate (**384**) was reduced with sodium borohydride in water solution to L-threono-γ-lactone (**385**) in 74% yield. Further reduction of the lactone grouping with bis(3-methyl-2-butyl) borane to the aldehyde stage proceeded smoothly and afforded L-threose in about 63% yield.

Application of tri-*n*-butyltin hydride to the reduction of the ester chloride **386** gave a good yield of methyl di-O-acetyl-L-threuronate (**387**).

Fuganti and co-workers[232] used 2R:3R-tartaric acid for the synthesis of 3-benzamido-2,3,6-trideoxy-L-*xylo*-hexose (**393**). The acid was converted according to a known procedure into 2,3-O-isopropylidene-4-O-*p*-toluenesulfonyl-L-threitol (**388**). Reduction of **388** with lithium aluminum hydride gave 4-deoxy-2,3-O-isopropylidene-L-threitol (**389**), which was oxidized to the corresponding threose (**390**). Condensation of **390** with ethoxycarbonylmethine-triphenylphosphorane yielded a mixture of E and Z unsaturated esters (**391**). Michael reaction type addition of ammonia to **391** followed by hydrolysis of the product with hydrochloric acid gave the α-lactone hydrochloride. This material was N-benzoylated, and after chromatographic separation the *xylo* stereoisomeric (**392**) was isolated in 60% yield. Half-reduction of the lactone grouping in **392** with DIBAH afforded **393**.

$$
\begin{array}{ccc}
\text{CO}_2\text{H} & & \text{CO} \longrightarrow \\
\text{H}-\text{C}-\text{OAc} & \xrightarrow[\text{H}_2\text{O}]{\text{NaBH}_4} & \text{H}-\text{C}-\text{OAc} \\
\text{AcO}-\text{C}-\text{H} & & \text{AcO}-\text{C}-\text{H} \\
\text{CO}_2\text{Me} & & \text{CH}_2 \longrightarrow \text{O}
\end{array}
$$

$$
\xrightarrow{(\text{C}_5\text{H}_{11})_2\text{BH}} \underline{\text{L}}\text{- Threose}
$$

384   **385**

$$
\begin{array}{ccc}
\text{COCl} & & \text{CHO} \\
\text{H}-\text{C}-\text{OAc} & \xrightarrow{n\text{-Bu}_3\text{SnH}} & \text{H}-\text{C}-\text{OAc} \\
\text{AcO}-\text{C}-\text{H} & & \text{AcO}-\text{C}-\text{H} \\
\text{CO}_2\text{Me} & & \text{CO}_2\text{Me}
\end{array}
$$

386   **387**

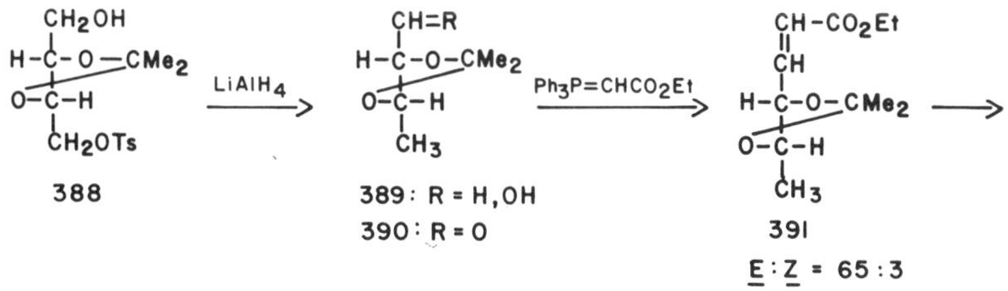
LiAlH4
Ph3P=CHCO2Et
388
389: R = H,OH
390: R = O
391
E:Z = 65:3

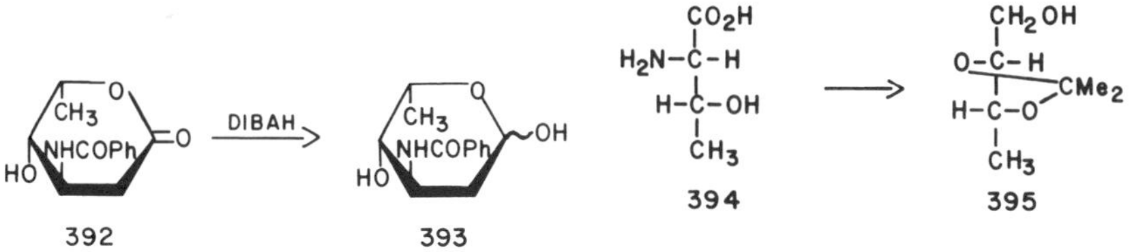
DIBAH
392
393
394
395

The D-enantiomer of **393** was obtained[232] in an identical sequence of reactions starting from 2,3-O-isopropylidene-4-deoxy-D-threitol (**395**). This compound was prepared from L-threonine (**394**) in the following way: the amino acid was deaminated to 2*S*:3*R*-dihydroxy-butyric acid. Esterification of the carboxyl group and protection of both hydroxyl groups with an isopropylidene grouping gave methyl 4-deoxy-2,3-O-isopropylidene-D-threonate. Reduction of the ester group afforded **395** smoothly.

In continuation of a study[233] on the synthesis of β-hydroxy-amino acids and 2-amino-2-deoxy-aldonic acids by base-catalyzed condensation of *N*-pyruvylideneglycinatocopper(II) complex (**396**) with aldehydes, its reactivity was compared[234] with that of bisglycinatocopper(II) (**397**). It was found in the reaction with 2,3-*O*-isopropylidene-L-glyceraldehyde that milder reaction conditions could be employed and the product, 2-amino-2-deoxy-4,5-*O*-isopropylidene-L-xylonic acid (**398**), was obtained in a better yield (70% vs. 31%).

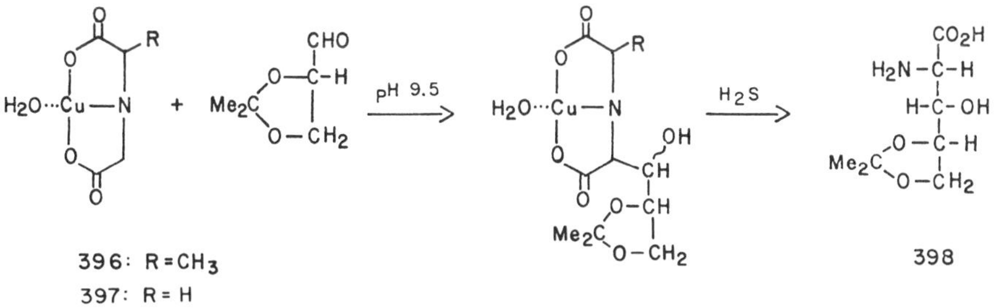
pH 9.5
H2S
396: R = CH3
397: R = H
398

It is known that deamination of some readily available amino acids can be performed with nearly quantitative retention of configuration. Based on this conversion, Yamada and co-workers[235] elaborated an ingenious synthesis of D-ribose and D-lyxose starting from L-glutamic acid. The preliminary communication of this work was reported in the first volume of this series.[236] Now the full details have been published.[235] In an extension of the application of the unsaturated intermediate (399) for the synthesis of other pentoses, all 2,3-anhydro derivatives (400-403) were prepared.[237] LAH-Reduction of β-D-*ribo* and α-D-*lyxo* epoxides (400 and 401) afforded, after acetylation, 3-deoxy-pentosides 404 and 405, whereas the same reaction conditions converted β-D-*lyxo* α-D-*ribo* epoxides (402 and 403) into the corresponding 2-deoxy-derivatives (406 and 407).

Alkaline hydrolysis of methyl 2,3-anhydro-β-D-ribofuranoside (400) and -α-D-*lyxo*furanoside (401) led to methyl furanosides of D-xylose and D-arabinose. Reaction of 400 with sodium benzyl mercaptide gave, after acetylation, methyl 2-O-acetyl-3,5-di-S,O-benzyl-3-thio-β-D-*xylo*furanoside (408). α-D-*lyxo* Epoxide (401) was also opened with ammonia at C-3 and gave, after acetylation, methyl 3-acetamido-2-O-acetyl-5-O-benzyl-α-D-*arabino*-furanoside (**409**) as the single product.

From L-glutamic acid, a branched-chain sugar derivative (**410**) was prepared[238] according to the scheme below:

Yamada and co-workers elaborated a general approach to 6-deoxy-L-hexoses starting from L-alanine.[239] Deamination of this amino acid in acetic acid gave 2S-acetoxy-propionic acid (**411**). After conversion to the acid chloride it was reacted with propiolaldehyde dimethylacetal magnesium bromide to afford the acetylene **412**. Half-reduction to the *cis*-olefin followed by deacetylation and heating with phosphoric acid gave both anomeric methyl 2,3,6-trideoxy-L-hex-2-enopyranosid-4-uloses (**413**) (cf. Section 5). From the α-anomer (which predominated in the mixture), a number of glycosides of antibiotic sugars have been prepared, including methyl α-L-amicetoside (**65**,R = Me), α-L-mycaminoside (**414**) and α-L-oleandroside (**415**).

In the presence of diethylamine-formic acid catalyst, 1-nitro-2-alkanols (**416**) are added[240,241] to the double bond of acrylaldehyde, forming, in 40–80% yield, 5-hydroxy-4-nitro-alkanals (**417**, occurring in the cyclic hemiacetal form). Under the reaction conditions employed, the formation of ether-type Michael reaction products **418** was not observed.

When R ≠ H, both stereoisomeric forms of **417** are expected from the addition reaction. It is remarkable that for R = Me, only the *erythro* stereoisomer was obtained. In case of R = Et, both isomers, *threo* and *erythro*, were isolated and characterized. Compounds **417** can be regarded as 2,3,4 . . . -polydeoxy-4-nitro-DL-aldoses. Methanolic hydrogen chloride readily converts them into mixtures of anomeric methyl pyranosides (**419**). Hydrogenation of the nitro group in **419** in the presence of Raney nickel leads to methyl 4-amino-2,3,4-polydeoxy-$\alpha\beta$-DL-aldopyranosides (**420**). In this way sugars of this class containing five, six, and seven carbon atoms in the chain were prepared. Most important is the hexose of *erythro* configuration; D-enantiomer (tolyposamine) is a constituent of the antibiotic tolypomycin.

The dimethylation[241] of the amino group in anomeric methyl 4-amino-2,3,6-

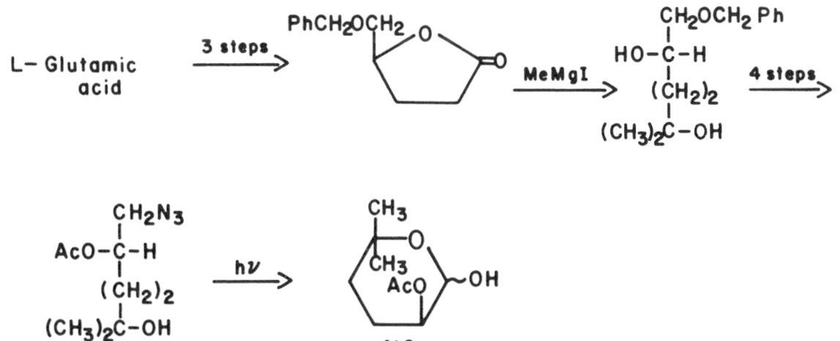

L- Alanine $\longrightarrow$ AcO–$\overset{\overset{\textstyle CO_2H}{|}}{\underset{\underset{\textstyle CH_3}{|}}{C}}$–H $\xrightarrow{\ (MeO)_2CH-C\equiv CMgBr\ }$

411

$$CH(OMe)_2$$

Structure **412**: $CH(OMe)_2$–C≡C–CO–$\overset{|}{C}$H(AcO)–CH_3

412

1. H_2, Lindlar cat.
2. NaOH
3. H_3PO_4

414 (NMe₂)   **415** (OMe)   **413**

trideoxy-DL-hexopyranosides (**421**) gave methyl glycosides of another antibiotic sugar, forosamine (**107**), a component of foromycins.

Sodium hydrogen carbonate-catalyzed condensation of sodium glyoxylate with 4-nitro-2-butanol gives[242] a 3.2:1 mixture of stereoisomeric 2-hydroxy-3-nitro-O-caprolactones of the *ribo* and *lyxo* configuration (**422** and **423**). The *lyxo* lactone (**423**) was isolated after conversion into the 2-O-tetrahydropyranyl derivative. Half-reduction of the lactone grouping gave a lactol, which was converted into a mixture of methyl glycosides (**424**) (α:β = 88:1). Equilibration of **424** with sodium hydrogen carbonate in aqueous methanol gave a mixture of glycosides from which methyl 3,4,6-trideoxy-3-nitro-α-DL-*xylo*-hexopyranoside (**425**) was isolated by thin layer chromatography (TLC). Hydrogenation of the nitro group in **425** to the amino group followed by Eschweiler–Clarke methylation gave methyl α-DL-desosaminide.

B. 1,2-Dihydropyridines

Natsume and Wada[243] found that photochemical addition of methanol to nicotinonitrile (**426**) leads to two dihydropyridines (**427** and **428**). Both compounds differed in stability; only the latter was relatively stable. After O,N -benzoylation, **429** could be isolated from the mixture as the sole product. Benzamide **429**, possessing six carbon atoms in the chain and two differently activated double bonds, appeared to be an interesting substrate for the synthesis of piperidinose-type sugars.

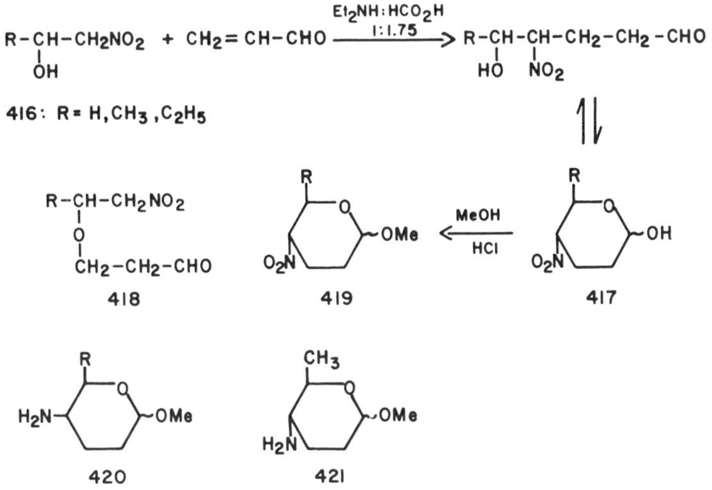

$$
\underset{\underset{OH}{|}}{R-CH-CH_2NO_2} + CH_2=CH-CHO \xrightarrow[\text{1:1.75}]{\text{Et}_2\text{NH:HCO}_2\text{H}} \underset{\underset{HO}{|}\ \underset{NO_2}{|}}{R-CH-CH-CH_2-CH_2-CHO}
$$

416: R = H, CH$_3$, C$_2$H$_5$

$$
\underset{\underset{\overset{O}{\underset{|}{|}}}{R-CH-CH_2NO_2}}{\underset{CH_2-CH_2-CHO}{}}
$$

418

419    $\xleftarrow[\text{HCl}]{\text{MeOH}}$ 417

420   421

The first goal was the synthesis of the antibiotic nojirimycin[244] (436, 5-amino-5-deoxy-L-glucose). *Cis*-hydroxylation of 429 followed by acetylation afforded a *ribo* enamine 430. Reaction of 430 with N -bromosuccinimide gave a methoxy-bromide (431). The attempt at substitution of the bromine atom by an acetate residue in reaction with tetrabutylammonium acetate resulted in elimination reaction leading to the unsaturated compound 432. *Cis*-hydroxylation of this material followed by acetylation gave a *manno* product 433. Conversion of the cyanohydrin acetate grouping in 433 to a keto group and its reduction to a hydroxyl group was accomplished by alkaline hydrolysis and subsequent treatment with sodium borohydride. The product was isolated in the form of N-

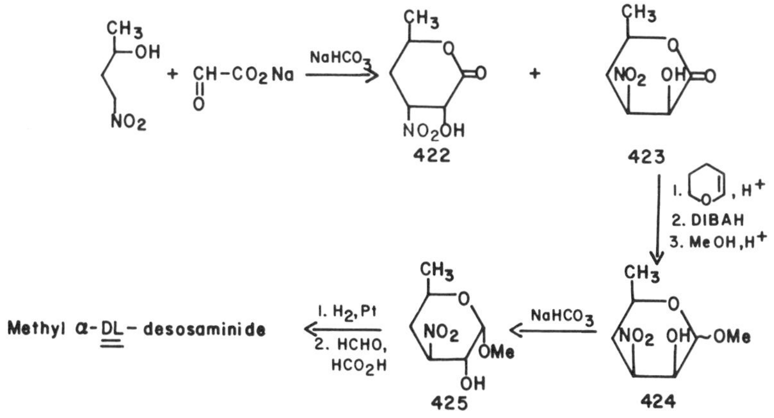

422   423

424

425

Methyl α-DL-desosaminide $\xleftarrow[\text{2. HCHO, HCO}_2\text{H}]{\text{1. H}_2\text{, Pt}}$ 425

benzoyl-2,3,4,6-tetra-O-acetate (**434**). Removal of acetyl groups afforded N-benzoyl-1-O-methyl-DL-nojirimycin (**435**) as the final product.

In a similar way, starting from N-methoxycarbonyl-5-cyano-1,2-dihydropyridine (**437**), peracetylated 5-methoxycarbonylamino-5-deoxy-DL-*xylo*pentopiperidinose (**438**) was obtained.[245]

Dihydropyridine **429** was next used[246] for the synthesis of methyl glycoside of 5-amino-5-deoxy-DL-*ido*-hexopiperidonose (**445**). In the first step, **429** was reacted with N-bromosuccinimide to afford two bromoacetates (**439** and **440**). The first bromoacetate (**429**) was deacetylated to the bromohydrin **441**. Reaction with silver oxide followed by acetylation gave a rearranged diacetate **442** (which was certainly preceded by an epoxide). The 1-O-acetyl group in **442** could be readily replaced by a methoxy group to form **443**. *Cis*-hydroxylation of the double bond followed by known degradation of the cyanohydrin grouping gave, after acetylation, methyl 2,3,4,6-tetra-O-acetyl-5-N-benzamido-5-deoxy-β-DL-*ido*-hexopiperidinoside (**444**). Alkaline deacetylation yielded the free glycoside **445**.

Dihydropyridines **429** and **446** afforded,[247] in a photochemical reaction with singlet oxygen, reactive *endo*-peroxides **447** and **448**. These products enabled novel ways of functionalization of the substrates directed towards carbohydrates. Thus, the reaction of **448** with thiols gave a mixture of anomeric 2,3-unsaturated thioglycosides (**449** and **450**). Warm acetic acid readily isomerized both compounds into the glycal-type derivative **451**. The boron trifluoride catalyzed addition of methanol to the double bond of **451** gave a sulfur-containing 2-deoxy sugar **452**. *Cis*-hydroxylation of **451** (R = CH₂Ph) followed by acetylation gave 1,2,4-tri-O-acetyl-3-S-benzyl-5-deoxy-5-methoxycarbamido-3-thio-α-DL-*xylo*-pentopiperidinose (**453**). Similarly, *cis*-hydroxylation of **449** and **450** gave thioglycosides of 5-deoxy-5-methoxycarbamido-α-DL-*lyxo*piperidinose (**454**) and

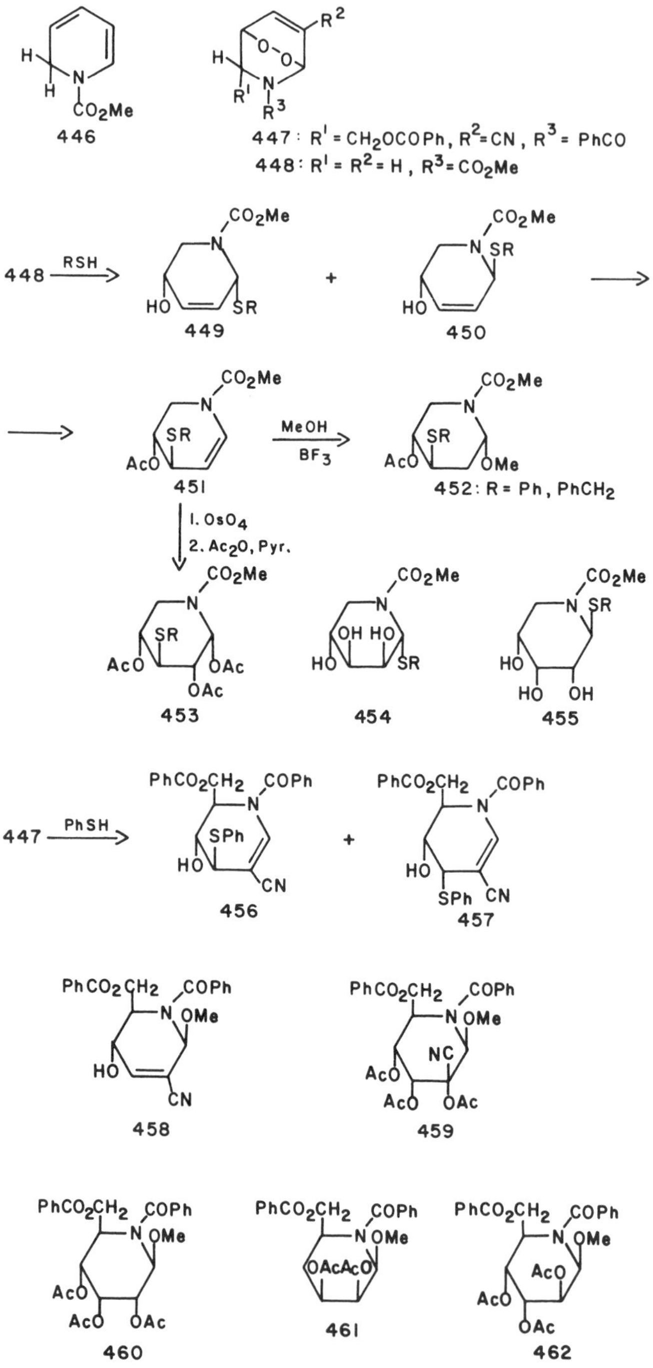

446

447: $R^1 = CH_2OCOPh$, $R^2 = CN$, $R^3 = PhCO$
448: $R^1 = R^2 = H$, $R^3 = CO_2Me$

$448 \xrightarrow{RSH}$ **449** + **450** $\longrightarrow$

$\longrightarrow$ **451** $\xrightarrow[BF_3]{MeOH}$ **452**: R = Ph, PhCH$_2$

451 ↓ 1. OsO$_4$ 2. Ac$_2$O, Pyr.

453   **454**   **455**

$447 \xrightarrow{PhSH}$ **456** + **457**

458   **459**

460   **461**   **462**

-β-DL-ribopiperidinose (455), respectively. Both thioglycosides could be converted into O-glycosides by a N-bromosuccinimide-silver nitrate-mediated reaction with methanol.

The reactions of *endo*-peroxide 447 were similar to those of 448.[248] Opening of the peroxide bridge with thiophenol immediately gave the products of allylic rearrangement, 456 and 457. However, reaction of 447 with dimethyl sulfide in methanol gave a 2,3-unsaturated product 458. *Cis*-hydroxylation of acetylated 458 furnished, after additional acetylation, a triacetate 459. The degradation of the cyanohydrin acetate grouping at C-2 to a keto group followed by reduction gave three products, which were separated and characterized as acetates 460–462.

C. Cyclic Hydroxysulfides

McCormick and McElhinney[249] devised a simple synthesis of glycosides and nucleosides containing sulfur in the sugar ring. Compounds of this type may possess interesting biological properties.

A cyclic hydroxysulfide is prepared either by reaction of a metal sulfide with an open-chain compound containing two leaving groups at both ends or by an intramolecular aldol condensation of a linear oxosulfide. In the next stage the hydroxysulfide is oxidized at sulfur atom to a hydroxysulfoxide. The key step is the reaction of hydroxysulfoxide with acetic anhydride (Pummerer rearrangement): α-acetoxysulfide is obtained having the anomeric center. The 1-O-acetyl group can be replaced under mild conditions by a number of nucleophiles to form glycosides, thioglycosides, and nucleosides.

Cis- and *trans* 3,4-thiolandiols (463 and 464) were obtained[249] by reaction of sodium sulfide with 1,4-dichloro-2,3-butanediol. Oxidation of both cyclic hydroxysulfides gave the corresponding sulfoxides, 465 (a mixture of two stereoisomers) and 466. After esterification the sulfoxides were subjected to Pummerer rearrangement. From 465 (both stereoisomers) all-*cis* triester sulfide (467) was obtained. However, cyclic phenylboronate ester gave under the same conditions a *trans* 1-O-acetate (468). The *trans* diester sulfoxide 466 yielded both of the possible rearranged products, 470 and 471, in roughly equal proportions.

Compound 468 readily replaced[250] 1-O-acetyl group for a number of alkoxy, aryloxy, and thioalkoxy groups in reactions with alcohols, phenols, or thiols in the presence of *p*-toluenesulfonic acid. In the case of alcohols and phenols, the reaction was stereospecific and β-anomers (472) were formed. With thiols, both anomeric thioglycosides (473) were obtained in about equal proportions.

When trifluoroacetic anhydride was used for the Pummerer rearrangement of phenylboronate ester 465, the 1-O-trifluoroacetyl derivative 469 was formed rapidly and could be used *in situ* for further reactions. The 1-O-trifluoroacetyl group could be replaced by bromide which, in turn, was exploited for glycoside synthesis.

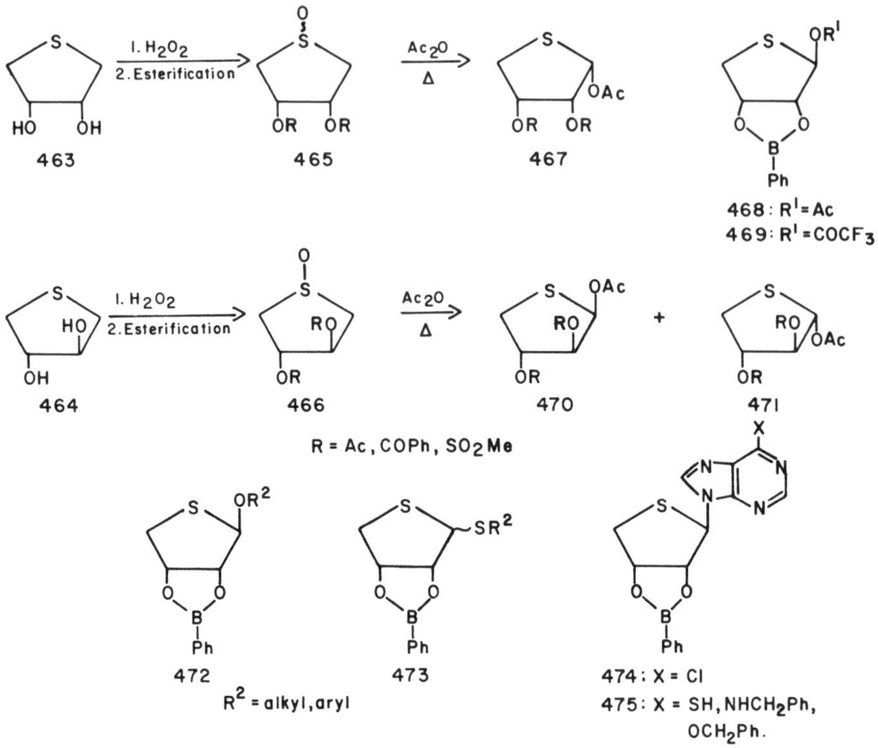
1. H2O2
2. Esterification
Ac2O
Δ
463
465
467
468: R1 = Ac
469: R1 = COCF3
464
466
470
471
R = Ac, COPh, SO2Me
472
473
R2 = alkyl, aryl
474: X = Cl
475: X = SH, NHCH2Ph, OCH2Ph.

In 4-thiothreose series the 1-O-acetyl group could be exchanged under similar conditions for bromide, *p*-nitrobenzyloxy, and thiouronium groups.

6-Chloropyrine nucleoside (**474**) was obtained[251] from 1-O-acetyl-4-thioerythrose phenylboronate.

This compound served as substrate for the preparation of a series of nucleosides (**475**) via substitution of 6-chlorine atom by various nucleophiles. Nucleosides also have been obtained in the 4-thiothreose series.

The substrate for 5-thiopentose was obtained[252] by intramolecular aldol condensation of diacetonyl sulfide (**476**). Reduction of the keto group in **477** afforded *trans* and *cis* diols (**478** and **479** 1.3:1) in 85% yield. Acetylation of **478**, oxidation to sulfoxide, and Pummerer rearrangement gave both regioisomeric 4-C-methyl and 2-C-methyl triacetates (**480** and **481**), with the latter distinctly prevailing. The same cycle of reactions applied to **479** gave the 5-thiopentose system (**482**) in low yield. Preparation of the 1,3-phenylboronate ester (**483**) followed by oxidation to a mixture of stereoisomeric sulfoxides and Pummerer rearrangement proved to be a better way: the yield of both regioisomeric 2-C-methyl and 4-C-methyl compounds (**484** and **485**) was 20%. Compound **485**

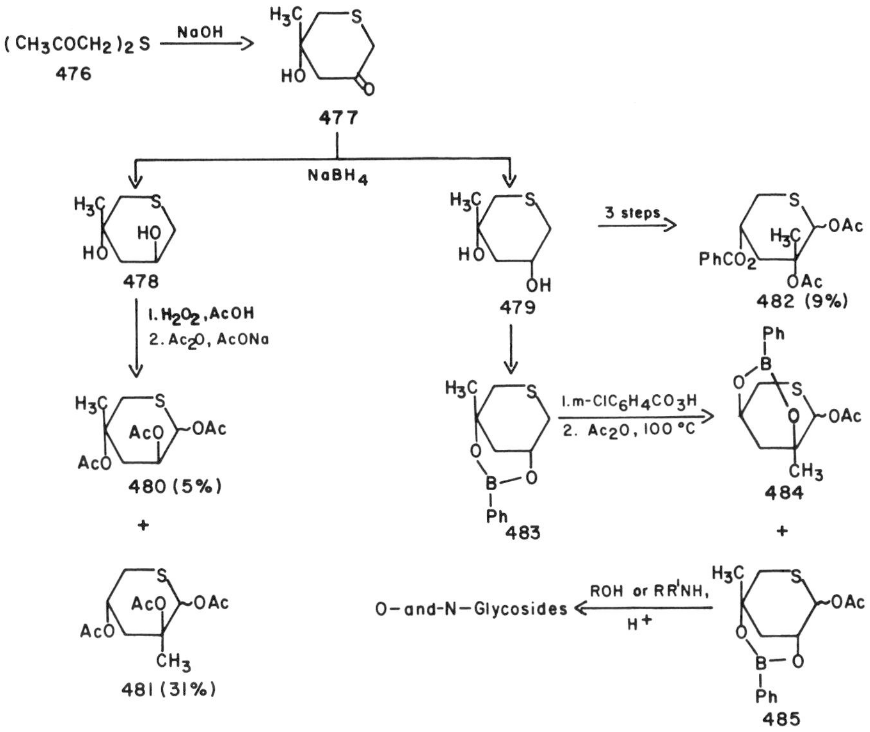

$(CH_3COCH_2)_2S$ $\xrightarrow{\text{NaOH}}$
476

477

NaBH$_4$

478

479   3 steps →   482 (9%)

PhCO$_2$

1. H$_2$O$_2$, AcOH
2. Ac$_2$O, AcONa

480 (5%)

+

481 (31%)

483   1. m-ClC$_6$H$_4$CO$_3$H   2. Ac$_2$O, 100 °C   484

+

O−and−N−Glycosides ← $\xrightarrow[\text{H}^+]{\text{ROH or RR'NH,}}$

485

underwent O- and N-glycosidation with alcohols and amines in the presence of p-toluenesulfonic acid, whereas 2-C-methyl esters **484** (and also **482**) remained unchanged under these conditions.

D. Other Substrates

Niemczura and co-workers[253] synthesized a modified daunosamine, 4-deoxy-DL-daunosamine (**491**), in a novel and original way. The approach was based on transformations of oxazolino-α-pyrone (**487**), which was regarded as a pyranose precursor substituted with blocked amino and hydroxyl group at positions 3 and 4. Such a substrate should serve for the synthesis of the whole series of 3-amino-2,3,6-tri-deoxy-DL-hexoses.

Oxazolino-α-pyrone (**487**) was obtained in five steps starting from 3-benzo-yloxy-2,4-pentanedione (**486**). Catalytic hydrogenation of **487** afforded two products with the same *threo* stereochemistry: 3-benzamido-5-caprolactone (**488**) and an analogous compound (**489**) having the phenyl ring perhydrogenated. With very active platinum catalyst, this last compound was the exclusive product of hydrogenation. It is interesting that no stereoisomeric *erythro* caprolactone was detected. Reduction of the lactone's carbonyl group with Red-al reagent gave *N*-cyclohexylcarbonyl-4-deoxy-DL-daunosamine (**490**). Hydrolysis of the amide with hydrobromic acid yielded 4-deoxy-DL-daunosamine hydrobromide (**491**). Methyl 4-deoxy-α-DL-daunosaminide was obtained also by Chmielewski[254]; the synthesis started from the aldehyde **137** and involved addition of hydrazoic acid to αβ-unsaturated carbonyl system followed by reduction of the azide to amino group and conversion of the ester grouping to the methyl group.

A new access to branched-chain sugars based on the dioxaphospholen condensation was elaborated by David and co-workers.[255]

Dioxaphospholen **492** was condensed with 2,3-O-isopropylidene-D-glyceraldehyde to give a single product **493**. Hydrolysis of **493** under controlled conditions removed the phosphate residue and isopropylidene grouping. Glycosidation of the product with methanol and an acidic resin gave three products (**494-496**), which were identified as different glycosides of the same sugar: 1-deoxy-3-C-methyl-D-*ribo*-hexose. Most remarkable in this synthesis is the efficiency of asymmetric induction: from a reagent containing a single center of chirality, a product with 3 centers of the *ribo*-stereochemistry was obtained.

8. SYNTHESIS OF CYCLITOLS

A route from furan-vinylene carbonate Diels–Alder adduct to inositols[1] has been reinvestigated.[256] *Endo* and *exo* isomers of oxabicyclo-[2,2,1]-hept-5-ene-2,3-diol carbonate (**7**) were separated and subjected to epoxidation or *cis*-hydroxylation. Successive alkaline and acidic hydrolysis of epoxide **497**, obtained from the *endo* adduct, produced a mixture of *allo* and *myo* inositols. *cis*-Hydroxylation of the *endo* adduct gave 1,4-anhydro-*allo*-inositol **498**, which afforded *neo*-inositol upon acidic hydrolysis. An analogous transformation of the *exo* adduct gave *allo* and *myo* inositols by way of the 1,4:2,3-dianhydro intermediate **499** and *epi*-inositol by hydrolysis of the *cis*-hydroxylation product **500**.

The Diels–Alder adduct of cyclopentadiene and β-bromo acrylic acid **501** has been exploited by Just as a convenient starting material for the synthesis of carbocyclic analogs of C-nucleosides. Hydroxylation of the double bond in **502**, followed by dehydrobromination and ozonolysis gave **503**, which after sequential hydride reduction, periodate oxidation, and oxidation with Collins's reagent gave the key intermediate: lactone **504**. An alternative route to **504** starting from norbornadiene was also described.[257] The condensation of **504** with amino-guanidine[258] gave a compound containing an amino-triazole ring attached to a carbocyclic analog of a ribofuranoside moiety. Carbocyclic analogs of pyrazomycin, showdomycin, and azapseudouridine were obtained from adduct **501** by way of silylated keto-ester derivative **505**.[259,260] Similar transformations of compounds resulting from epoxidation[261] or hydroboration[262] of **501** have been described.

The known 7-*endo*-oxabicyclo [2.2.1]-hept-5-ene-2-carboxylic acid **506**, obtained by the condensation of acrylic acid with furan, was used as a substrate for synthesis of validamine, (**510**) a component of validamycin antibiotics. The treatment of **506** with hydrogen peroxide in formic acid gave the tricyclic compound **507**, which after reduction and hydrolysis afforded cyclitol. Treatment of **508** with 2,2-dimethoxy-propane in the presence of an acid gave a mixture of diisopropylidene derivatives in which compound **509** was predominant. Introduction of an amino group, by way of a tosyl ester and azide displacement, followed by hydrogenation and hydrolysis, completed the synthesis of DL-validamine.[263] An alternative route to validamine and its analogs from **510** involved its bromolactonization, or opening of 1,4-anhydro ring in the bicyclic intermediate by action of hydrogen bromide in acetic acid.[264]

An interesting approach to the synthesis of highly oxygenated cyclohexane derivatives has been developed by Ichihara. Benzoquinone epoxides and epoxycyclohexenones were obtained by the thermal retro Diels–Alder reaction of epoxide **511**, obtained from quinones and dimethylfulvene without isolation of the primary adduct.[265] Intermediate **512** (R = Ac), obtained in this way from hydroxymethyl-benzoquinone, was converted into the natural compound senepoxide (**514**) as shown in Scheme 6. Reductive, regioselective cleavage of the diepoxide **513** was achieved by treatment with hydrazine hydrate.[266] The stereoselective total synthesis of crotoepoxide (**515**)[267] from cyclohexenone **512** (R = Ms), and of other natural compounds with related structure, was described.[268]

Only a few syntheses aimed at cyclic polyols started from acyclic precursors as shown in the following examples. Ethyl crotonate and ethyl acetoacetate served

506   507   508

509   510

as substrates in the synthesis of 3-hydroxy-3-(hydroxymethyl)-5-methyl-1,2-cy-clohexanedione (**516**) compounds related to a metabolite leucogenenol.[269] The Dieckmann condensation of diethyl oxalate and diethyl glutarate, resulting in the formation of diethyl 2,3-cyclopentanedione-1,4-dicarboxylate (**517**), was employed for the synthesis of carbocyclic ribonucleoside analogs according to Scheme 6.[270,271]

Bisdiazoketone **518**, obtained by a conventional diazomethane chain elongation reaction from xylaric acid dichloride, underwent unexpected ring closure when decomposed thermally in acetic acid containing cupric acetate. Apart from the desired heptadiulose pentaacetate, compound **519** was obtained in 32% yield. Reduction of the keto group in **519** followed by acetylation gave DL-2-C-acetoxymethyl-1,3,4,5,6-penta-O-acetyl-*epi*-inositol.[272,273] Although the synthesis started from D-xylose it is mentioned here because it proceeds through achiral intermediates, and consequently results in formation of a racemic product.

511   512 R= Ac, Ms   513

513 $\longrightarrow$ 514   515

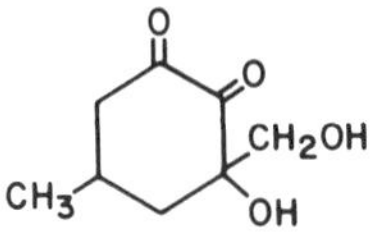

516

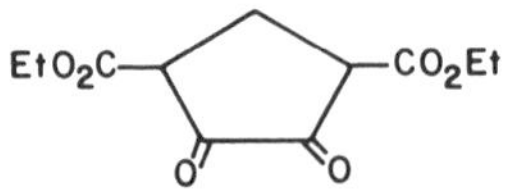

517

Among the derivatives of cycloalkanes, trioxatris-σ-homobenzenes (3,6,9-trioxatetracyclo [6.1.0^{2,4}.0^{5,7}]nonanes) are of special interest in the synthesis of cyclitols in view of their versatile reactivity. Prinzback has shown that radical bromination of epoxide **520** affords a mixture of dibromides **521–523**; individual compounds can be isolated in pure form, although they are prone to equilibration.[274] Cis-trioxatris-σ-homobenzene **524**, which was prepared from the dibromide **521**, was converted selectively into 1,2:3,4-dianhydro-epi-inositol **525**, and further, through a sulfonate ester into trans-trioxatris-σ-homobenzene **526**,[275] (trans-3,6,9-trioxatetrocyclo [6.1.0.0^{2,4}.0^{5,7}]nonane) previously prepared from benzene monoxide.[276] Reaction of triepoxide **524** with hydrazine gave compound **527**, which was converted by the action of phthalic anhydride into **528**. The alkaline hydrolysis of derivative **528** followed by treatment with Raney nickel yielded streptamine **529** (R = H). In a similar way the total syntheses of 2-deoxystreptamine **530** and hyosamine **531** was achieved, starting from mono-bromoepoxycyclohexenes.[277] An improved method of synthesis of trans-trioxatris-σ-homobenzene **526** from dibromide **523** by way of a ditosylate **532** was described.[278]

A new route to cyclitols from cyclohexene was elaborated by Hasegawa and co-workers. 1-O-Benzylcyclohex-2-enol (**533**) was converted into a mixture of

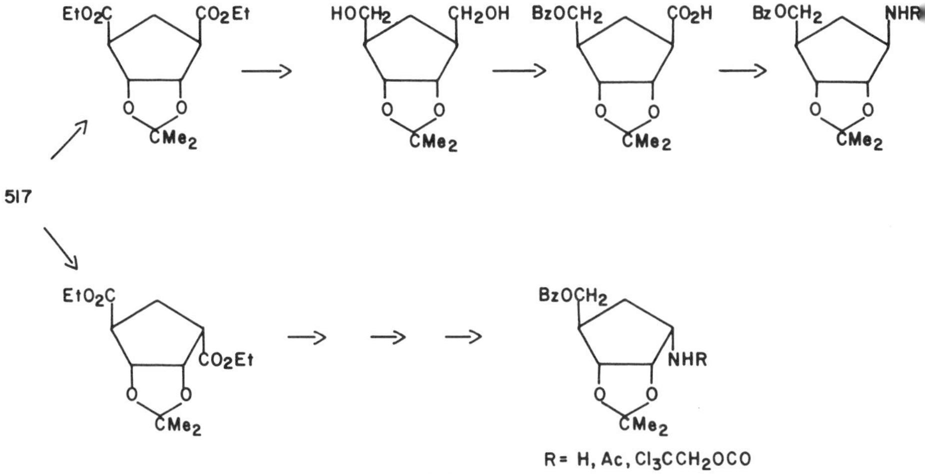

517

R = H, Ac, Cl$_3$CCH$_2$OCO

Scheme 6

```
CHN2
 |
 CO
 |
H-C-OAc
 |
AcO-C-H
 |
 H-C-OAc
 |
 CO
 |
CHN2
 518
```

519

compounds **535-537** via dibromide **534**. Further transformations of the dibenzyl derivative **535**, the main component of the mixture, are presented in Scheme 7.[279]

Treatment of dibromide **534** with sodium azide gave a mixture of allylic azides, which were reduced and separated as the corresponding acetamido derivatives. DL-*trans*-2-acetamido-1-O-benzylcyclohex-3-enol **538** was obtained in 62% yield and the minor components were identified as *trans*- and *cis*-4 acetamido-1-O-benzylcyclohex-2-enols. Further functionalization of the double bond in **538** was described.[280] The dibromide obtained from **535** gave on treatment with sodium azide a mixture of DL-(1,3/2)-3-azido-1,2-di-O-benzylcyclohex-4-enediol **532** and DL-(1,3/4)-1-azido-3,4-di-O-benzylcyclohex-5-enediol **540**. The corresponding acetamido derivatives **541** and **542** were converted into a variety of deoxyinosamine and deoxyinosadiamine derivatives.[281]

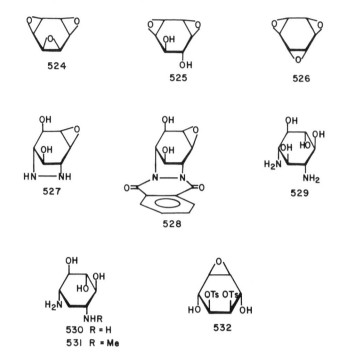

In continuation of his studies in the field of cyclitols, Posternak completed the synthesis of all-*cis* cyclopentane 1,2,3- and 1,3,4-triols, 1,2,3,4-tetraol, and 1,2,3,4,5-pentaol. Their preparations from cyclohex-3-enol or cyclohex-2-en-1,4-diol involved the epoxidation of the double bond followed by hydride reduction of *cis*-epoxy-ol arrangement, or hydrolytic displacement of bromine in the presence of a *trans*-neighboring benzoyloxy group. The configuration of the cyclitols obtained was confirmed by ¹H nmr spectra.[282] A number of keto derivatives of cyclopentanols were obtained, starting from cyclopentenone dimethylketal, by way of allylic bromination, followed by *cis*-hydroxylation.[283] The preparation of 2,3-dihydroxycyclopentanone and 2,3-dihydroxycyclopentenone,

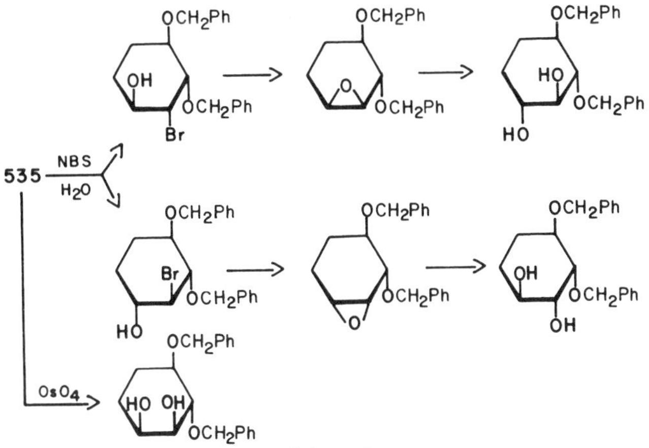

Scheme 7

and a number of their derivatives, were also described.[284] An investigation of the reaction of penta-O-acetyl-myo-inosose-2 (543) with diazoalkanes revealed that diazomethane reacted with formation of the spiroepoxide 544 only, whereas higher diazoalkanes afforded a mixture of the corresponding spiroepoxide and the ring expansion products 545. The configuration of the spiroepoxides was established, and a number of their reactions were described.[285] Deacetylation of compound 545 led to the bicyclic hemiacetal 546.[286]

A long series of papers by Shvets and co-workers is devoted to synthesis of asymmetrically substituted myo-inositols[287,288] and, in particular, inositolphosphatides.[289] Their route to optically active compounds from myo-inositol consists of the formation of diastereoisomeric orthoesters in the reaction of a suitably protected racemic myo-inositol derivative with the orthoacetate of D-mannose, followed by the separation of the diastereoisomers and hydrolysis. Naturally occurring $(+)$ and $(-)$ bornesitols (3- and 1-O-methyl-sn-myo-inositols) were obtained in this way, among many other enantiomerically pure chiral myo-inositol derivatives.[288]

The Suami and Ogawa group has been particularly active in the field of carbocyclic polyols. The main line of their interest was confined to the synthesis of aminocyclitols, constituents of aminoglycoside antibiotics and their analogs. Besides, inosamines[290] and aminocyclopentanepolyols[291,292] obtained in their laboratory were utilized as substrates in synthesis of nucleoside analogs. A

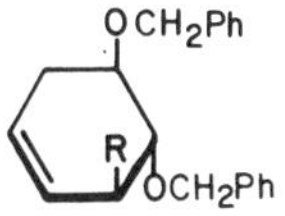

539 R = N₃
541 R = NHAc

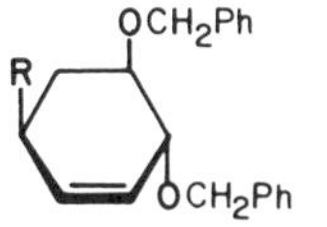

540 R = N₃
542 R = NHAc

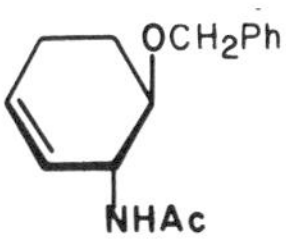

538

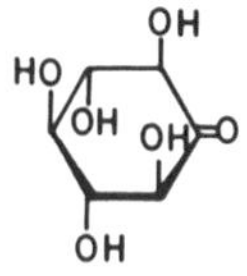

543

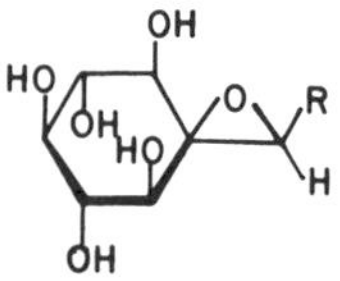

544 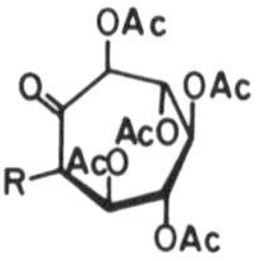   545   546

R = alkyl, H

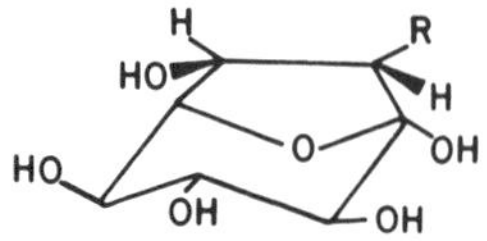

number of "pseudosugar"-monosaccharide analogs, in which the ring oxygen is replaced by methylene group, were also obtained by the Japanese group. The synthesis of pseudo-β-DL-galacto-pyranose DL-(1,3,4,5/2)-5-hydroxymethyl-1,2,3,4-cyclohexanepentaol and pseudo-α-DL-altropyranose DL-(1,3,4,5/2)-5-hydroxymethyl-1,2,3,4-cyclohexanepentaol from *myo*-inositol involved the treatment of suitably protected deoxyinosose **547** with diazomethane, followed by the opening of the spiroepoxide thus formed with hydroiodic acid and by dehydroiodination with zinc. Hydroboration of olefine **548**, after acetylation, gave cyclitols **549–551** in 13, 17, and 13% yields respectively.[293]

Validatol **552** and its deoxy derivative **553**—cyclitols obtained by hydrogen-olysis of the antibiotic validomycin A—were synthesized by conventional transformations of tri-O-acetyl(1,3/2,4,6)-4-bromo-6-bromomethyl-1,2,3-cyclohex-anetriol.[294,295]

Six isomeric 5-hydroxymethyl-1,2,3,4-cyclohexanetetraols and a number of their derivatives were prepared from *endo*-3-acetoxy-*endo*-5-acetoxymethyl-7-oxabicyclo[2.2.1]-heptanes bearing *exo*-2 substituent, by way of the hydrolytic

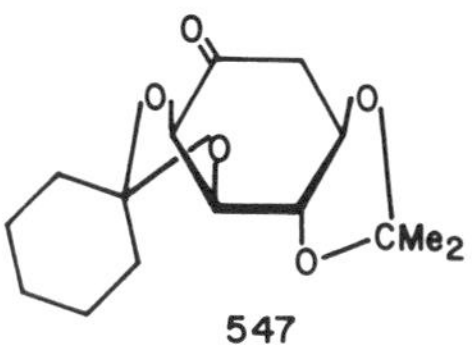

547 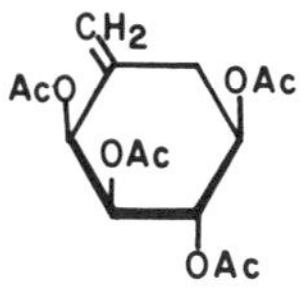   548

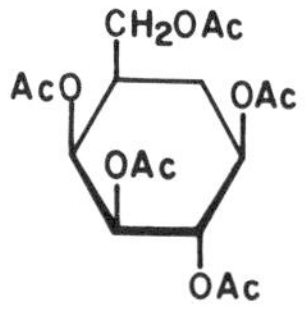

549 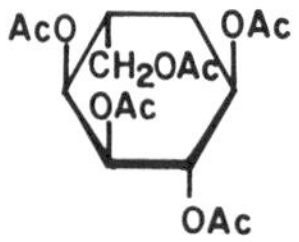   550 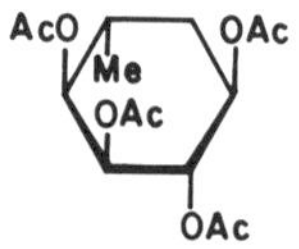   551

552 R = OH
553 R = H

cleavage of the 1,4-anhydro ring.[296]

Cyclization of cis-3,4-cyclohexylidenedioxy-2,5-dihydroxytetrahydrofuran with nitromethane was applied to the synthesis of aminocyclopentanetetraols.[297,298] The reactions and configurational relations of these compounds were described in some detail.[299] Diaminocyclopentanetriols were obtained from appropriately protected aminocyclopentanetetraols by way of periodate oxidation and cyclization of the resulting dialdehyde with nitromethane.[300] In a similar way a number of inosadiamines-1,4 were synthesized.[301,302] The general approach to the syntheses of deoxyaminocyclitols from aminocyclitol derivatives, consisted of chlorination with sulfuryl chloride followed by dechlorination with Raney nickel or with tributyltinhydride was described. 2,4-Dideoxy- and 2,4,5-trideoxystreptamines were prepared from 2-deoxystreptamine by these procedures.[303] A convenient synthesis of DL-proto-quercitol from myo-inositol was described.[304] A number of partially blocked myo-inositol derivatives, potential substrates in synthesis of various cyclitols, were prepared. Their tosylation and transformation into derivatives of chiro-inositol and muco-inositol was reported.[305] All isomers of 1,2:3,4- and 1,2:4,5-dianhydrocyclohexanehexaols (5 racemates and 6 meso compounds) were synthesized from appropriate inositol sulfonate diesters, by treatment with sodium methoxide.[306] Labeled compounds: 1-O-methyl-^{14}C-DL-myo-inositol (methyl-^{14}C-bornesitol) and 5-O-methyl-^{14}C-myo-inositol (methyl-^{14}C-sequoitol) were synthetized in connection with research on inositol metabolism by the methylation of benzylated myo-inositol derivatives with methyl-^{14}C iodide, followed by hydrogenolysis.[307]

REFERENCES

1. J. K. N. Jones and W. A. Szarek, in *The Total Synthesis of Natural Products,* Vol. 1 (J. ApSimon, Ed.), Wiley-Interscience, New York, 1973, p. 1.

2. A. H. Weiss and J. Shapira, "Manufacture of Sugars," *CEP Symposium Series,* **67**(108), 137 (1971).

3. F. Hoyle, A. H. Olavesen, and N. C. Wickramasinghe, *Nature,* **271,** 229 (1978).

4. N. W. Gabel and C. Ponamperuma, *Nature,* **216** 453 (1967).

5. A. G. Cairns-Smith, P. Ingram, and G. L. Walker, *J. Theor. Biol.,* **35,** 601 (1972).

6. W. Heidmann, P. Decker, and R. Pohlmann, *Origin of Life* (H. Noda, Ed.), Proceedings of the 2nd ISSCL Meeting, Bns, Cent. Acad. Soc. Japan, Tokyo, 1977, p. 625; *Chem. Abstr.,* **90,** 68,101x (1979).

7. T. Mizuno, K. Kawai, K. Muramatsu, and K. Banba, *Nippon Nogei Kagaku Kaishi*, **46,** 73 (1972); *Chem. Abstr.*, **77,** 70,865z (1972).

8. T. Mizuno, T. Naiki, Y. Yasuhara, Y. Shima, and Y. Asada, *Shizuoka Daigaku Nogakubu Kenkyu Hokoku* **23,** 75 (1973); *Chem. Abstr.*, **83,** 203,644z (1975).

9. K. Sasajima, *Kagaku To Seibutsu*, **12,** 595 (1974); *Chem. Abstr.*, **83,** 23,463f (1975).

10. T. Mizuno, C. Nakano, Y. Yamada, and K. Kondo, *Shizuoka Daigaku Nogakubu Kenkyu Hokoku*, **27,** 65 (1977); *Chem. Abstr.*, **90,** 83,421v (1979).

11. H. B. Chermiside, A. Gill, A. Furst, and J. Shapira, *Proc. West. Pharmacol. Soc.*, **14,** 112 (1971).

12. T. Mizuno, J. Nishigaki, K. Tachenchi, and M. Matsuka, *Nippon Nogei Kagaku Kaishi*, **47,** 327 (1973); *Chem. Abstr.*, **79,** 123,980h (1973).

13. A. H. Weiss, R. D. Partridge, H. Tambawala, and J. A. Shapira, *Polyols from Formaldehyde*, Dechema—Monographien Nos. 1264–1291, Vol. 68, Verlag Chemie, Weinheim/Bergstrasse (1971), p. 239.

14. T. Mizuno and A. H. Weiss, in *Advances in Carbohydrate Chemistry and Biochemistry* (R. S. Tipson and D. Horton, Eds.), Vol. 29, Academic Press, New York, 1974, p. 173.

15. Y. Uchida, *Kagaku To Kogyo (Tokyo)*, **28,** 77 (1975); *Chem. Abstr.*, **84,** 135,948v (1976).

16. T. Mizuno, *Kagaku No Ryoiki*, **26,** 58 (1973).

17. A. H. Weiss, R. B. La-Pierre, and J. Shapira, *J. Catal.*, **16,** 332 (1970).

18. H. Tambavala and A. H. Weiss, *J. Catal.*, **26,** 388 (1972).

19. A. H. Weiss and T. John, *J. Catal.*, **32,** 216 (1974).

20. T. I. Khomenko, O. A. Golovina, M. H. Sakharov, O. V. Krylov, R. D. Partridge, and A. H. Weiss, *J. Catal.*, **48,** 354 (1976).

21. A. H. Weiss, V. A. Seleznov, M. M. Sakharov, O. V. Krylov, Ya. B. Gorokhovatsky, and N. P. Evmenenko, *J. Catal.*, **48,** 354 (1977).

22. S. B. Ziemecki, R. B. La Pierre, A. H. Weiss, and M. M. Sakharov, *J. Catal.*, **50,** 455 (1977).

23. K. Fujino, J. Kobayashi, and I. Higuchi, *Nippon Kagaku Kaishi*, **12,** 2292 (1972); *Chem. Abstr.*, **78,** 72,472p (1973).

24. V. A. Likholobov, A. H. Weiss, and M. M. Sakharov, *React. Kinet. Katal. Lett.*, **8,** 155 (1978).

25. R. D. Partridge, T. I. Khomenko, C. A. Golovina, M. M. Sakarov, A. H. Weiss, and O. V. Krylov, *Kinet. Katal.*, **18,** 557 (1977).

26. O. A. Golovina and T. I. Khomenko, *Kinet. Katal.*, **18,** 556 (1977).

27. Ya. B. Gorokhovatsky, N. P. Evmenenko, and V. F. Gaevsky, *Kinet. Katal.*, **18,** 558 (1977).

28. A. H. Weiss, *Kinet. Katal.*, **18,** 539 (1977).

29. A. A. Morozov, O. E. Levanevskii, T. K. Tamashaeva, and N. N. Ikonnikova, *Kinet. Katal.*, **20,** 1580 (1979).

30. T. Mizuno, Y. Harada, and Y. Okamura, *Shizuoka Daigaku Nogakubu Kenkyu Hokoku*, **26,** 45 (1976); *Chem. Abstr.*, **87,** 184,809w (1977).

31. H. Bauer, A. R. Hermanto, and W. Voelter, Abstracts, 10th International Symposium on Carbohydrate Chemistry, Sydney, Australia, July, 1980, F-17.

32. T. Matsuura, Y. Shigemasa, and C. Sakazawa, *Chem. Lett.*, 713 (1974).

33. Y. Shigemasa, M. Shimao, C. Sakazawa, and T. Matsuura, *Bull. Chem. Soc. Japan*, **48,** 2099 (1975).

34. Y. Shigemasa, T. Fujitani, C. Sakazawa, and T. Matsuura, *Bull. Chem. Soc. Japan*, **50**, 1527 (1977).

35. Y. Shigemasa, M. Shimao, C. Sakazawa, and T. Matsuura, *Bull. Chem. Soc. Japan*, **50**, 2138 (1977).

36. Y. Shigemasa, O. Nagae, C. Sakazawa, R. Nakashima, and T. Matsuura, *J. Am. Chem. Soc.*, **100**, 1309 (1978).

37. Y. Shigemasa, Y. Matsuda, C. Sakazawa, R. Nakashima, and T. Matsuura, *Bull. Chem. Soc. Japan*, **52**, 1091 (1979).

38. Y.Shigemasa, T. Tayi, C. Sakazawa, R. Nakashima, and T. Matsuura, *J. Catal.*, **58**, 296 (1979).

39. Y. Shigemasa, S. Akagi, R. Nakashima, and S. Saito, *Carbohydr. Res.*, **80**, C1 (1980).

40. Y.Shigemasa, Y. Matsuda, C. Sakazawa, and T. Matsuura, *Bull. Chem. Soc. Japan*, **50**, 222 (1977).

41. Y. Shigemasa, H. Komaki, E. Waki, and R. Nakashima, *Tottori Daigaku Kogakubu Kenkyu Hokoku*, **9**, 116 (1978); *Chem. Abstr.*, **91**, 57,347b (1979).

42. S. Morgenlie, *Carbohydr. Res.*, **80**, 215 (1980).

43. T. Mizuno, S. Goda, K. Kanemitsu, and K. Utumi, *Shizuoka Daigaku Nogakubu Kenkyu Hokoku*, **27**, 91 (1977); *Chem. Abstr.*, **90**, 138,108f (1979).

44. R. R. Schmidt and A. Lieberknecht, *Angew. Chem.*, **90**, 821 (1978).

45. Y. Araki, J. Nagasawa, and Y. Ishido, *Carbohydr. Res.*, **58**, C4 (1977).

46. T. Tamura, T. Kunieda, and T. Takizawa, *Tetrahedron Lett.*, **1972**, 2219.

47. T. Kunieda and T. Takizawa, *Yuki Gosei Kagaku Kyokai Shi*, **33**, 560 (1975); *Chem. Abstr.*, **83**, 177,527s (1975).

48. T. Tamura, T. Kunieda, and T. Takizawa, *J. Org. Chem.*, **39**, 38 (1974).

49. K. Hosoda, T. Kunieda, and T. Takizawa, *Chem. Pharm. Bull. (Tokyo)*, **24**, 2927 (1976).

50. T. Kunieda and T. Takizawa, *Heterocycles*, **8**, 661 (1977).

51. T. Matsuura, T. Kunieda, and T. Takizawa, *Chem. Pharm. Bull. (Tokyo)*, **25**, 239 (1977).

52. K. Hosoda, T. Kunieda, and T. Takizawa, *Chem. Pharm. Bull. (Tokyo)*, **24**, 2927 (1976).

53. N. Mitsuo, T. Kunieda, and T. Takizawa, *J. Org. Chem.*, **38**, 2255 (1973).

54. T. Kunieda, T. Tamura, and T. Takizawa, *J. Chem. Soc., Chem. Commun.*, 885 (1972).

55. T. Kunieda, T. Tamura, and T. Takizawa, *Chem. Pharm. Bull. (Tokyo)*, **25**, 1749 (1977).

56. H. Takahata, T. Kunieda, and T. Takizawa, *Chem. Pharm. Bull. (Tokyo)*, **23**, 3017 (1975).

57. Y. Nii, T. Kunieda, and T. Takizawa, *Tetrahedron Lett.*, **1976**, 2323.

58. Y. Nii, T. Kunieda, and T. Takizawa, *Chem. Pharm. Bull. (Tokyo)*, **26**, 1999 (1978).

59. N. Mitsuo, T. Kunieda, and T. Takizawa, *Chem. Pharm. Bull. (Tokyo)*, **25**, 231 (1977).

60. N. Mitsuo, T. Kunieda, and T. Takizawa, *Chem. Pharm. Bull. (Tokyo)*, **26**, 1993 (1978).

61. T. Matsuura, T. Kunieda, and T. Takizawa, *Chem. Pharm. Bull. (Tokyo)*, **25**, 1225 (1977).

62. T. Kunieda, Y. Abe, S. Sami, and T. Takizawa, *Heterocycles*, **12**, 183 (1979).

63. Y. Abe. and T. Kunieda, *Tetrahedron Lett.*, 5007 (1979).

64. T. Ando, S. Shioi, and M. Nakagawa, *Bull. Chem. Soc. Japan*, **45**, 2611 (1972).

65. K. Sonogashira and M. Nakagawa, *Bull. Chem. Soc. Japan*, **45**, 2616 (1972).

66. K. Mimaki, M. Masunari, G. Nakaminami, and M. Nakagawa, *Bull. Chem. Soc. Japan*, **45**, 2620 (1972).

67. G. Schneider, T. Horvath, and P. Sohár, *Carbohydr. Res.*, **56**, 43 (1977).

68. C. Fuganti and P. Grasselli, *J. Chem. Soc., Chem. Commun.*, 299 (1978).

69. C. Fuganti, P. Grasselli, and G. Marinoni, *Tetrahedron Lett.*, 1161 (1979).

70. I. Dyong and R. Wiemann, *Angew. Chem.*, **90**, 728 (1978).

71. I. Dyong and H. Friege, *Chem. Ber.*, **112**, 3273 (1979).

72. I. Iwataki, Y. Nakamura, K. Takahashi, and T. Matsumoto, *Bull. Chem. Soc. Japan*, **52**, 2731 (1979).

73. S. M. Makin, Y. E. Rayfeld, O. V. Limanova, and R. M. Arshova, *Zh. Org. Khim.*, **15**, 1843 (1979).

74. P. Chautemps, *Compt. Rend.*, **284C**, 807 (1977).

75. G. Nakaminami, S. Shioi, Y. Sugiyama, S. Isemura, M. Shibuya, and M. Nakagawa, *Bull. Chem. Soc. Japan*, **45**, 2624 (1972).

76. I. Dyong and D. Glittenberg, *Chem. Ber.*, **110**, 2721 (1977).

77. H. Grisebach, W. Hofheinz, and W. Doerr, *Chem. Ber.*, **96**, 1823 (1963).

78. I. Dyong and N. Jersch, *Chem. Ber.*, **112**, 1849 (1979).

79. I. Dyong and N. Jersch, *Chem. Ber.*, **109**, 896 (1976).

80. I. Dyong, R. Knollmann, and N. Jersch, *Angew. Chem.*, **88**, 301 (1976).

81. I. Dyong, R. Knollmann, N. Jersch, and H. Luftmann, *Chem. Ber.*, **111**, 559 (1978).

82. I. Dyong and H. Bendlin, *Chem. Ber.*, **111**, 1677 (1978).

83. I. Dyong and H. Bendlin, *Chem. Ber.*, **112**, 717 (1979).

84. Reference 1, p. 21.

85. K. Kefurt, Z. Kefurtová, and J. Jarý, *Collect. Czech. Chem. Commun.*, **37**, 1035 (1972).

86. K. Kefurt, K. Capek, J. Capková, Z. Kefurtová, and J. Jarý, *Collect. Czech. Chem. Commun.*, **37**, 2985 (1972).

87. K. Torssell and M. P. Tyagi, *Acta Chem. Scand.*, **B31**, 7 (1977).

88. K. Torssell and M. P. Tyagi, *Acta Chem. Scand.*, **B31**, 297 (1977).

89. J. E. Jensen and K. Torssell, *Acta Chem. Scand.*, **B32**, 457 (1978).

90. M. Chmielewski, *Polish J. Chem.*, **54**, 1913 (1980).

91. M. Chmielewski, *Tetrahedron*, **36**, 2345 (1980).

92. M. Chmielewski, *Tetrahedron*, **35**, 2067 (1979).

93. M. Chmielewski, *Carbohydr. Res.*, **68**, 144 (1979).

94. M. Chmielewski, J. Jurczak, and A. Zamojski, *Tetrahedron*, **34**, 2977 (1978).

95. S. Shibahara, S. Kondo, K. Maeda, H. Umezawa, and M. Ohno, *J. Am. Chem. Soc.*, **94**, 4353 (1972).

96. N. Elming, *Adv. Org. Chem.*, **2**, 67 (1960).

97. O. Achmatowicz Jr., P. Bukowski, G. Grynkiewicz, B. Szechner, A. Zamojski, and Z. Zwierzchowska, *Rocz. Chem.*, **46**, 879 (1972).

98. J. Šrogl, M. Janda, J. Stibor, and J. Kucera, *Collect. Czech. Chem. Commun.*, **38**, 455 (1973).

99. G. Büchi, E. Demole, and A. F. Thomas, *J. Org. Chem.*, **38**, 123 (1973).

100. J. Šrogl and J. Pavlikova, *Collect. Czech. Chem. Commun.*, **33**, 1954 (1968).

101. L. N. Kralinina, R. I. Kruglikova, and V. I. Bogomoleva, *Khim. Geterotcikl. Soedin.*, 229 (1970).

102. J. Šrogl, M. Janda, and I. Stibor, *Collect. Czech. Chem. Commun.*, **38**, 3666 (1973).

103. J. Šrogl, M. Janda, and I. Stibor, *Collect. Czech. Chem. Commun.*, **39**, 185 (1974).

104. I. Stibor, J. Šrogl, and M. Janda, *J. Chromatogr.*, **91**, 767 (1974).

105. T. Kinoshita, K. Miyano, and T. Miwa, *Bull. Chem. Soc. Japan*, **48**, 1865 (1975).

106. T. Kinoshita and T. Miwa, *Carbohydr. Res.*, **28**, 175 (1973).

107. T. Kinoshita and T. Miwa, *Bull. Chem. Soc. Japan*, **51**, 225 (1978).

108. J. Buddrus and H. Herzog, *Chem. Ber.*, **112**, 1260 (1979).

109. Y. Lefebvre, *Tetrahedron Lett.*, 133 (1972).

110. R. Laliberté, G. Médawar, and Y. Lefebvre, *J. Med. Chem.*, **16**, 1084 (1973).

111. G. Piancatelli, A. Scetri, and M. D'Auria, *Tetrahedron Lett.*, 2199 (1977).

112. O. Achmatowicz Jr., P. Bukowski, B. Szechner, Z. Zwierzchowska, and A. Zamojski, *Tetrahedron*, **27**, 1973 (1971).

113. P. D. Weeks, D. E. Kuhla, R. P. Allingham, H. A. Watson Jr., and B. Wlodecki, *Carbohydr. Res.*, **56**, 195 (1977).

114. G. Grynkiewicz, B. Barszczak, and A. Zamojski, *Synthesis*, 364 (1979).

115. G. Grynkiewicz and A. Zamojski, *Synth. Commun.*, **8**, 491 (1978).

116. O. Achmatowicz Jr. and P. Bukowski, *Rocz. Chem.*, **47**, 99 (1973).

117. O. Achmatowicz Jr. and M. H. Burzyńska, *Polish J. Chem.*, **53**, 265 (1979).

118. O. Achmatowicz Jr. and B. Szechner, *Tetrahedron Lett.*, 1205 (1972).

119. O. Achmatowicz Jr. and P. Bukowski, *Can. J. Chem.*, **53**, 2524 (1975).

120. O. Achmatowicz Jr. and G. Grynkiewicz, *Carbohydr. Res.*, **54**, 193 (1977).

121. O. Achmatowicz Jr. and B. Szechner, *Rocz. Chem.*, **49**, 1715 (1975).

122. O. Achmatowicz Jr. and B. Szechner, *Carbohydr. Res.*, **50**, 23 (1976).

123. O. Achmatowicz Jr. and B. Szechner, *Rocz. Chem.*, **50**, 729 (1976).

124. M. H. Burzyńska, Ph.D. Thesis, Institute of Organic Chemistry, Polish Academy of Sciences, Warsaw, 1976.

125. G. Grynkiewicz and M. H. Burzyńska, *Tetrahedron*, **32**, 2109 (1976).

126. O. Achmatowicz Jr. and A. Zamojski, *Rocz. Chem.*, **42**, 453 (1968).

127. O. Achmatowicz Jr., R. Bielski, and P. Bukowski, *Rocz. Chem.*, **50**, 1535 (1976).

128. O. Achmatowicz Jr. and P. Bukowski, *Bull. Acad. Pol. Sci., Ser. Sci. Chim.*, **19**, 305 (1971).

129. O. Achmatowicz Jr. and R. Bielski, *Carbohydr. Res.*, **55**, 165 (1977).

130. O. Achmatowicz Jr. and R. Bielski, *Rocz. Chem.*, **51**, 1389 (1977).

131. O. Achmatowicz Jr., G. Grynkiewicz, and B. Szechner, *Tetrahedron*, **32**, 1051 (1976).

132. O. Achmatowicz Jr. and G. Grynkiewicz, *Rocz. Chem.*, **50**, 719 (1976).

133. G. Grynkiewicz, O. Achmatowicz Jr. and H. Bartoń, *Rocz. Chem.*, **50**, 719 (1976).

134. T. Shono and Y. Matsumura, *Tetrahedron Lett.*, 1363 (1976).

135. G. Kuświk and G. Grynkiewicz, *Polish J. Chem.*, **54**, 1319 (1980).

136. M. Yamashita, M. Yoshikane, T. Ogata, and S. Inokawa, *Tetrahedron*, **35**, 741 (1979).

137. G. Grynkiewicz, *Carbohydr. Res.*, **80**, 53 (1980).

138. G. Grynkiewicz and A. Zamojski, *Z. Naturforsch.*, **35b**, 1024 (1980).

139. G. Just and A. Martel, *Tetrahedron Lett.*, 1517 (1973).

140. G. Just and K. Grozinger, *Tetrahedron Lett.*, 4165 (1974).

141. G. Just and K. Grozinger, *Can. J. Chem.*, **53**, 2701 (1975).

142. G. Just, A. Martel, K. Grozinger, and M. Ramjeesingh, *Can. J. Chem.*, **53**, 131 (1975).

143. G. Just, M. Ramjeesingh, and T. Liak, *Can. J. Chem.*, **54**, 2940 (1976).

144. G. Just and M. Ramjeesingh, *Tetrahedron Lett.*, 985 (1975).

145. M. Lim, Ph.D. Thesis, McGill University, Montreal, 1976.

146. G. Just and M. Lim, *Can. J. Chem.*, **55**, 2993 (1977).

147. A. P. Kozikowski, W. C. Floyd, and M. P. Kuniak, *J. Chem. Soc., Chem. Commun.*, 582 (1977).

148. A. P. Kozikowski and W. C. Floyd, *Tetrahedron Lett.*, 19 (1978).

149. R. Noyori, *Acc. Chem. Res.*, **12**, 61 (1979).

150. R. Noyori, S. Makino, T. Okita, and Y. Hayakawa, *J. Org. Chem.*, **40**, 806 (1975).

151. R. Noyori, T. Sato, and Y. Hayakawa, *J. Am. Chem. Soc.*, **100**, 2561 (1978).

152. T. Sato, R. Ito, Y. Hayakawa, and R. Noyori, *Tetrahedron Lett.*, 1829 (1978).

153. T. Sato, M. Watanabe, and R. Noyori, *Tetrahedron Lett.*, 4403 (1978).

154. T. Sato, M. Watanabe, and R. Noyori, *Tetrahedron Lett.*, 2897 (1979).

155. T. Sato, K. Marunouchi, and R. Noyori, *Tetrahedron Lett.*, 3669 (1979).

156. T. Sato and R. Noyori, *Bull. Chem. Soc. Japan*, **53**, 1195 (1980).

157. A. Zamojski and T. Koźluk, *J. Org. Chem.*, **42**, 1089 (1977).

158. T. Koźluk, Ph.D. Thesis, Institute of Organic Chemistry, Polish Academy of Sciences, Warsaw, 1978.

159. R. Bognár and P. Herczegh, *Magy. Kem. Foly.*, **83**, 298 (1977); *Chem. Abstr.*, **87**, 184,814u (1977).

160. R. Bognár and P. Herczegh, *Carbohydr. Res.*, **52**, 11 (1976).

161. R. Bognár and P. Herczegh, *Carbohydr. Res.*, **54**, 292 (1977).

162. Reference 1, p. 32.

163. D. Anker, D. Descours, and H. Pacheco, *Compt. Rend.*, **277C**, 215 (1973).

164. D. Descours, D. Anker, H. Pacheco, M. Chareire, and G. Carret, *Eur. J. Med. Chem., Chim. Ther.*, **12**, 313 (1977).

165. J. E. Roff and R. K. Brown, *Can. J. Chem.*, **51**, 3354 (1973).

166. A. Saroli, D. Descours, D. Anker, H. Pacheco, and M. Chareire, *J. Heterocycl. Chem.*, **15**, 765 (1978).

167. Y. Suhara, F. Sasaki, G. Koyama, K. Maeda, H. Umezawa, and M. Ohno, *J. Am. Chem. Soc.*, **94**, 6501 (1972).

168. S. Yasuda and T. Matsumoto, *Tetrahedron*, **29**, 4087 (1973).

169. Reference 1, page 49.

170. Reference 1, page 35.

171. R. R. Schmidt and R. Angerbauer, *Angew. Chem.*, **89**, 822 (1977).

172. V. B. Mochalin and I. S. Varpakhovskaya, *Zh. Org. Khim.*, **12**, 2257 (1976).

173. S. David, J. Eustache, and A. Lubineau, *Compt. Rend.* **276C**, 146 (1973).

174. S. David, J. Eustache, and A. Lubineau, *J. Chem. Soc., Perkin I*, 2274 (1974).

175. S. David, A. Lubineau, and A. Thieffry, *Tetrahedron*, **34**, 299 (1978).

176. S. David, J. Eustache, and A. Lubineau, *J. Chem. Soc., Perkin I*, 1795 (1979).

177. S. David and J. Eustache, *J. Chem. Soc., Perkin I*, 2230 (1979).

178. S. David and J. Eustache, *J. Chem. Soc., Perkin I*, 2521 (1979).

179. R. M. Srivastava, F. Sweet, and R. K. Brown, *J. Org. Chem.*, **37**, 190 (1972).

180. G. F. Woods and S. C. Temin, *J. Am. Chem. Soc.*, **72**, 139 (1950).

181. Reference 1, p. 43.

182. K. Ranganayakulu and R. K. Brown, *J. Org. Chem.*, **39**, 3941 (1974).

183. V. B. Mochalin and I. S. Varpakhovskaya, *Zh. Org. Khim.*, **12**, 2626 (1976).

184. V. B. Mochalin, A. N. Kornilov, A. N. Vulfson, and I. S. Varpakhovskaya, *Zh. Geterocykl. Soed.*, **2**, 167 (1975).

185. W. Priebe and A. Zamojski, *Tetrahedron*, **36**, 287 (1980).

186. F. Sweet and R. K. Brown, *Can. J. Chem.*, **46**, 1543 (1968).

187. A. Zamojski, M. Chmielewski, and A. Konował, *Tetrahedron*, **26**, 183 (1970).

188. M. Chmielewski, J. Jurczak, and A. Zamojski, *Rocz. Chem.*, **46**, 627 (1972).

189. J. Jurczak, M. Chmielewski, and A. Zamojski, *Polish J. Chem.*, **52**, 743 (1978).

190. W. Streicher, H. Reinshagen, and F. Turnovsky, *J. Antibiot.*, **31**, 725 (1978).

191. M. Chmielewski, J. Jurczak, and A. Zamojski, *Tetrahedron*, **34**, 2977 (1978).

192. A. Konował, J. Jurczak, and A. Zamojski, *Tetrahedron*, **32**, 2957 (1976).

193. A. Konował, K. Belniak, J. Jurczak, M. Chmielewski, O. Achmatowicz Jr., and A. Zamojski, *Rocz. Chem.*, **50**, 505 (1976).

194. K. Belniak and A. Zamojski, *Rocz. Chem.*, **51**, 1545 (1977).

195. D. M. Vyas and G. W. Hay, *J. Chem. Soc., Perkin I*, **1975**, 180.

196. D. M. Vyas and G. W. Hay, *Can. J. Chem.*, **53**, 1362 (1975).

197. V. B. Mochalin, A. N. Kornilov, I. S. Varpakhovskaya, and A. N. Vulfson, *Zh. Org. Khim.*, **12**, 58 (1976).

198. M. Chmielewski and A. Zamojski, *Bull. Acad. Pol. Sci., Ser. Sci. Chim.*, **20**, 751 (1972).

199. V. B. Mochalin and A. N. Kornilov, *Zh. Obshch. Khim.*, **43**, 218 (1973).

200. V. B. Mochalin and A. N. Kornilov, *Zh. Obshch. Khim.*, **44**, 2334 (1974).

201. A. Banaszek and A. Zamojski, *Polish J. Chem.*, **53**, 287 (1979).

202. D. Descours, D. Anker, and H. Pacheco, *Compt. Rend.*, **283C**, 691 (1976).

203. D. Descours, D. Anker, J.-Y. Sollaud, J. Legheand, H. Pacheco, and M. Chareire, *Eur. J. Med. Chem., Chim. Ther.*, **14**, 67 (1979).

204. V. B. Mochalin, J. N. Porshniev, and G. I. Samokvalov, *Zh. Obshch. Khim.*, **38**, 85 (1968).

205. J. Mieczkowski and A. Zamojski, *Bull. Acad. Pol. Sci., Ser. Sci. Chim.*, **23**, 581 (1975).

206. A. Banaszek and A. Zamojski, *Carbohydr. Res.*, **51**, 276 (1976).

207. Reference 1, page 44.

208. K. Ranganayakulu, U. P. Singh, T. P. Murray, and R. K. Brown, *Can. J. Chem.*, **52**, 988 (1974).

209. M. Chmielewski and A. Zamojski, *Rocz. Chem.*, **46**, 1767 (1972).

210. A. Banaszek and A. Zamojski, *Carbohydr. Res.*, **25**, 453 (1972).

211. J. Mieczkowski and A. Zamojski, *Carbohydr. Res.*, **55**, 177 (1977).

212. S. David, A. Lubineau, and J.-M. Vatele, *J. Chem. Soc., Perkin I*, **1976**, 1831.

213. M. Chmielewski and A. Zamojski, *Rocz. Chem.*, **46**, 2223 (1972).

214. A. Banaszek, *Bull. Acad. Pol. Sci., Ser. Sci. Chim.*, **20**, 925 (1972).

215. A. Banaszek, *Bull. Acad. Pol. Sci., Ser. Sci. Chim.*, **23**, 585 (1975).

216. S. David, A. Lubineau, and J.-M. Vatele, *J. Chem. Soc., Chem. Commun.*, 701 (1975).

217. S. David, A. Lubineau, and J.-M. Vatele, submitted for publication.

218. A. Thieffry, Thesis, University of Paris South, 1977.

219. S. David, A. Lubineau, and J.-M. Vatele, *J. Chem. Soc., Chem. Commun.*, 535 (1978).

220. A. Konowaľ, O. Achmatowicz Jr., and A. Zamojski, *Rocz. Chem.*, **50**, 879 (1976).

221. M. Chmielewski and A. Zamojski, *Rocz. Chem.*, **46**, 2039 (1972).

222. A. Banaszek, *Bull. Acad. Pol. Sci., Ser. Sci. Chim.*, **20**, 935 (1972).

223. E. Langstaff, R. Y. Moir, R. A. B. Bannard, and A. A. Casselman, *Can. J. Chem.*, **46**, 3649 (1968).

224. A. Banaszek, *Bull. Acad. Pol. Sci., Ser. Sci. Chim.*, **22**, 79 (1974).

225. M. Chmielewski and A. Zamojski, *Bull. Acad. Pol. Sci., Ser. Sci. Chim.*, **20**, 755 (1972).

226. A. Banaszek, *Bull. Acad. Pol. Sci., Ser. Sci. Chim.*, **23**, 633 (1975).

227. A. Banaszek, *Bull. Acad. Pol. Sci., Ser. Sci. Chim.*, **22**, 1045 (1974).

228. S. David and A. Lubineau, *Nouv. J. Chim.*, **1**, 375 (1977).

229. M. Chmielewski, A. Konowaľ, and A. Zamojski, *Carbohydr. Res.*, **70**, 275 (1979).

230. Reference 1, p. 19.

231. G. Nakaminami, H. Edo, and M. Nakagawa, *Bull. Chem. Soc. Japan*, **46**, 266 (1973).

232. G. Fronza, C. Fuganti, P. Grasselli, and G. Marinoni, *Tetrahedron Lett.*, 3883 (1979).

233. T. Ichikawa, S. Maeda, T. Okamoto, Y. Araki, and Y. Ishido, *Bull. Chem. Soc. Japan*, **44**, 2779 (1971), and references cited therein.

234. S. Ohdan, T. Okamoto, S. Maeda, T. Ichikawa, Y. Araki, and Y. Ishido, *Bull. Chem. Soc. Japan*, **46**, 981 (1973).

235. M. Taniguchi, K. Koga, and S. Yamada, *Tetrahedron*, **30**, 3547 (1974).

236. Reference 1, p. 23.

237. M. Taniguchi, K. Koga, and S. Yamada, *Chem. Pharm. Bull.*, **22**, 2318 (1974).

238. W. A. Szarek, D. M. Vyas, and L. Chen, *Carbohydr. Res.*, **53**, C1 (1977).

239. K. Koga, S. Yamada, M. Yoh, and T. Mizoguchi, *Carbohydr. Res.*, **36**, C9 (1974).

240. S. Zen, E. Kaji, and H. Kohno, *Chem. Lett.*, 1029 (1974).

241. E. Kaji, H. Kohno, and S. Zen, *Bull. Chem. Soc. Japan*, **50**, 928 (1977).

242. T. Kinoshita, Y. Kawashima, K. Hayashi, and T. Miwa, *J. Chem. Soc., Chem. Commun.*, 766 (1979).

243. M. Natsume and M. Wada, *Tetrahedron Lett.*, 4503 (1971).

244. M. Natsume and M. Wada, *Chem. Pharm. Bull.*, **23**, 2567 (1975).

245. M. Natsume and M. Wada, *Chem. Pharm. Bull.*, **24**, 2651 (1976).

246. M. Natsume and M. Wada, *Chem. Pharm. Bull.*, **24**, 2657 (1976).

247. M. Natsume, Y. Sekine, and H. Soyagimi, *Chem. Pharm. Bull.*, **26**, 2188 (1978).

248. M. Natsume, M. Wada, and M. Ogawa, *Chem. Pharm. Bull.*, **26**, 3364 (1978).

249. J. E. McCormick and R. S. McElhinney, *J. Chem. Soc., Chem. Commun.*, 171 (1969); *J. Chem. Soc., Perkin I*, 2533 (1976).

250. J. E. McCormick and R. S. McElhinney, *J. Chem. Soc., Perkin I*, 64 (1978).

251. J. E. McCormick and R. S. McElhinney, *J. Chem. Soc., Perkin I*, 500 (1978).

252. J. E. McCormick and R. S. McElhinney, *J. Chem. Res. (S)*, 52 (1979).

253. H.-K. Hung, H-Y. Lam, W. Niemczura, M.-Ch. Wang, and Ch.-M. Wong, *Can. J. Chem.*, **56**, 638 (1978).

254. M. Chmielewski, *Polish J. Chem.*, **54**, 1197 (1980).

255. S. David, M. Ch. Lepine, G. Aranda, and G. Vass, *J. Chem. Soc., Chem. Commun.*, 747 (1976).

256. C. R. Kowarski and S. Sarel, *J. Org. Chem.*, **38**, 117 (1973).

257. G. Just, G. Reader, and B. Chalard-Faure, *Can. J. Chem.*, **54**, 849 (1976).

258. G. Just and B. Chalard-Faure, *Can. J. Chem.*, **54**, 861 (1976).

259. G. Just and S. Kim, *Tetrahedron Lett.*, 1063 (1976).

260. G. Just and S. Kim, *Can. J. Chem.*, **54**, 2935 (1976).

261. G. Just and R. Ouellet, *Can. J. Chem.*, **54**, 2925 (1976).

262. G. Just and S. Kim, *Can. J. Chem.*, **54**, 2935 (1976).

263. T. Suami, S. Ogawa, K. Nakamoto, and I. Kasahara, *Carbohydr. Res.*, **58**, 240 (1977).

264. S. Ogawa, I. Kasahara, and T. Suami, *Bull. Chem. Soc. Japan*, **52**, 118 (1979).

265. A. Ichihara, M. Kobayashi, K. Oda, and S. Sakamura, *Tetrahedron Lett.*, 4231 (1974).

266. A. Ichihara, K. Oda, M. Kobayshi, and S. Sakamura, *Tetrahedron Lett.*, 4235 (1974).

267. K. Oda, A. Ichihara, and S. Sakamura, *Tetrahedron Lett.*, 3187 (1975).

268. A. Ichihara, K. Oda, and S. Sakamura, *Tetrahedron Lett.*, 5105 (1972).

269. F. A. H. Rice, *Carbohydr. Res.*, **21**, 65 (1972).

270. A. Holy, *Collect. Czech. Chem. Commun.*, **41**, 647 (1976).

271. A. Holy, *Collect. Czech. Chem. Commun.*, **41**, 2096 (1976).

272. D. E. Kiely and C. E. Cantrell, *Carbohydr. Res.*, **23**, 155 (1972).

273. C. E. Cantrell, D. E. Kiely, G. J. Abruscato, and J. M. Riordan, *J. Org. Chem.*, **42**, 3562 (1977).

274. R. Schwesinger and H. Prinzbach, *Angew. Chem.*, **87**, 625 (1975).

275. H. Prinzbach and H. W. Schneider, *Tetrahedron Lett.*, 3073 (1975).

276. C. H. Foster and G. A. Berthold, *J. Amer. Chem. Soc.*, **94**, 7939 (1972).

277. H. Prinzbach, R. Keller, and R. Schwesinger, *Angew. Chem.*, **87**, 626 (1975).

278. H. Prinzbach, R. Keller, and R. Schwesinger, *Angew. Chem.*, **87**, 627 (1975).

279. A. Hasegawa, T. Kobayashi, and M. Kiso, *Agric. Biol. Chem.*, **44**, 165 (1980).

280. M. Kiso, T. Kobayashi, and A. Hasegawa, *Agric. Biol. Chem.*, **44**, 169 (1980).

281. M. Kiso, T. Kobayashi, and A. Hasegawa, *Agric. Biol. Chem.*, **44**, 419 (1980).

282. F. G. Cocu and T. Posternak, *Helv. Chim. Acta*, **54**, 1676 (1971).

283. F. G. Cocu and T. Posternak, *Helv. Chim. Acta*, **55**, 2828 (1972).

284. F. G. Cocu and T. Posternak, *Helv. Chim. Acta*, **55**, 2838 (1972).

285. A. Giddey, F. G. Cocu, B. Pochelon, and T. Posternak, *Helv. Chim. Acta*, **57**, 1963 (1974).

286. F. G. Cocu, B. Pochelon, A. Giddey, and T. Posternak, *Helv. Chim. Acta*, **57**, 1974 (1974).

287. V. N. Krylova, N. I. Kobel'kova, G. F. Oleynik, and V. I. Shvets, *Zh. Org. Khim.*, **16**, 62 (1980), and the references cited therein.

288. V. I. Shvets, B. A. Klyashvhitskii, A. E. Stepanov, and R. P. Evstigneeva, *Tetrahedron*, **29**, 331 (1973).

289. A. E. Stepanov and V. I. Shvets, *Chem. Phys. Lipids*, **25**, 247 (1979).

290. T. Suami, Y. Fukai, Y. Sakota, M. Karimoto, N. Takoi, and Y. Tsukamoto, *Bull. Chem. Soc. Japan*, **44**, 1695 (1971).

291. T. Suami, S. Nishiyama, K. Tadano, and F. W. Lichtenthaler, *Bull. Chem. Soc. Japan*, **46**, 2562 (1973).

292. K. Tadano, Y. Emori, M. Ayabe, and T. Suami, *Bull. Chem. Soc. Japan*, **50**, 1572 (1977).

293. T. Suami, S. Ogawa, T. Ishibashi, and J. Kasahara, *Bull. Chem. Soc. Japan*, **49**, 1388 (1976).

294. S. Ogawa, K. Nakamoto, M. Takahara, Y. Tanno, N. Chida, and T. Suami, *Bull. Chem. Soc. Japan*, **52**, 1174 (1979).

295. S. Ogawa, T. Toyokuni, M. Omata, N. Chida, and T. Suami, *Bull. Chem. Soc. Japan*, **53**, 455 (1980).

296. S. Ogawa, M. Ara, T. Kondoh, M. Saitoh, R. Masuda, T. Toyokuni, and T. Suami, *Bull. Chem. Soc. Japan*, **53**, 1121 (1980).

297. R. Ahluwalia, S. J. Angyal, and B. M. Luttrell, *Aust. J. Chem.*, **23**, 1819 (1970).

298. T. Suami, Y. Sakota, K. Tadano, and S. Nishiyama, *Bull. Chem. Soc. Japan*, **44**, 2222 (1971).

299. T. Suami, K. Todano, S. Nishiyama, and F. W. Lichtenthaler, *J. Org. Chem.*, **38**, 3691 (1973).

300. K. Tadano, Y. Emori, M. Ayabe, and T. Suami, *Bull. Chem. Soc. Japan*, **49**, 1108 (1976).

301. T. Suami, K. Tadano, and S. Horiuchi, *Bull. Chem. Soc. Japan*, **48**, 2895 (1975).

302. K. Tadano, T. Shiratori, and T. Suami, *Bull. Chem. Soc. Japan*, **49**, 3193 (1976).

303. S. Ogawa, Y. Hongo, H. Fujimori, K. Iwata, A. Kasuga, and T. Suami, *Bull. Chem. Soc. Japan*, **51**, 2957 (1978).

304. T. Suami, S. Ogawa, T. Ueda, and H. Uchino, *Bull. Chem. Soc. Japan*, **45**, 3226 (1972).

305. S. Ogawa, S. Oki, H. Kunimoto, and T. Suami, *Bull. Chem. Soc. Japan*, **50**, 1867 (1977).

306. S. Ogawa, S. Oki, and T. Suami, *Bull. Chem. Soc. Japan*, **52**, 1095 (1979).

307. R. H. Shad and F. Loewus, *J. Label. Compounds*, **6**, 333 (1970).

SUGGESTED READING

S. David and A. Lubineau, "Synthesis of a protected glycoside of α-D-purpurosamine C by cycloaddition," *J. Org. Chem.*, **44**, 4986 (1979).

R. R. Schmidt and R. Angerbauer, "A convenient preparation of 2,3-unsaturated N-galactosyl derivatives," *Carbohydr. Res.*, **72**, 272 (1979).

S. David and M.-Ch. Lépine, "Preparation of sugars with branched chains, a methylene bridge, or C-1 phenyl substituents," *J. Chem. Soc., Perkin I*, 1262 (1980).

A. P. Kozikowski, K. L. Sorgi, and R. J. Schmiesing, "Stereochemistry of the alkoxyselenation of substituted 3,4-dihydropyrans: A useful process for the construction of 2-alkoxy-5,6-dihydro-2H-pyran," *Chem. Commun.*, 477 (1980).

S. David, A. Lubineau, and J. M. Vatele, "The synthesis by the cycloaddition method of the trisaccharide antigenic determinant of the A human blood group system, and six related trisaccharides, one of them active in the "acquired B" system," *Nouv. J. Chim.*, **4**, 547 (1980).

G. Berti, S. Magi, G. Catelani, and L. Monti, "A total synthesis of the racemic 3,6-dideoxyhexoses," *Gazz. Chim. Ital.*, **110**, 173 (1980).

R. R. Schmidt and R. Angerbauer, "A short synthesis of racemic uronic acids and 2,3-anhydrouronic acids," *Carbohydr. Res.*, **89**, 159 (1981).

R. Angerbauer and R. R. Schmidt, "Synthesis and glycosidation reactions of acetylated racemic pseudoglycals," *Carbohydr. Res.*, **89**, 193 (1981).

W. Abele and R. R. Schmidt, "De novo-Synthese von verzweigten dl-Pseudoglycalen," *Tetrahedron Lett.*, 4807 (1981).

G. Fronza, C. Fuganti, P. Grasselli, and G. Pedrocchi-Fantoni, "Synthesis of the *N*-benzoyl derivatives of L-arabino, L-xylo and L-lyxo (L-vancosamine) isomers of 2,3,6-trideoxy-3C-methyl-3-aminohexose from a non-carbohydrate precursor," *Tetrahedron Lett.*, 5073 (1981).

I. Dyong, H. Friege, H. Luftmann, and H. Merten, "Synthese biologisch wichtiger Kohlenhydrate. 27. Total-synthese und Konfigurationsbestimmung Me-$\overset{|}{C}{}^3$-NHR-verzweigter 2,3,6-Trideoxy-hexosen," *Chem. Ber.*, **114**, 2669 (1981).

Y. Araki, J. Nagasawa, and Y. Ishido, "Synthesis of DL-apiose derivatives by photochemical cycloaddition of 1,3-dihydroxypropanone-2," *J. Chem. Soc., Perkin I*, 12 (1981).

M. Yamaguchi and T. Mukaiyama, "The stereoselective synthesis of D- and L-ribose," *Chem. Lett.*, 1005 (1981).

T. Harada and T. Mukaiyama, "A convenient synthesis of 2-deoxy-D-ribose," *Chem. Lett.*, 1109 (1981).

T. Mukaiyama, M. Yamaguchi, and J. Kato, "A novel method for the synthesis of 2,2,2-tribromo-ethanols from aldehydes and carbon tetrabromide in the presence of stannous fluoride: A synthesis of diacetyl-D-erythronolactone," *Chem. Lett.*, 1505 (1981).

K. Suzuki, Y. Yuki, and T. Mukaiyama, "The stereoselective synthesis of D-ribulose," *Chem. Lett.*, 1529 (1981).

P. M. Wovkulich and M. R. Uskokovic, "A chiral synthesis of L-acosamine and L-daunosamine via an enantioselective intramolecular [3 + 2] cycloaddition," *J. Am. Chem. Soc.*, **103**, 3956 (1981).

Y. Shigemasa, T. Hamada, M. Hirabayashi, E. Waki, R. Nakashima, K. Harada, N. Takeda, and M. Suzuki, "A selective synthesis of 3-C-hydroxymethylpentofuranose in the formose reaction," *Chem. Lett.*, 899 (1981).

G. Kresze and E. Kysela, "Polyhydroxyamines via Diels–Alder synthesis with nitroso compounds. IV. Preparation and stereochemistry of dideoxyinosamine derivatives," *Liebigs Ann. Chem.*, 202 (1981).

G. Kresze, E. Kysela, and W. Dittel, "Polyhydroxyamines via Diels–Alder synthesis with nitroso compounds. V. 1,3-Cyclohexadiene-5,6-dicarboxylic acid derivatives as reactants," *Liebigs Ann. Chem.*, 210 (1981).

G. Kresze, W. Dittel, and H. Melzer, "Polyhydroxyamines via Diels–Alder synthesis with nitroso compounds. VI. Synthesis of inosamine derivatives," *Liebigs Ann. Chem.*, 224 (1981).

G. Kresze and W. Dittel, "Polyhydroxyamines via Diels–Alder synthesis with nitroso compounds. VII. Synthesis of Conduramine-F1," *Liebigs Ann. Chem.*, 610 (1981).

F. M. Hause and R. P. Rhep "A Brief Total Synthesis of N-Benzoyl-DL-daunosamine" J. Org. Chem. **46,** 227 (1981).

T. Mukaiyama, T. Miwa, and T. Nakatsuka, "A stereoselective synthesis of 2-amino-2-deoxy-D-arabinose and D-ribose," *Chem. Lett.*, 145 (1982).

M. Murakami and T. Mukaiyama, "A new method for the generation of a boron enolate of an ester. A new synthesis of 2-deoxy-D-ribose," *Chem. Lett.*, 241 (1982).

M. Mukaiyama, K. Suzuki, and T. Yamada, "4-O-Benzyl-2,3-O -isopropylidene-L-threose: A new and useful building block for the synthesis of L-sugars," *Chem. Lett.*, 929 (1982).

T. Mukaiyama, Y. Yuki, and K. Suzuki, "The stereoselective synthesis of L-tagatose: An application of Zn (II) mediated highly stereoselective addition of 2-furyllithium to polyoxygenated aldehyde," *Chem. Lett.*, 1169 (1982).

M. Murakami and T. Mukaiyama, "A stereoselective synthesis of 2-amino-2-deoxy-D-ribose," *Chem. Lett.*, 1271 (1982).

T. Mukaiyama, T. Sugaya, S. Marui, and T. Nakatsuka, "A new method for the synthesis of α-L-threofuranosides from acyclic precursor," *Chem. Lett.*, 1555 (1982).

L. F. Tietze, K. H. Glüsenkamp, K. Harms, G. Remberg, and G. M. Sheldrick, "Diels–Alder reactions of malondialdehyde derivatives with reversed electron demand," *Tetrahedron Lett.*, 1147 (1982).

R. R. Schmidt and M. Maier, "Hetero-Diels-Alder reactions of α-methoxymethylene substituted 1,3-dicarbonyl compounds with enol and enediol ethers," *Tetrahedron Lett.*, 1789 (1982).

K. Kobayashi, M. Kai and T. Hiyama, "A new approach to amino sugars," Abstract, 14th International Conference on Organic Synthesis, Tokyo, Japan, 1982, No. C-I-7302.

N. Minami, S. S. Ko, and Y. Kishi, "Stereocontrolled synthesis of D-pentitols, 2-amino-2-deoxy-D-pentitols and 2-deoxy-D-pentitols from D-glyceraldehyde acetonide," *J. Am. Chem. Soc.*, **104**, 1109 (1982).

G. Fronza, C. Fuganti, P. Grasselli, L. Majori, G. Pedrocchi-Fantoni, and F. Spreafico, "Synthesis of enantiomerically pure forms of N-acyl derivatives of C-methyl analogues of the aminodeoxy sugar L-acosamine from non-carbohydrate precursor," *J. Org. Chem.*, **47**, 3289 (1982).

T. Kunieda, Y. Abe, Y. Iitaka, and M. Hirobe, "Highly stereo- and regioselective formation of 2-oxazolone telomers, potential synthetic intermediates for amino sugars," *Org. Chem.*, **47**, 4291 (1982).

G. Berti, G. Catelani, F. Colonna, and L. Monti, "A highly diastereoselective synthesis of DL-oleandrose," *Tetrahedron*, **38**, 3067 (1982).

D. Descours, D. Picq, D. Anker, and H. Pacheco, "Synthèse et étude conformationelle des quatre méthyl-2-amino-2,4-didésoxy-DL-pentopyranosides," *Carbohydr. Res.*, **105**, 9 (1982).

K. Dziewiszek, M. Chmielewski, and A. Zamojski, "A new synthesis of D-glycero-D-mannoheptose," *Carbohydr. Res.*, **104**, C1 (1982).

W. R. Roush and R. J. Brown, "Total synthesis of carbohydrates: Stereoselective syntheses of 2,6-dideoxy-D-arabinohexose and 2,6-dideoxy-D-ribo-hexose," *J. Org. Chem.*, **47**, 1373 (1982).

T. Katsuki, A. W. M. Lee, P. Ma, V. S. Martin, S. Masamune, K. B. Sharpless, D. Tuddenham, and F. J. Walker, "Synthesis of saccharides and related polyhydroxylated natural products. 1. Simple alditols," *J. Org. Chem.*, **47**, 1373 (1982).

P. Ma, V. S. Martin, S. Masamune, K. B. Sharpless, and S. M. Viti, "Synthesis of saccharides and related polyhydroxylated natural products. 2. Simple deoxyalditols," *J. Org. Chem.*, **47**, 1378 (1982).

A. W. M. Lee, V. S. Martin, S. Masamune, K. B. Sharpless, and F. J. Walker, "Synthesis of saccharides and related polyhydroxylated natural products. 3. Efficient conversion of 2,3-erythro-aldoses to 2,3-threo-aldoses," *J. Am. Chem. Soc.*, **104**, 3515 (1982); cf. also *ibid.*, **104**, 6167 (1982).

S. Danishefsky and J. F. Kerwin Jr., "A simple synthesis of DL-chalcose," *J. Org. Chem.*, **47**, 1597 (1982).

S. Danishefsky, J. F. Kerwin Jr., and S. Kobayashi, "Lewis acid catalyzed cyclocondensations of functionalized dienes with aldehydes," *J. Am. Chem. Soc.*, **104**, 358 (1982).

S. Danishefsky, N. Kato, D. Askin, and J. F. Kerwin Jr., "Stereochemical consequences of the Lewis acid catalyzed cyclocondensation of oxygenated dienes with aldehydes: A rapid and stereoselective entry to various natural products derived from propionate," *J. Am. Chem. Soc.*, **104**, 360 (1982).

A. P. Kozikowski and A. K. Ghosh, "Diastereoselection in intermolecular nitrile oxide cycloaddition (NOC) reactions: Confirmation of the "antiperiplanar effect" through a simple synthesis of 2-deoxy-D-ribose," *J. Am. Chem. Soc.,* **104,** 5788 (1982).

O. Achmatowicz Jr. and M. H. Burzyńska, "Stereospecific synthesis of methyl DL-hex-2-ulopyranosides from furan compounds," *Tetrahedron,* **38,** 3507 (1982).

T. Koźluk and A. Zamojski, "The synthesis of 3-deoxy-DL-streptose," *Tetrahedron,* **39,** 805 (1983).

T. Koźluk and A. Zamojski, "Hydroxylation of 6-substituted 2,7-dioxabicyclo[3.2.0]hept-3-enes. The synthesis of analogs of 3-deoxy-DL-streptose," *Collect. Czech. Chem. Commun.,* **48,** 1659 (1983).

A. Jaworska and A. Zamojski, "A new method of oligosaccharide synthesis: rhamnobioses," *Carbohydr. Res.,* in press.

A. Jaworska and A. Zamojski, "A new method of oligosaccharide synthesis: Rhamnotrioses," *Carbohydr. Res.,* in press.

K. Harada, N. Takeda, M. Suzuki, Y. Shigemasa, and R. Nakashima, "Structural characterization of formose by chemical ionization mass spectrometry," *Nippon Kagaku Kaishi,* 1617 (1982); *Chem. Abstr.,* **98,** 54304u (1982).

Y. Shigemasa, H. Sakai, and R. Nakashima, "Formose reactions. XVII. Some factors affecting the selective formation of 2,4-bis (hydroxymethyl)-3-pentulose," *Nippon Kagaku Kaishi,* 1926 (1982); *Chem. Abstr.,* **98,** 54305v (1982).

T. Matsumoto and S. Inoue, "Formose reactions, Part 3: Selective formose reaction catalyzed by organic bases," *J. Chem. Soc., Perkins I,* 1975 (1983).

Y. E. Raifeld, L. L. Zilberg, B. M. Arshanova, O. V. Limanova, M. B. Levinskii, and S. M. Makin, "Chemistry of enol ethers. LIX. Synthesis and separation of stereoisomeric acetals of 3-O-methyl ethers of 2,6-dideoxy-DL-hexoses," *Zh. Org. Khim.,* **18,** 1870 (1982).

R. R. Schmidt, Ch. Beitzke, and A. K. Forrest, "Synthesis of 5-deoxy-DL-ribo-hexa-furanuronate derivatives via 7-oxa-norbornanones," *J. Chem. Soc., Chem. Commun.,* 909 (1982).

The Total Synthesis
of Pyrrole Pigments
1973–1980

A. H. JACKSON

Department of Chemistry,
University College,
Cardiff, Wales

K. M. SMITH

Department of Chemistry,
University of California,
Davis, California

1. INTRODUCTION

In our earlier article[1] we reviewed the main methods then available for the synthesis of naturally occurring porphyrins and bile pigments, and the structurally and biosynthetically related vitamin B_{12} (cobalamin). The review essentially covered material published up to the middle of 1971. In the present article we describe new developments during the subsequent nine years. In order to preserve continuity with the earlier article and for ease of cross-referencing we have used a similar format and order of topics.

Much of the recent synthetic work has depended upon the methods developed during the 1960s. Only those syntheses which differ appreciably from those reported earlier, or which concern porphyrins of considerable chemical or biochemical significance, will be considered in any detail for reasons of economy in space.

Since our original article was written, a regular series of reviews of porphyrin chemistry has begun[2]; two books on the chemistry of pyrroles,[3,4] a second edition of *Porphyrins and Metalloporphyrins*[5] (based on Falk's original book), and a small handbook[6] concerned with laboratory methods and spectroscopic data have now appeared. A seven-volume treatise[7] on the chemistry and biochemistry of porphyrins has also recently been published; volumes 1 and 6 are particularly relevant to the subject matter of this article.

2. NOMENCLATURE

A similar system of nomenclature will be used as in the earlier article. Approaches to the further systematization of trivial names have been reviewed elsewhere,[8] and the numbering system of the International Union of Pure and Applied Chemistry (IUPAC) is slowly gaining ground. The Fischer system[9] is used where necessary in this article in referring to porphyrins to preserve continuity with the earlier article and with the enormous volume of the earlier literature on porphyrins; however, the IUPAC system[10] will be used as before for vitamin B_{12} and corrinoid compounds because of its clear advantages in this field.

3. STRATEGY OF PORPHYRIN SYNTHESIS

Interest in porphyrin synthesis continues unabated. The first examples of a truly stepwise procedure, in which four pyrrole residues are linked together one by one, have now been described (see Section 7). Methods such as this are particularly useful for the preparation of isotopically labeled compounds for biosynthetic studies, where it is of considerable advantage both experimentally and economically to introduce the label as late as possible in the synthetic sequence. It is still, however, often necessary to carry out modification of side chains, or introduce labels, at the porphyrin stage because of the lability of the desired side chains under the conditions required to construct the polypyrrolic precursors and to cyclize them to porphyrins.

A number of improvements have also been described in the other methods used for synthesizing unsymmetrical porphyrins from open chain tetrapyrroles. However, the direct coupling of two dipyrrolic precursors to give porphyrins is still often the method of choice; as a "convergent" procedure it minimizes the number of stages involved and is usually only limited by the need for one of the dipyrrolic intermediates to be symmetrical (to avoid the formation of more than one porphyrin).

Progress in the synthesis of chlorophylls and bacteriochlorophylls (di- and tetra-hydro porphyrins) has been relatively slow, but the recent isolation of new reduced porphyrins, such as the physiologically active marine pigment bonellin[11] and sirohydrochlorin[12] (an intermediate in vitamin B_{12} biosynthesis) is likely to stimulate attempts to synthesize these compounds directly from partially reduced open chain precursors rather than by reduction of preformed porphyrins (see Section 11). Considerable progress has already been made in the development of new methods for synthesizing the partially reduced bile pigments (Section 9).

4. PYRROLES

A. Ring Synthesis

Methods for synthesizing pyrroles have been extensively reviewed recently[3,4,13] both in connection with their preparation from acyclic precursors and by modification of the side chains of preformed pyrroles. The Knorr synthesis and its variations are still the most widely used procedures for preparing pyrroles with the appropriate side chains required for the synthesis of naturally occurring porphyrins.

Symmetrical 3,4-di-alkylpyrroles (2) may be prepared by heating aliphatic (or aromatic) azines (1) in the presence of nickel or cobalt chlorides.[14] Unsymmetrical 3,4-disubstituted pyrroles (e.g., 5) are available by a versatile new route[15] involving 1,3-dipolar addition of p-toluene sulfonylmethylisocyanide (3) to suitably activated alkanes, for example, 4. Improvements in the synthesis of precursors for Knorr syntheses leading to 3,4-dialkylpyrroles have also been described—for example, boron trifluoride catalyzed condensation of 2-pentanone with acetic anhydride led to the 1,3-diketone **6**, which on condensation with diethyl oximino malonate gave exclusively the pyrrole **7** (uncontaminated by isomeric products).

B. Introduction and Modifications of Pyrrole Side Chains

2,4-Disubstituted pyrroles lacking substituents at the 3- and 5-positions are often required in porphyrin synthesis, and these can now be prepared from pyrrole itself. Thus, trichloroacetylation or trifluoroacetylation of pyrrole affords the 2-trihaloacetylpyrrole **8**, which can be converted into the corresponding esters **9**,[16] carboxylic acids,[17] or amides.[18] The amides have been acetylated, or formylated, at the 4-position specifically,[19] and the resulting acyl pyrroles (**10**) have been reduced to the alkyl pyrroles (**11**). This general procedure has now been improved[20,21] by direct Friedel–Crafts formylation of the trichloroacetylpyrrole (**8a**) with dichloromethyl methyl ether (aluminium chloride) to the diacylpyrroles (**12**) followed by transformation of the trichloroacetyl group to the carboxylic acid, ester, or amide. In a similar fashion, 2-cyanopyrrole has also been acetylated at the 4-position, and in some instances the 2-cyano group was converted into the corresponding 2-aldehyde; removal of the latter then gave 3-monosubstituted pyrroles in good yield.[22]

Cyclic (pyrrole) acetals derived from 2,2-dimethylpropanediol have recently been used successfully in protection of the aldehyde group and in some base catalyzed reactions,[23] and the parent acetal (**13**) and some derivatives have now been prepared.[24] A useful new formylation procedure involves treatment of appropriate pyrroles with triethylorthoformate and trifluoroacetic acid.[25]

C. Porphobilinogen

In further studies[26] of the synthesis of porphobilinogen and isoporphobilinogen, α-free pyrroles (prepared by Knorr type syntheses) were aminomethylated by the Tscherniak–Einhorn reaction, for example, treatment with N-hydroxy-methylchloroacetamide in hot ethanolic hydrogen chloride. The resulting chloro-acetylamino derivatives (**14**) could be reduced catalytically to the acetylamino-methyl pyrroles (**15**). In related studies an acetylaminomethyl pyrrole (**16b**) was also prepared[26] by treatment of 2,4-dimethylpyrrole with acetyl thiocyanate followed by Raney nickel reduction of the intermediate acetyl thiocarbamide (**16a**).

An effective new route[27] to porphobilinogen (**19**) involves as a key step the

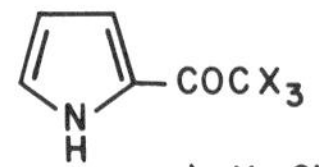

COCX_3

(8) a) X = Cl
    b) X = F

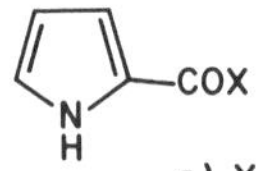

COX

(9) a) X = OH
    b) X = O alkyl
    c) X = NR_2

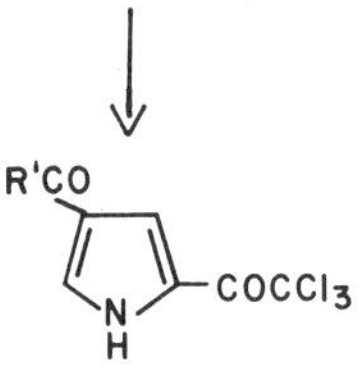

R'CO — COCCl$_3$
(12)

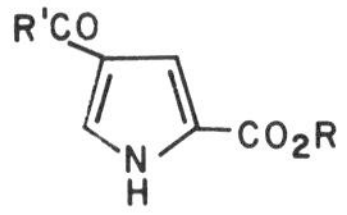

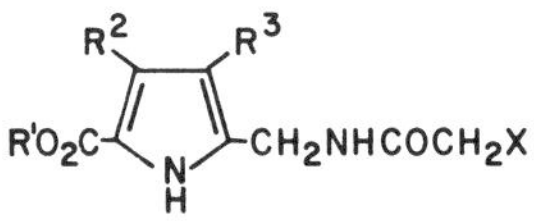

R'CO — CO$_2$R
(10)

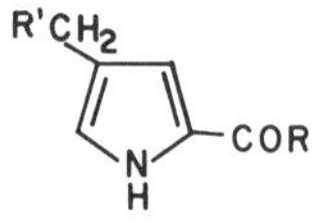

R'CH$_2$ — COR
(11)

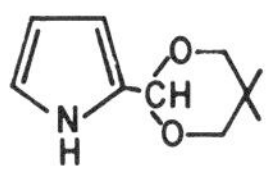

(13) CH〈O,O〉X

R'O$_2$C — CH$_2$NHCOCH$_2$X
R^2 R^3
(14) X = Cl
(15) X = H

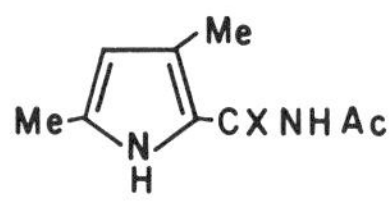

Me — Me — CXNHAc
(16) a) X = O
     b) X = H$_2$

R' = Et, CH$_2$Ph

R^2, R^3 = CH$_2$CO$_2$Et, CH$_2$CH$_2$CO$_2$Et

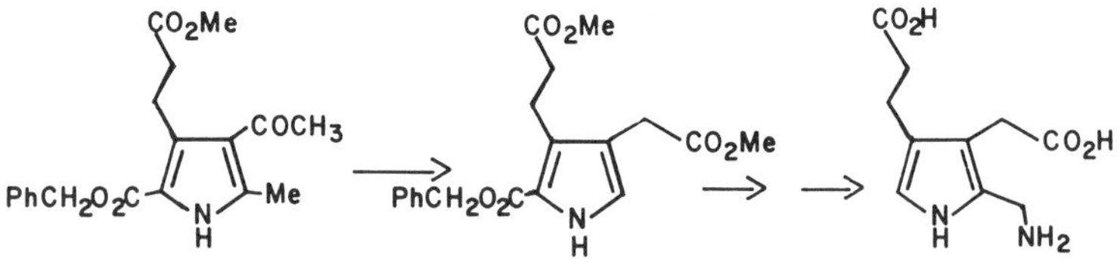

CO$_2$Me

PhCH$_2$O$_2$C — COCH$_3$ — Me
(17)

CO$_2$Me

PhCH$_2$O$_2$C — CH$_2$CO$_2$Me
(18)

CO$_2$H

CO$_2$H
NH$_2$
(19)

thallium (III) nitrate oxidative rearrangement of the acetyl pyrrole (17) (prepared by a Knorr type synthesis) to the corresponding methoxycarbonyl methyl pyrrole (18); this route has been adapted to the synthesis of ^{14}C-labeled material by using a labeled acetyl group. It also provides a useful alternative general method for introducing acetic acid side chains into pyrroles.

[13]C-labeled porphobilinogen, required for biosynthetic studies,[28] has been prepared via reductive C-methylation using ^{13}C-formaldehyde (derived from ^{13}C-methanol). Further syntheses of 2-aminomethyl pyrroles and related lactams have been described,[29] utilizing the azaindole route, and this method has also been adapted to the preparation of porphobilinogen labeled with ^{14}C in the propionic acid side chain.[30]

An entirely new approach to porphobilinogen synthesis has been described

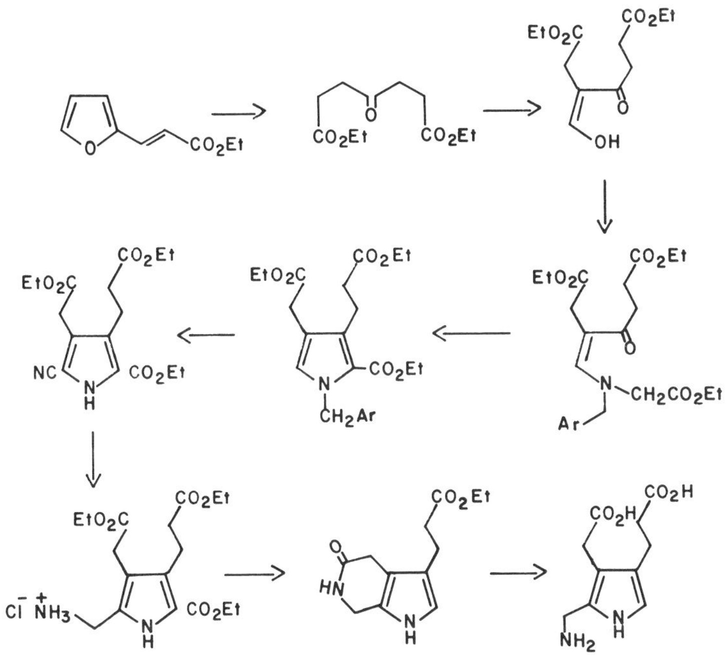

Scheme 1

by Evans and co-workers[31] and is shown in Scheme 1; improvements in the synthesis of one of the intermediates have also been reported.[32] Although the method is fairly specific it is clearly capable of adaptation for the preparation of certain types of intermediates for porphyrin synthesis.

5.   DI-, TRI-, AND TETRAPYRROLIC COMPOUNDS

A significant improvement in the synthesis of unsymmetrical pyrromethanes in good yields involves the coupling of 2-acetoxymethyl pyrroles with 2-unsubstituted pyrroles in methanol containing a catalytic amount of toluene-p-sulfonic acid.[33] More recently the use of tin(IV) chloride has been described[34] as a catalyst for similar reactions, and this is likely to be of particular utility with less reactive pyrroles. A variety of aminomethyl pyrromethanes, and the related cyclic lactams, have been prepared by standard routes for use in biosynthetic studies.[35]

The use of thioacetals as protecting groups for formyl pyrroles has enabled certain otherwise inaccessible pyrromethanes to be prepared.[36] Similarly, MacDonald and co-workers have utilized Girard derivatives of formyl pyrroles to synthesize certain di-, tri-, and tetrapyrroles for biosynthetic studies; [37] recent studies in Cardiff, however, have shown that protection of formyl groups is not essential in pyrromethane synthesis.[38]

Bromination of appropriate 5-acetoxymethyl pyrrole-2-carboxylic acids in ethanol-free chloroform at $\leqslant 10°$ affords previously inaccessible 5-bromo-5'-bromomethylpyrromethanes[39] with alternating acetic acid (or methyl) and propionic acid side chains useful for copro- and uroporphyrin syntheses.

Several new tripyrrenes (and biladienes) have been described[40,41] by condensation of symmetrical pyrromethanes with one mole (or two) of a 2-formyl-5-methyl pyrrole. Tripyrrenes may be prepared either by reduction of tripyrrenes[37] or by coupling of aminomethyl pyrroles with pyrromethanes.[42] Several syntheses of acetate and propionate bearing α-amino-methylbilanes and tripyrranes, of interest in relation to uroporphyrinogen biosynthesis, can also now be prepared by rational routes[35,37,43-45] usually involving reduction of intermediate b-bilenes and protection of the aminomethyl group as a cyclic lactam. Four 1-methylbilanes (labeled with ^{14}C in the methyl groups) have been synthesized in a similar manner and found (as expected) not to be incorporated into vitamin B_{12}.[46] All these new aminomethyl-, -tripyrranes, and -bilanes are relatively unstable compounds as they do not contain any stabilizing groups and must be handled with care, so that although vital for biosynthetic studies they are probably less useful at present as synthetic intermediates.

6. PORPHYRINS FROM MONO- AND DIPYRROLIC PRECURSORS

A. From Monopyrroles

As indicated in our earlier article, the direct polymerization of monopyrroles to porphyrins is of little synthetic value as far as the unsymmetrical natural pigments are concerned. However, this might provide a feasible approach at least to the copro- and uroporphyrins if sufficiently sophisticated methods could be developed for their chromatographic separation on a preparative basis. This has already been achieved analytically by high performance liquid chromatography (hplc).[47]

B. From Pyrromethanes

A large number of new syntheses of porphyrins from pyrromethanes have been reported in recent years by the MacDonald method.[71] As mentioned previously, this is generally limited by the need for one of the pyrromethanes (the diformyl component, or the di-α-free pyrromethane or its αα'-dicarboxylic analog) to be symmetrical or otherwise two or more porphyrins may be formed. Notable examples of the use of this method have included syntheses of deuteriated derivatives[48] of protoporphyrin IX (**20a**) required for nmr studies of haem proteins, diacetyl deuteroporphyrin IX (**20b**),[49] coproporphyrin III (**20c**)[50] [α,γ-[14]C$_2$]uroporphyrin III (**21a**),[51] harderoporphyrin (**20d**)[52] ester, the "S411"porphyrin (the 2-acrylic acid analog of coproporphyrin III) (**20e**),[53] isochlorocruoroporphyrin (**20f**),[54] isopemptoporphyrin (**20g**),[54] protoporphyrin XIII,[55] a range of hepta, hexa, and pentacarboxylic porphyrins (**21b**)[56] related to biosynthetic intermediates between uroporphyrinogen III and coproporphyrinogen III, and a

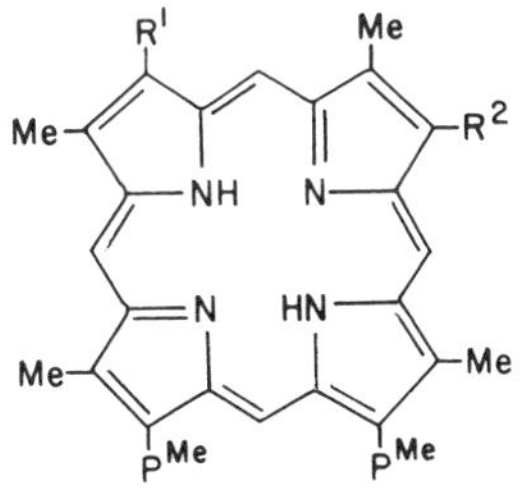

(20)

P^{Me}= CH$_2$CH$_2$CO$_2$Me

A^{Me}= CH$_2$CO$_2$Me

a) R^1= R^2= CH=CH$_2$
b) R^1= R^2= COCH$_3$
c) R^1= R^2= P^{Me}
d) R^1= CH=CH$_2$, R^2= P^{Me}
e) R^1= CH=CHCO$_2$Me, R^2= P^{Me}
f) R^1= CH=CH$_2$, R^2= CHO
g) R = CH=CH$_2$, R^2= H
h) R^1= H, R^2= CH=CH$_2$
i) R^1= CHO, R^2= CH=CH$_2$

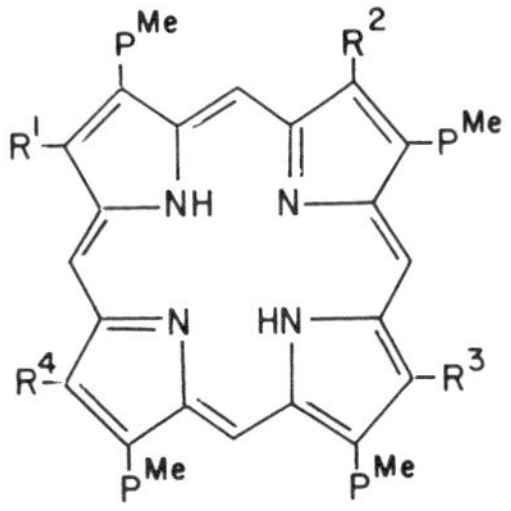

(21)

a) R^1= R^2= R^3= R^4= A^{Me}
b) R^1, R^2, R^3, R^4= Me or A^{Me}
c) R^1= R^2= R^3= A^{Me}, R^4= Me

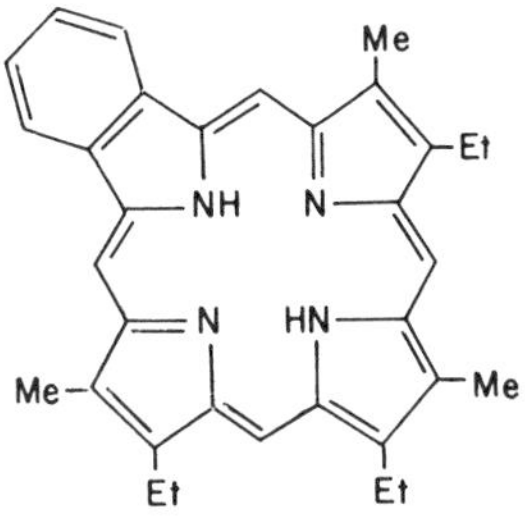

(22)

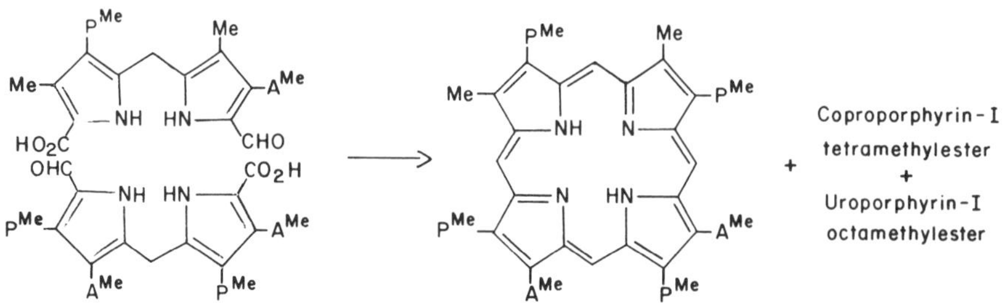

Coproporphyrin-I
tetramethylester
+
Uroporphyrin-I
octamethylester

(23)

Scheme 2

benzoporphyrin (22)[57] related to a petroporphyrin. Useful practical improvements in many of the foregoing syntheses involve the use of pyrromethane α,α'-dicarboxylic acids (rather than the α,α'-unsubstituted pyrromethanes), carrying out the condensations in the presence of p-toluene sulfonic acid, and then adding zinc acetate before aeration and oxidation to porphyrin.[52]

In an extension[58] of the MacDonald synthesis, an unsymmetrical pyrromethane bearing an electron withdrawing carboxylic substituent has been coupled with the α,α'-diformyl analog. MacDonald found that the two porphyrins produced (one having negative groups on adjacent rings, the other on opposite rings) could be separated easily by crystallization, as they did not form mixed crystals, (a problem often bedeviling researchers working with mixtures of porphyrins).

A further extension of the MacDonald method is to condense two α-free-α'-formylpyrromethanes (or the corresponding α'-carboxylic acids) together under mildly acidic conditions.[38] Self-condensation may afford type I porphyrins, for example (cf. Scheme 2), whereas if two different pyrromethanes are used mixtures of porphyrins result. However, with the advent of hplc methods, these can be easily separated if the synthesis can be designed in such a way that the porphyrins produced have different numbers of ester side chains. A range of porphyrins related to copro- and uroporphyrins I and III have now been synthesized[39] in this way, including the type I hexacarboxylic porphyrin (23) and phyriaporphyrin (21c) (cf. Scheme 2), thus providing a more rapid route for their synthesis than via open chain tetrapyrroles.

C. From Pyrromethenes

Improved yields of coproporphyrin I and aetioporphyrin I have been obtained in Fischer-type syntheses by heating the appropriate pyrromethene perbromide (or the hydrobromide with one equivalent of bromine) in formic acid[59]; in a similar manner 5,5'-dimethylpyrromethenes and 5,5'-dibromopyrromethenes have been converted by bromine in boiling formic acid to other, less symmetrical porphyrins in good yields.[60]

D. From Pyrroketones

Clezy's modification of the MacDonald synthesis using a 5,5'-diformyl-pyrroketone instead of a diformyl-pyrromethane has been extended to the synthesis of the α-oxy-derivative of mesoporphyrin IX[61] and to the γ-oxy-derivative of protoporphyrin IX[62] (Scheme 3). The iron complex of the latter undergoes oxidative ring opening in pyridine solution to form biliverdin IX-γ (25), the parent compound of a series of three green pigments found in certain types of butterflies and caterpillars[63]; this provides a model for their probable mode of biosynthesis.

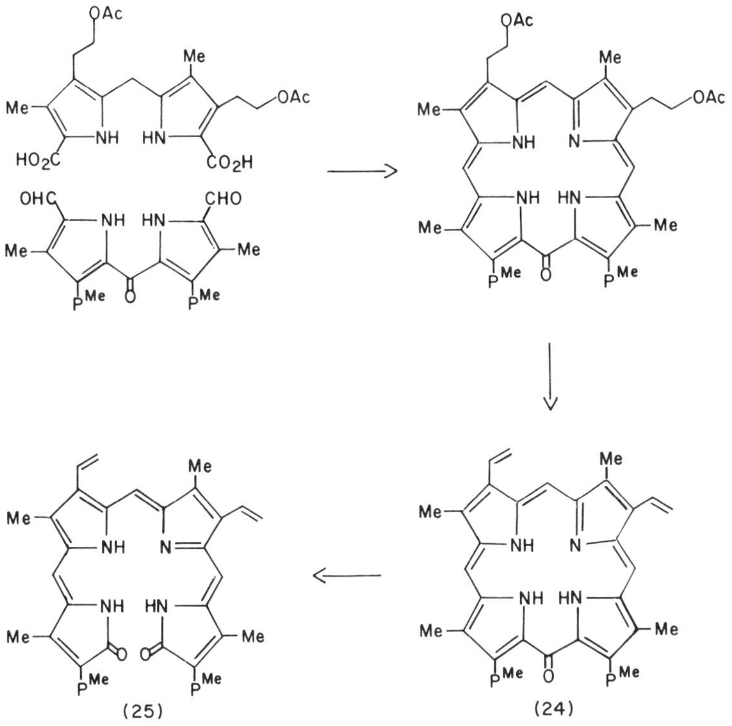

Scheme 3

7. PORPHYRIN SYNTHESIS FROM OPEN-CHAIN TETRAPYRROLIC INTERMEDIATES

A. From *a,c*-Biladienes and *b*-Bilenes

Many further applications of the *a,c*-biladiene route have now appeared; salient examples include the synthesis of [β,γ, and δ-^{13}C-] protoporphyrin IX (**20a**)[64] isocoproporphyrin (**26**),[65] *meso*-substituted porphyrins,[66] protoporphyrins III,[67] IX,[68] and XIII,[67] and pemptoporphyrin (**20h**). The ^{13}C-protoporphyrin was prepared by cyclization of an *a,c*-biladiene with ^{13}C-formaldehyde,[64] and the *meso*-substituted porphyrin was prepared by condensation of nickel *ac*-biladienes with aldehydes.[66] Further improvements in the copper salt catalyzed cyclization of the *ac*-biladienes and *b*-bilenes have been reported[69] and subsequently utilized in many of these syntheses; this enables even tetrapyrroles with electronegative groups in the terminal rings to be cyclized to porphyrins of otherwise difficult accessibility. An acrylic ester substituted porphyrin has also been prepared[70] directly from an acrylic pyrrole by the *ac*-biladiene route and copper catalyzed cyclization (hitherto these had been prepared by Knovenagel type condensations of formyl porphyrins).

Perhaps the most important new development in the *ac*-biladiene route has been the synthesis of discrete unsymmetrical tripyrrene intermediates.[65] Russian workers had shown[71] that α,α'-disubstituted pyrromethanes could be condensed successively with one and two formyl pyrrole units to form first a tripyrrene and then a symmetrical *ac*-biladiene, which could be cyclized to porphyrin. The Liverpool group,[65] however, showed that by use of unsymmetrical differentially protected pyrromethane diesters, condensation could be effected with two different formyl pyrroles to yield first tripyrrenes and then unsymmetrical *ac*-biladienes. Cyclization of the latter afforded unsymmetrically substituted porphyrins. This procedure represents the first truly stepwise rational porphyrin synthesis, as exemplified by the synthesis of isocoproporphyin (**26**) (cf. Scheme 4).

The original *b*-bilene route has now been modified, and the cyclization of bilenes with terminal methyl groups can be effected by treatment with copper acetate and pyridine (as in the *ac*-biladiene route). Examples of the use of the *b*-bilene method include pemptoporphyrin (**20h**),[72] mesoporphyrins III and IX,[73] an isomer of coproporphyrin III[74] [shown not to be identical with isocoproporphyrin (**26**)], the methyl esters of magnesium-free chlorophylls -c_1 and -c_2[75] (**27**), chlorocruoroporphyrin (**20i**),[76] harderoporphyrin (**20d**),[77] S-411 porphyrin (**20e**),[77] and isocoproporphyrin (**25**).[78] The monohydroxyethyl and monoethyl analogs of protoporphyrin IX have also been synthesized[79] by the *b*-bilene method via appropriate precursors bearing acetoxyethyl, acetyl, and ethyl side chains; the acetoxyethyl groups were converted into vinyl groups by methods described

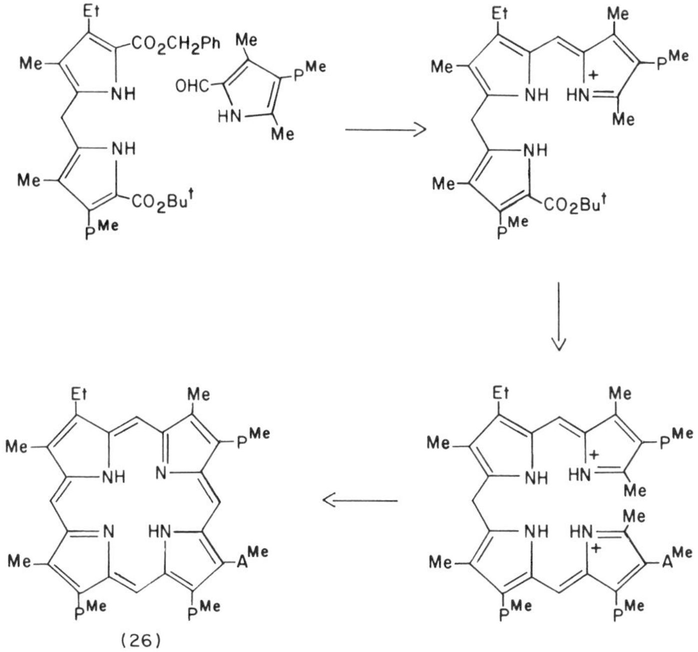

(26)

Scheme 4

previously, and the acetyl groups were reduced with borohydride to hydroxyethyl groups. Limitations to the *b*-bilene route caused by electron withdrawing substituents have also been discussed.[80]

Perhaps the most striking example of the *b*-bilene route in recent years has been its use in the synthesis of porphyrin-*a* (**31**) and its hexa-hydro analog. The key intermediate[81] was the formyl porphyrin ester (**28**) (Scheme 5), which was converted to the corresponding acid chloride and used to acylate the magnesium

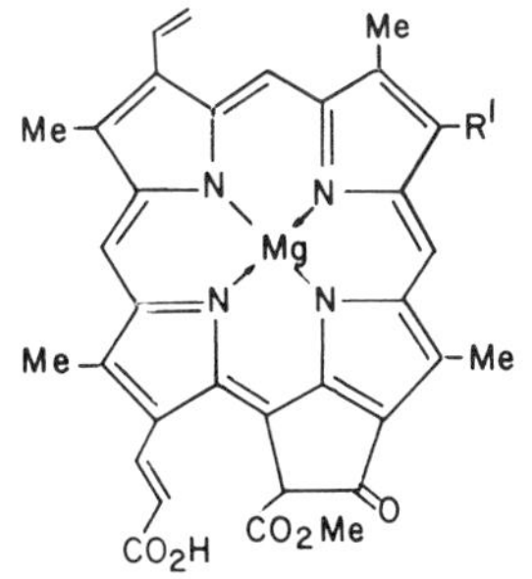

(27) a) R = Et , chlorophyll- $\underline{c}_1$

b) R = CH=CH$_2$ chlorophyll-$\underline{c}_2$

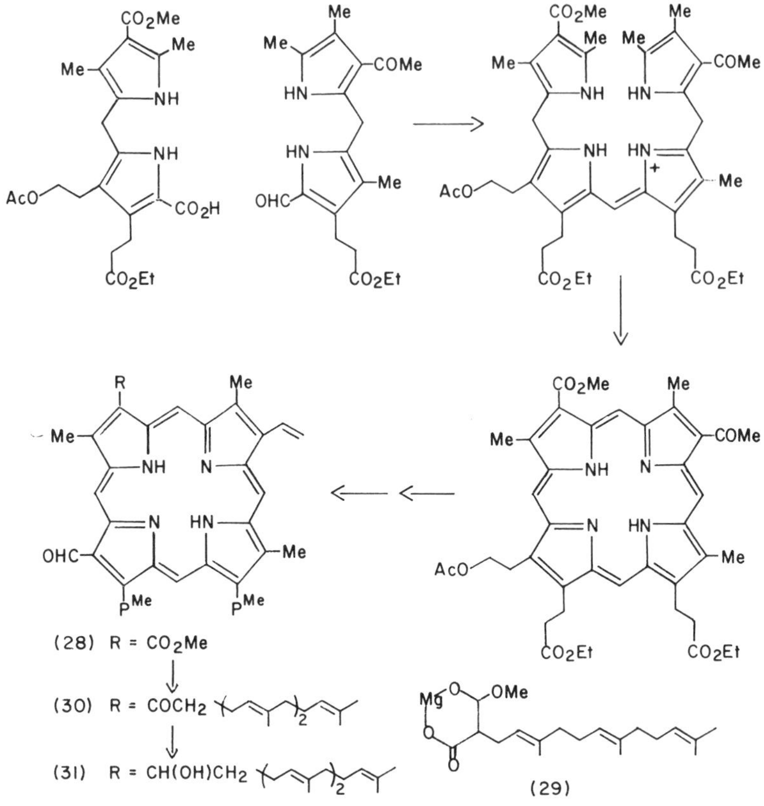

Scheme 5

enolate (29) prepared from *trans-trans*-farnesyl bromide and malonic ester. De-methylation and decarboxylation of the resulting β-keto ester with lithium iodide then gave the corresponding ketone (30), which was reduced with borohydride to the alcohol (the 8-formyl group being protected as an acetal). Hydrolysis of the acetal and re-esterification then gave (RS)-porphyrin-*a*-dimethyl ester (31),[82] identical with the natural product by uv, nmr, and hplc. A closely isomeric porphyrin was also synthesized and shown to be easily distinguishable chro-matographically and spectroscopically from porphyrin-*a* dimethyl ester.

An interesting alternative synthesis of the benzoporphyrin (22) described above has also been accomplished by the *b*-bilene route,[57] utilizing the conden-sation of a pyrromethane imine salt with a formyl pyrromethane containing a fused cyclohexanone ring. The use of the imine salt is reminiscent of Rapoport's synthesis of deoxophylloerythroaetio porphyrin described earlier,[1] but in this instance cyclization of the terminal-substituted methyl and iminomethyl bilene was effected by copper acetate, and the resulting porphyrin was subsequently converted into the desired benzoporphyrin (**22**).

B.   From Oxobilanes

Coproporphyrins III (**20c**) and IV have been synthesized by the *a*-oxobilane route.[83] A range of other porphyrins have also been synthesized by the *b*-oxobilane route including protoporphyrin-I,[84] pemptoporphyrin,[23] isopemptoporphyrin,[23] chlorocruoroporphyrin,[23] rhodoporphyrin,[85] and deuteriomethyl protoporphyrin IX[48] (required for nmr studies of haemoproteins), S-411 porphyrin,[53] harderoporphyrin,[54] and hepta-, hexa-, and pentacarboxylic porphyrins[56] related to uroporphyrins III and I.

Recently, isocoproporphyrin and its dehydro analog have also been prepared by the *b*-oxobilane route,[86] and the corresponding porphyrinogens were shown to be metabolized by chicken haemolysates to dihydroprotoporphyrin IX and protoporphyrin IX, respectively.[87]

C.   Modification of Porphyrin Side Chains

Several procedures required for the preparation of porphyrins with labile side chains have already been referred to above or in the earlier article, for example, the conversion of acetoxyethyl or aminoethyl groups (introduced at the pyrrole stage) into vinyl groups. The preparation of porphyrin β-keto esters has now been improved[88] by the use of imidazolides (rather than acid chlorides) in condensations with magnesium methyl hydrogen malonate. The porphyrin acyl imidazolides have also been converted into acrylic esters[89] by the sequence porCo-imidazole $\rightarrow$ $-CH_2OH$ $\rightarrow$ CHO $\rightarrow$ $CH=CHCO_2Me$. Alternatively the porphyrin ketoesters may be reduced with borohydride to hydroxyesters and dehydrated to acrylic esters.[77]

A useful synthesis of coproporphyrin III and related compounds has now been achieved by modification[90] of protoporphyrin IX; the key reaction in this process is the terminal oxidation of the vinyl residues by thallium(III) trifluoroacetate (Scheme 6); this has now been utilized for the preparation of a specifically ^{14}C-labeled coproporphyrin III required for biosynthetic studies.[91] Further adaptation of this process has allowed the preparation of a range of other intermediates and analogs[92] between copro- and protoporphyrin, including hardero-, pempto-, and chlorocruoroporphyrin and their isomers, as well as dihydroprotoporphyrin.[92]

Oxidation of protoporphyrin IX dimethyl ester with permanganate[93] affords a mixture of the corresponding 2,4-diformyl derivative, chlorocruoro (**20i**), and

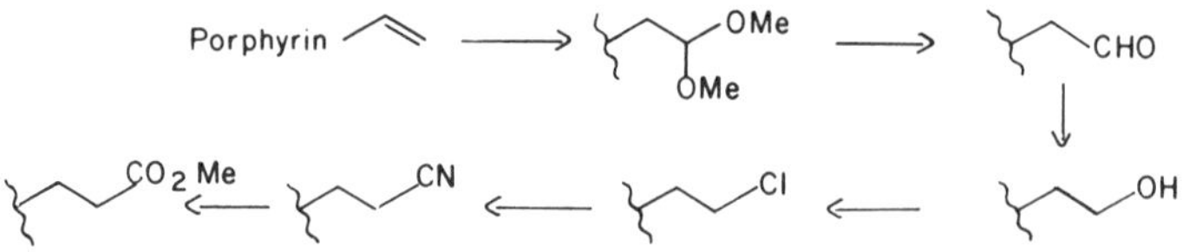

Scheme 6

isochlorocruoro (**20f**) porphyrins. Dipolar addition of *p*-nitrophenylazide to the vinyl groups of protoporphyrin followed by loss of diazomethane also affords the same three porphyrins in good overall yield[94] (no by-products being formed), and this provides an excellent route for their preparation as they can be readily separated by preparative hplc.

The oxidative ring closure of a magnesium complex of porphyrin β-ketoesters (**32a**) onto the neighboring *meso*-position by iodine in methanol affords a methoxy phaeoporphyrin ester (**33a**) and provides a model for the biosynthesis of the isocyclic ring of chlorophyll.[95] A more efficient process was the use of thallium (III) trifluoroacetate as oxidizing agent, which led to the first formal total synthesis of phaeoporphyrin a_5 dimethyl ester (**33b**) and its monovinyl analog **33c**.[96]

Meso-methyl porphyrins have been synthesized by *meso*-formylation of metalloporphyrins followed by reduction of the formyl residue in two stages.[97]

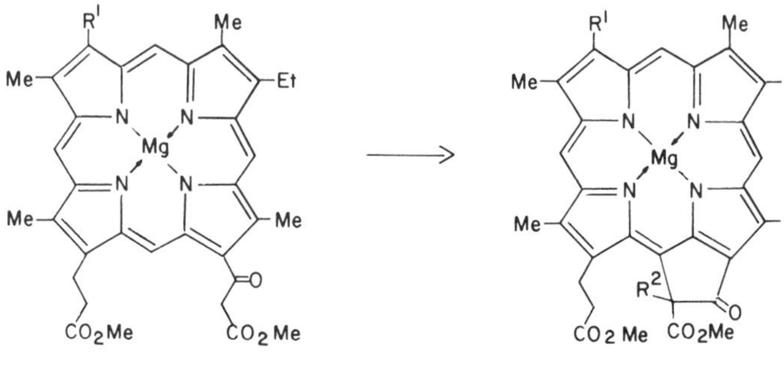

(32) a) R = Et
      b) R = V

(33) a) R^1 = Et , R^2 = OMe
      b) R^1 = Et , R^2 = H
      c) R^1 = V , R^2 = H

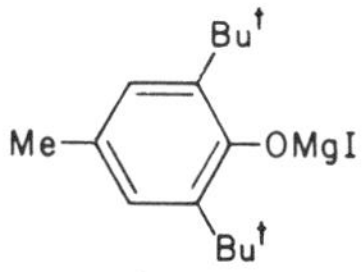

(34)

8. CHLORINS AND OTHER PARTIALLY REDUCED PORPHYRINS

The plant chlorophylls have a *trans*-configuration in the D ring but it has recently been shown that phaeophobides with the unnatural *cis*-configuration may be prepared by photochemical reduction of zinc porphyrins.[98]

Phaeophytins *a* and *b* can now be prepared on a large scale by utilizing Girard T reagent and subsequent chromatography.[99] Preparative hplc is now also a very useful method for separation of both phaeophytins and phaeophorbides.[100] Rhodoporphyrin XV and its vinyl analogs can now be prepared from phaeophytin *a* in good overall yield by an efficient new method.[99]

Magnesium has been introduced into both porphyrins and chlorins via magnesium alkoxides, but more recently magnesium pyridine complexes have been utilized. Also very effective is the use of an iodomagnesium phenolate (**34**). The latter was used as a mild procedure to introduce magnesium into bacteriophaeophytin *a*.[101] Other methods for introducing metals into chlorins have been described, including the use of magnesium acetate in acetone and dimethylsulfoxide.[102]

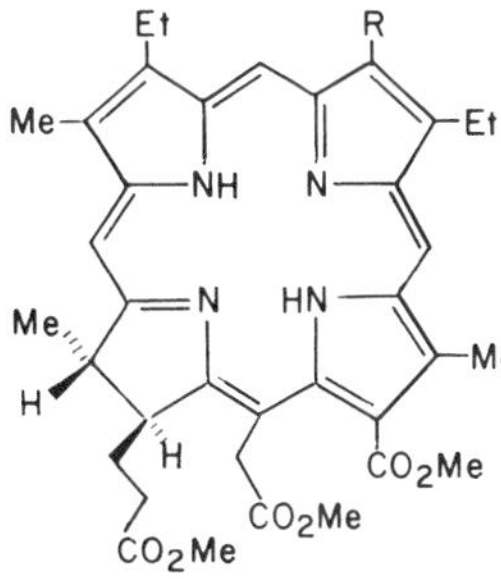

(35) a) R = Me
     b) R = CHO

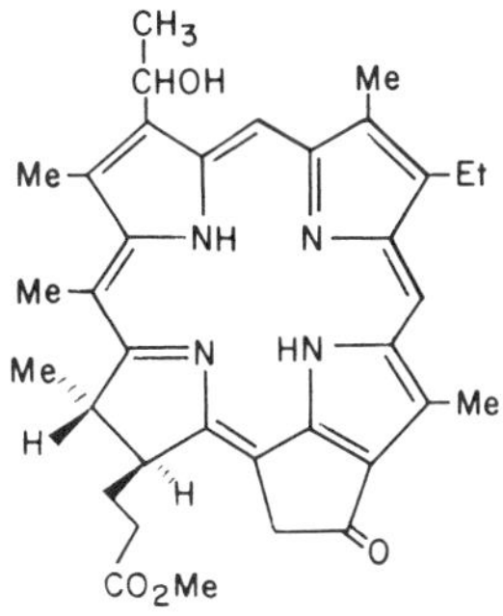

(36)

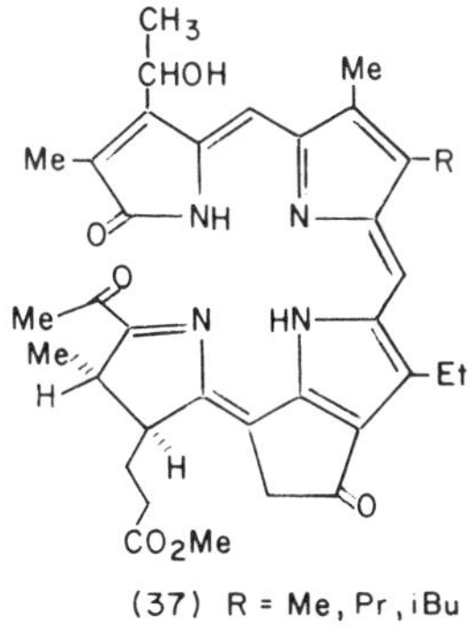

(37) R = Me, Pr, iBu

Very recently the photochemical conversion of the chlorophyll *a* derivative, chlorin e_6 trimethyl ester (**35a**), into the related rhodin g_7 (**35b**) of the *b*-series in low yield in carbon tetrachloride has been reported by Jurgens and Brockmann[103]; this is thought to be due to photochlorination of the 3-methyl group. A partial synthesis of a bacteriophaeophorbide *c* (**36**) (related to chlorobium chlorophyll 660) has also been described by the same group, starting from chlorin e_6 trimethyl ester (**35a**) and introducing the δ-*meso*-methyl group via *meso*-formylation followed by reduction.[104]

9. BILE PIGMENTS

As we mentioned in our earlier article, most natural bile pigments probably arise in nature by oxidative ring opening of haem at the α-position. The precise mechanism of this process is still obscure, although there is strong circumstantial evidence that the first intermediate is the α-*meso*-hydroxy heme, which subsequently undergoes cleavage by molecular oxygen to form biliverdin IXα. Exceptions to this general rule are biliverdin IXγ (**25**) and other related pigments found in certain types of butterflies, probably formed by oxidative ring opening of γ-oxyhaem (cf. Scheme 3 above).

Oxidative ring opening of unsymmetrical porphyrin metal complexes (e.g., haem) usually affords mixtures of bile pigments, and so it is not of preparative usefulness. However, the Buenos Aires group[104] has recently shown that in contrast to haem itself, the isomeric but more symmetrical iron complexes of protoporphyrins III and XIII undergo oxidative ring opening specifically at the α-position. Certain chlorophyll derivatives also undergo photooxidation specifically at the δ-*meso*-position—because of the higher reactivity of this position to electrophilic attack—to form dihydro-analogs of the bile pigments, for example, **37**.[105,106]

A useful reagent for oxidation of porphyrins (and chlorins) to *meso*-oxygenated derivatives is thallium trifluoroacetate,[107] which is more efficient than reagents previously used (cf. Ref. 1), but again, mixtures of products are formed with unsymmetrical porphyrins. Chlorins, however, are attacked at the *meso*-position adjacent to the reduced ring, and the products on hydrolysis undergo oxidative ring opening.[107]

For the systematic synthesis of naturally occurring bile pigments, the most versatile method is still the stepwise coupling of pyrrole—or partially reduced analogs—to form dipyrrolic units, cf.[108] which can be coupled together. ^{14}C-labeled bilirubin IXα (**40**), for example, has now been synthesized[109] by condensation of an α-unsubstituted oxopyrromethene (**38**) with an oxopyrromethene Mannich base (**39**); the label was incorporated in the dimethylaminomethyl res-

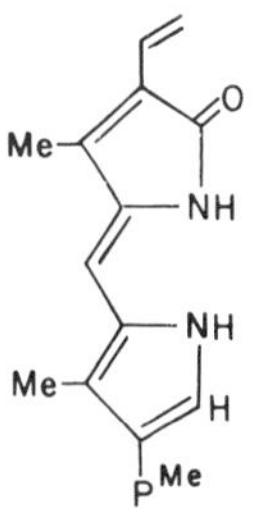

(38) 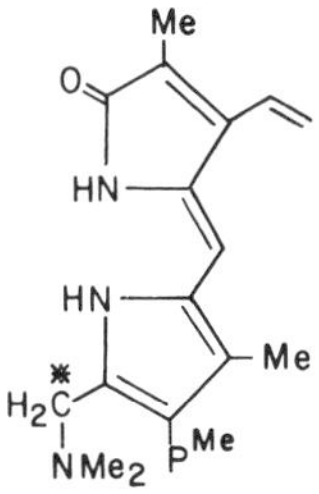   (39) 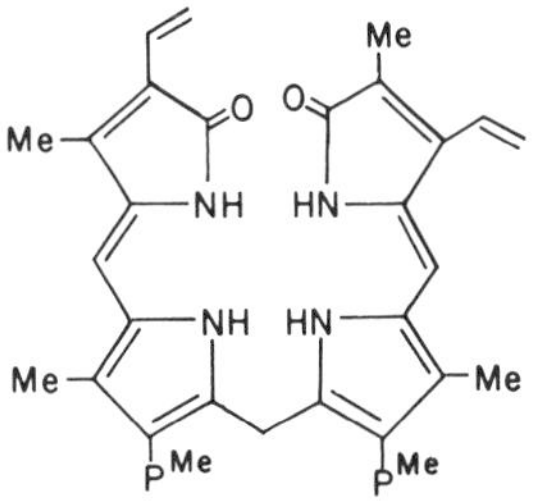   (40)

Scheme 7

idue of the latter via a Vilsmeier type formylation with ^{14}C-dimethylformamide (Scheme 7).

Total synthesis of several of the reduced bile pigments have been described during the past few years, including phycocyanobilin (**41**),[110] mesobilirhodin (**42**),[111] analogs of mesourobilin,[112] and phycoerythrobilin dimethyl ester (**43**).[113]

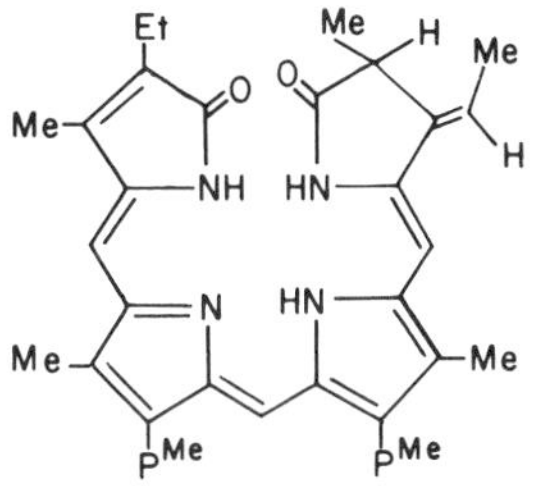

(41) 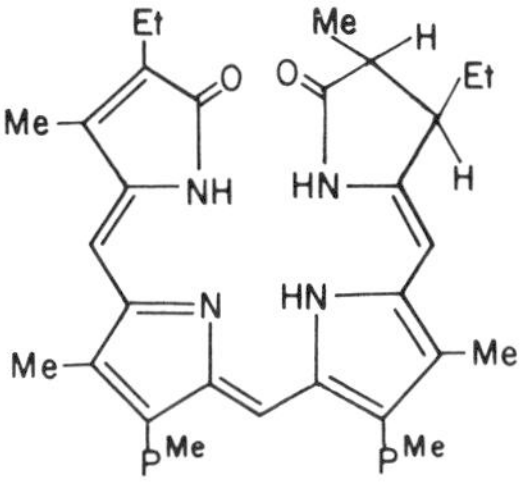   (42)

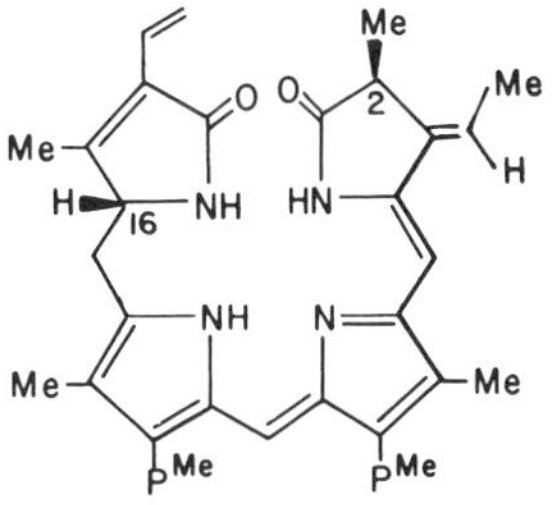

(2R, I6R form shown)

(43)

The overall strategy has been similar to that adopted in previous bile pigment syntheses, but a number of novel procedures have been devised by Gossauer's group, who have made the major contributions in this area (cf. Refs. 110-113).

In the phycocyanobilin syntheses, for example, one of the two oxopyrromethenes (46) required was prepared by a "sulfur contraction" route between a thiosuccinimide (44) and an α-bromomethyl pyrrole ester derivative (45a) or the corresponding phosphorous ylid (45b) (Scheme 8). Coupling of this oxopyrromethene (46) with the formyl oxopyrromethene (47) then gave phycocyanobilin dimethyl ester (41).

In another procedure a diazoketone (48) (derived from the corresponding acid chloride by treatment with diazomethane) was coupled with an α-free pyrrole (49) to give an adduct (50) that was subsequently cyclized with ammonium acetate to give the oxopyrromethene (51) required for bile pigment synthesis.

An interesting stereoselective synthesis of two new bile pigments[113a] (phycoerythrobilins) utilized the coupling of the optically active formyl oxopyrromethene (52) with the racemic oxopyrromethene (53) under acidic conditions; the mixture of diastereoisomeric reaction products was separated chromatograph-

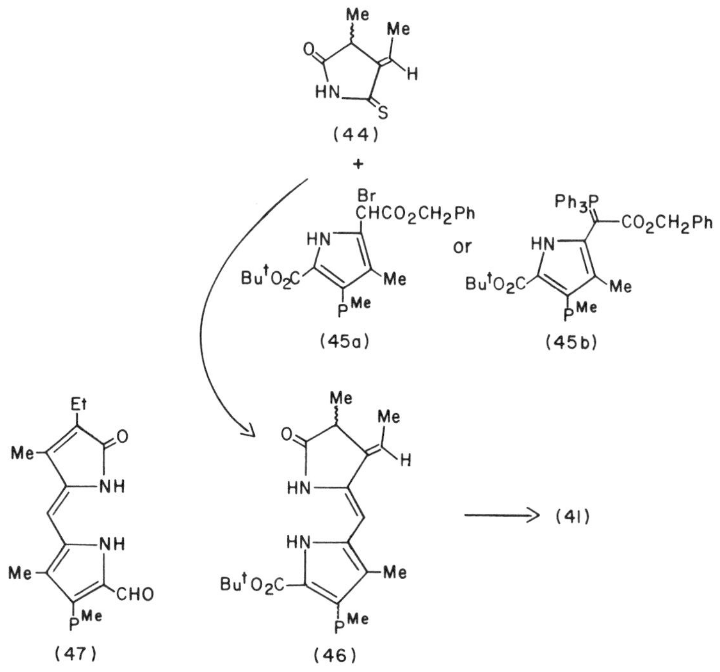

Scheme 8

a) R = CO₂ α − methylfenchyl
b) R = CHO

ically to give $(+)$-$(2R,16R)-$ and $(-)$-$(2S,16R)$-phycoerythrobilin dimethyl ester (43). The optically active 52b was prepared through the related fenchyl ester 52a (which behaves as chemically equivalent to a t-butyl ester group). The absolute configuration of the phycoerythrobilins was inferred from the use of the same optically active intermediate in synthesis of an optically active urobilin, and by oxidative degradations to optically active products. From the results it was deduced that the chromophore of the native chromoprotein has the R-configuration at both C-2 and C-16.

Racemic phycoerythrobilin dimethyl ester (cf. 43) was prepared[113b] from the thioacetal 54 via 52b and the oxopyrromethene 53; this gave a mixture of two racemic bile pigments, one of which, after chromatographic separation, was shown to be identical with phycoerythrobilin dimethyl ester obtained from R-phycoerythrin by treatment with boiling methanol (Scheme 9).

Isomerically pure biliverdin can be prepared from pure bilirubin IXα by oxidation with chloranil[114]; acidic reagents cause some isomerization with formation of the III and XIII isomers. A variety of synthetic and spectroscopic studies have been carried out by the Vienna group on model systems[115] related to the reduced bile pigments, in connection with studies of the conformation and

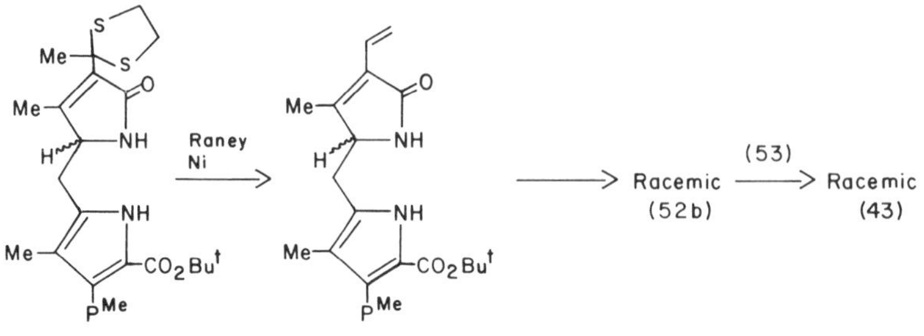

Scheme 9

photochemical isomerization of the pigments when associated with protein *in vivo*.

10. PRODIGIOSINS AND RELATED COMPOUNDS

Full details of the synthesis of metacycloprodigiosin (described in our original article[1] have now appeared.[116] Some 5-aryl-2,2'-pyrromethene analogs of prodigiosin[117] and a phenyl analog of metacycloprodigiosin have also been prepared.[118a,119] O-alkylation of 3-hydroxypyrroles and 3-hydroxy pyrromethenones can be effected by triethyloxonium tetrafluoroborate, whereas other alkylating reagents attack nitrogen; this was a key factor in the synthesis of a pyrromethane analog of metacycloprodigiosin.[118b]

11. CORRINS AND VITAMIN B$_{12}$

As mentioned in our earlier article,[1] the total synthesis of vitamin B$_{12}$ (55) has been completed following two convergent approaches. The first of these, termed the "Woodward–Eschenmoser Approach,"[1] involved construction of the corrin macrocycle (56) using classical chemistry in a joint approach by the Harvard and ETH Zurich groups.[120] In a quite separate endeavor,[121] the Swiss workers fashioned the corrin (56) by the "new road," involving an orbital-symmetry-allowed photocyclization of a seco-corrin.

In the time since the leaders of the Swiss and American groups described their successful syntheses, there has nevertheless been a great deal of progress, particularly with regard to developing an understanding of the characteristics of Eschenmoser's novel photocyclization, and in fabrication of model corrins by the isoxazole approach.

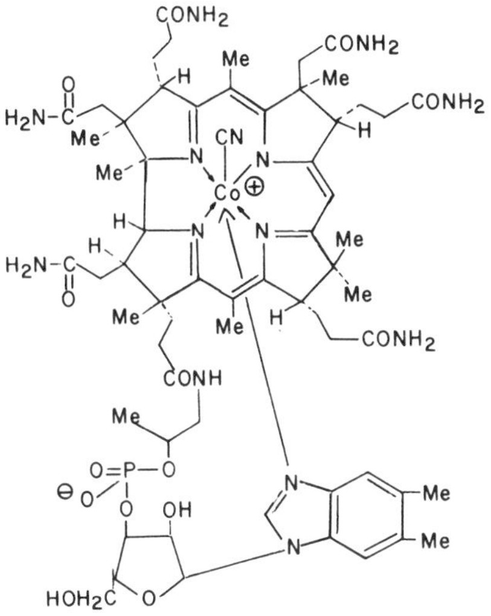

(55)

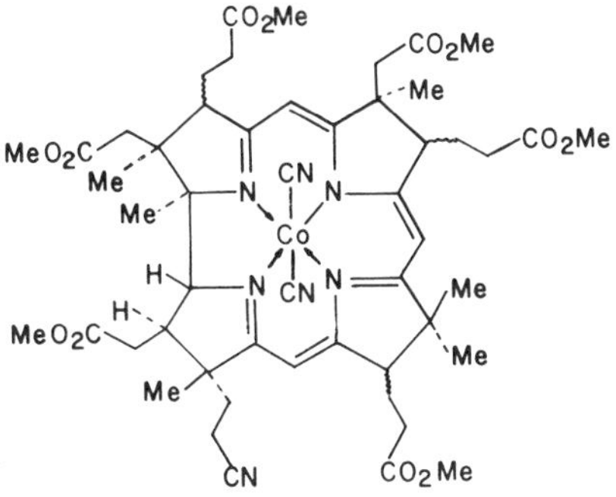

(56)

A. The Isoxazole Approach to Corrins

In our earlier review[1] we outlined a synthetic approach to corrins, devised by Cornforth, in which it was proposed to transform a model triisoxazole into a corrin. Stevens[122] had conceived, quite independently, a similar approach that has been strikingly successful. Following Stevens' route, the isoxazole rings

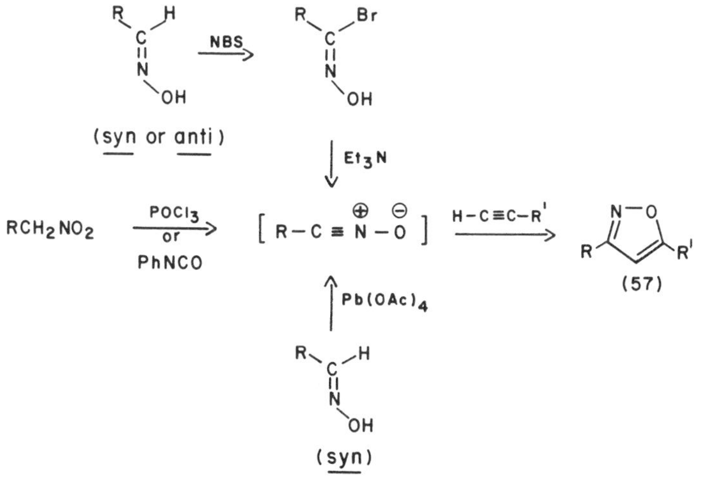

Scheme 10

were synthesized by one of the three routes outlined in Scheme 10, whereby primary nitrocompounds were transformed into nitrile oxides by dehydration with either phosphoryl chloride or phenyl isocyanate, or else the same oxides were formed by dehydrogenation with lead tetraacetate (*syn*) or N-bromosuccinimide (*syn* and *anti*). The unstable nitrile oxides were then reacted *in situ* with an appropriately substituted alkyne to give the isoxazole (**57**).

The metal-free corrin nucleus can nominally be synthesized by stepwise connection of isoxazole nuclei in either a clockwise or counterclockwise direction, Scheme 11. Depending on the mode of construction, the isomeric triisoxazoles **58** and **59** could be synthesized. At present most information is available for the counterclockwise approach to **59**, and since the basic principles for each case are very similar, only this route will be outlined.[122]

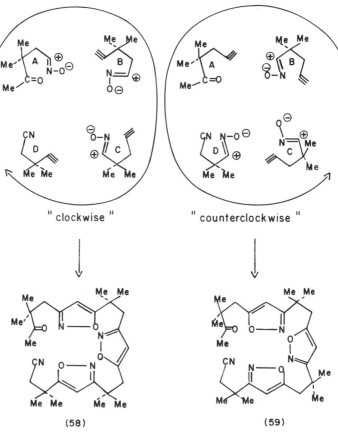

" clockwise "   " counterclockwise "

(58)   (59)

Scheme 11

The nitroester (**60**) was transformed into its nitrile oxide (Scheme 10) in the presence of the alkyne (**61**) to give the isoxazole (**62**). Treatment with acid gave the corresponding aldehyde, which was converted into the oxime (**63**) using hydroxylamine hydrochloride. Dehydrogenation (Scheme 10) and cyclo-addition with a second mole of the alkyne (**61**) afforded the bisisoxazole (**64**). Repetition of this same sequence of operations next gave the triisoxazole (**65**) in an overall yield of 40% through a sequence that was applicable to large-scale synthesis. The ketone carbonyl in compound (**65**) was protected as its ethylene ketal and the ester was reduced with lithium aluminum hydride to give the alcohol (**66**), which was transformed into the tosylate (**67**) in the usual way. Isoxazoles can

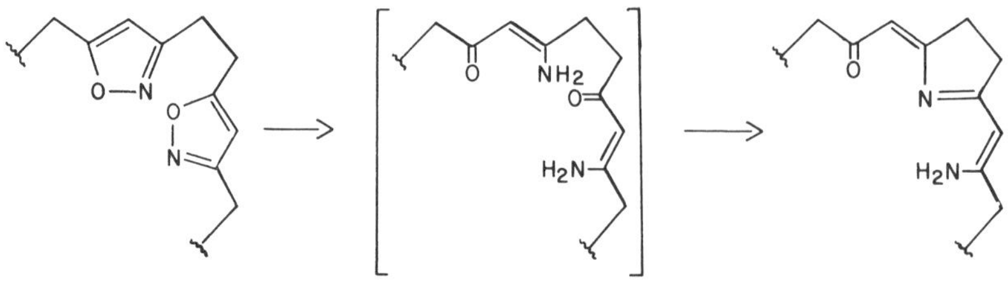

Scheme 12

be converted into β-enamino-ketones (Scheme 12) by catalytic hydrogenation, and the whole principle upon which the isoxazole approach was developed by Cornforth and Stevens depended upon the fact that (Scheme 12) if two or more isoxazoles in close proximity to each other are opened then they will cyclize in the reverse sense to afford pyrrolenine rings. Thus, catalytic hydrogenation of the triisoxazole (67) and recyclization with triethylamine gave a tricyclic ligand, which was chelated with nickel perchlorate to give 68. The fourth nitrogen atom was introduced by treatment of 68 with ammonium acetate in methanol to afford 69. With an excess of cyanide, compound 69 was demetalated and then recomplexed with zinc ions to give 70. The reason for this metal exchange was twofold. First, zinc(II) corrins can be demetalated relatively easily (compared with nickel corrins), but second and more important, Eschenmoser's photochemical cyclization of seco-corrin complexes does not proceed with nickel chelates, whereas it affords corrins from zinc(II) seco-corrins in virtually quantitative yield.[123] Dehydration of 70 gave the complex 71, which smoothly cyclized to give the *trans* corrin 72 by way of a photochemically allowed antarafacial (1,16)-sigmatropic shift of a hydrogen from ring D to the methylene carbon of Ring A; the product of this would be the system (73) which would then undergo a thermally allowed antarafacial 1,15-($\sigma - \pi$) electrocyclic ring closure to 72.

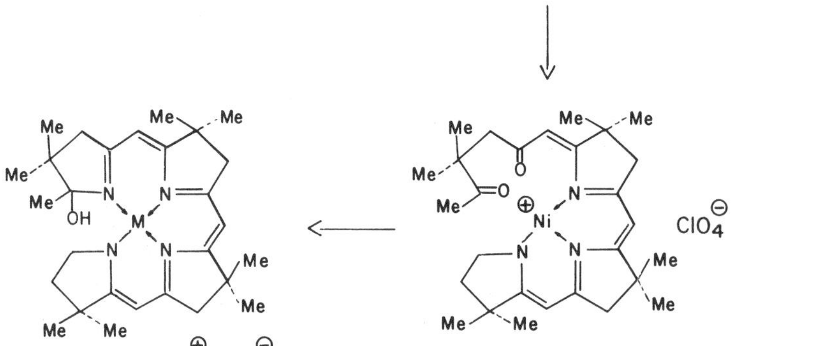

(69) M = Ni$^{\oplus}$ ClO$_4^{\ominus}$

(70) M = Zn$^{\oplus}$ Cl$^{\ominus}$

(68)

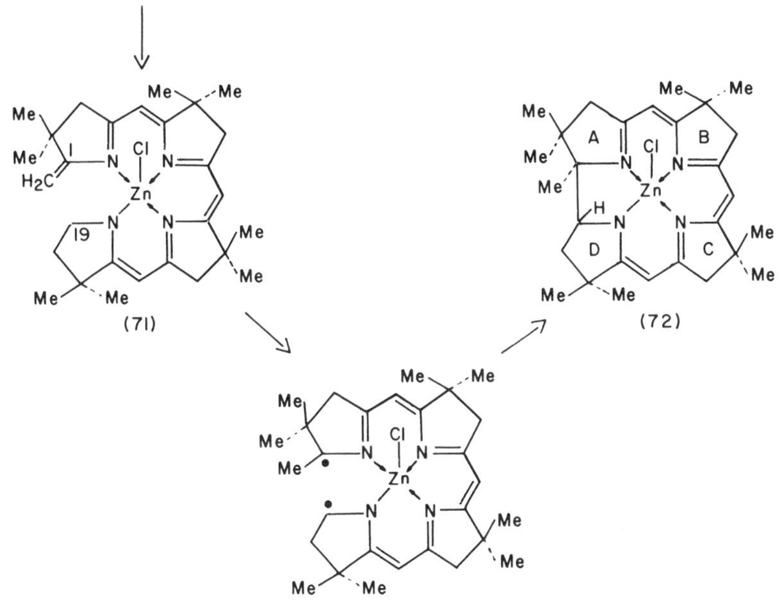

(71)

(72)

(73)

B. Recent Studies on the Cyclization of Seco-corrins

Fueled by the possibility that nature might use a procedure like seco-corrin cyclization for the biosynthesis of vitamin B$_{12}$, Eschenmoser[123] and his colleagues have carried out extensive studies of seco-corrin chemistry since the completion of the total syntheses for which the procedure of photocyclization was elegantly conceived and executed. Previous work had shown[123] that photocyclization of the chelate **74** proceeds with M = Li, MgCl, ZnCl, CdCl, Pd$^+$, or Pt$^+$, but not with M = H, Cu$^+$, Ni$^+$, Co(CN)$_2$, or MnCl. Moreover, the 19,19-dideuterium analog of **74** (M = Pd) cyclizes slower than **74** (M = Pd) and one deuterium migrates to the methylene carbon. Finally, the cyclization of the zinc and cadmium analogs of **74** is strongly sensitized by the cyclization product, but oxygen

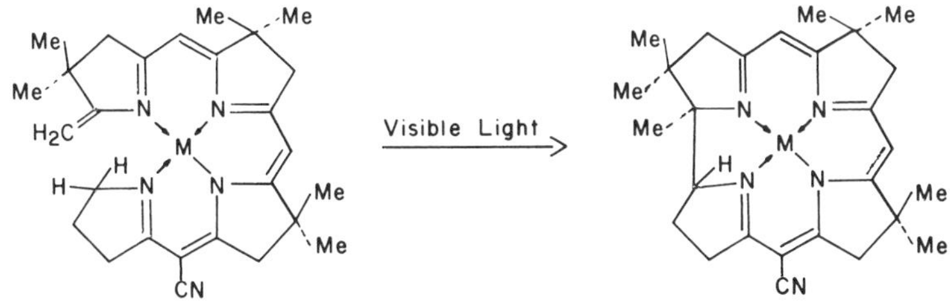

(74)   Visible Light

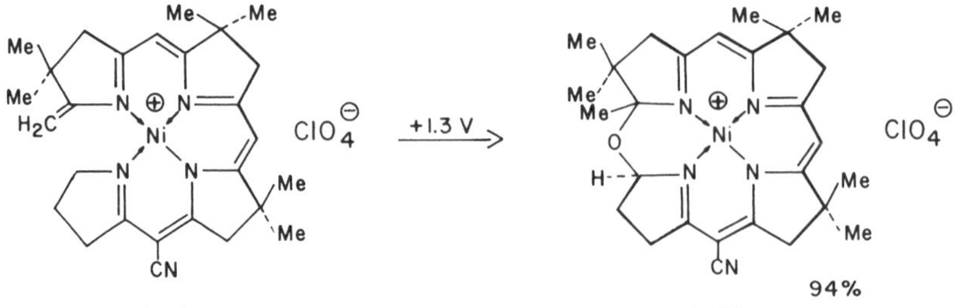

(75)   (76)

and specific triplet quenchers prevent the cyclization. Observations of these types led Eschenmoser and his colleagues to investigate the possibility of (electrochemical) redox simulation of the photochemical seco-corrin cyclization.[123]

Electrochemical oxidation of the nickel seco-corrin (75) in acetonitrile containing water gave the oxide (76) in a two-electron oxidation process; the oxygen atom was shown to come from water, but neither of the water protons were incorporated into the molecule. Instead, in the rate-determining step after initial one-electron oxidation, one of the hydrogens of the ring D methylene group migrates to the ring A methylene. The proposed mechanism[123] for formation of the seco-corrin oxide (76) is shown in Scheme 13. A reversible dimerization of

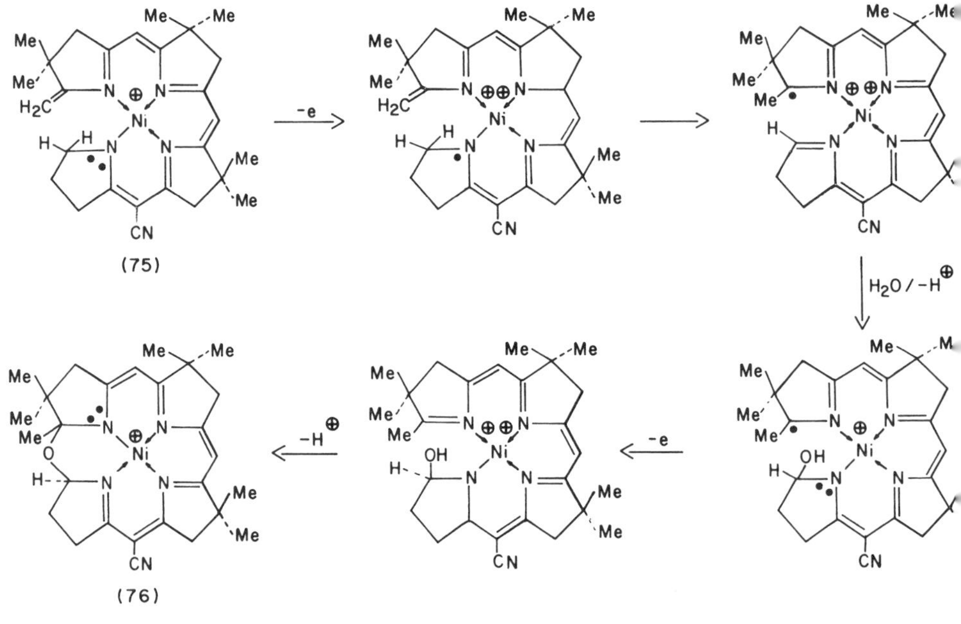

Scheme 13

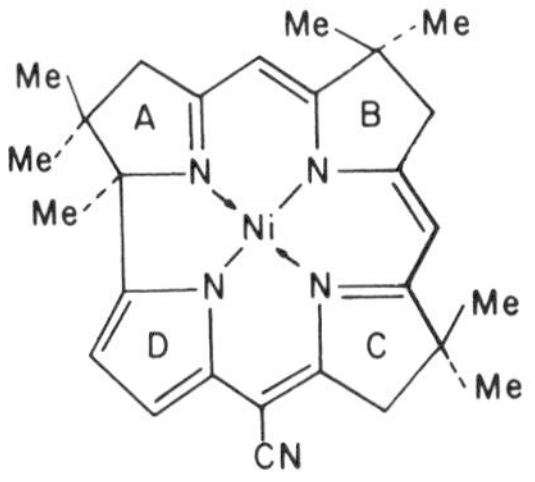

(77)

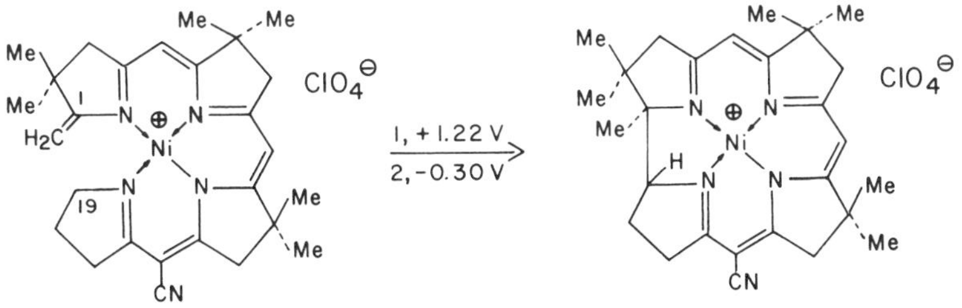

$$\xrightarrow[2,-0.30\,\text{V}]{1,+1.22\,\text{V}}$$

(75)   (78)

the cation-radical seems to be implicated by additional cyclic voltammetric evidence. The same seco-corrin oxide was also formed by oxidizing with trisphenanthrolineiron(III) perchlorate ($E_{1/2} = 0.99$ V) in place of electrolysis. Unfortunately, the oxide (76) proved to be inert towards acid catalyzed ring opening for transformation into a nickel-corrin, though it can be thermolyzed to yield the novel D-pyrrolocorrin salt (77) in 55% yield.

Electrolysis of the nickel(II) seco-corrin (75) under strictly anhydrous conditions gave the corrin complex (78), thereby demonstrating that the electrochemical simulation of the photocyclization is feasible; moreover, the nickel(II) seco-corrin could *not* be cyclized photochemically! (See Section 11A.)

In a variant of the sulfide contraction method,[124] Eschenmoser and his col-

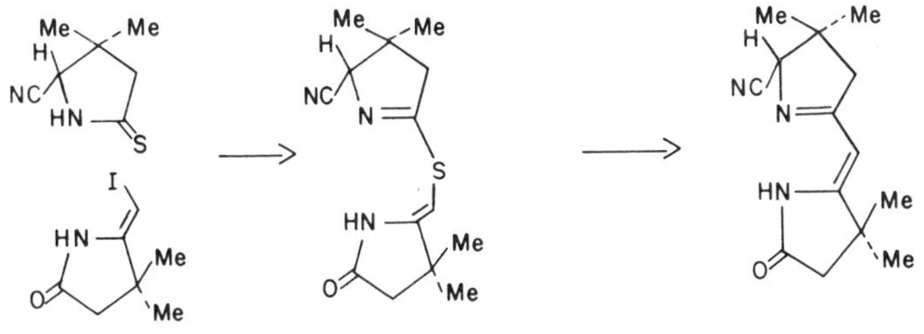

(79)   (80)

Scheme 14

1. Meerwein's reagent

2. $\begin{array}{c}Ph\\Ph\end{array}$N–NH$_2$

+ Thiolactam derivative of (80)

CdCl$_2$/ Ph$_3$P

leagues have also completed a synthesis of a Δ^{18}-dehydro-A/D-seco-corrin, which has been successfully cyclized to give a corrin.[125] If iodomethylene pyrrolidones are used (Scheme 14) with thiopyrrolidones in the presence of strong base, then the thiolactam sulfur displaces the vinylic iodine to give[125] the species **79**, which can be cleanly contracted to give a semicorrin (**80**) by heating in triethyl phosphite. Scheme 15 outlines the route to a Δ^{18}-dehydro-A/D-seco-corrin (**81**) using this new approach, with the added novelty that *trans*-2,3-diphenyl-1-aminoaziridine was used to achieve reduction of the lactam in ring D.[125] Irradiation of the cadmium complex (**82**) in methanol led to elimination of nitrogen and stilbene, to give the 19-methoxy-1,19-seco-corrin nickel(II) complex (**81**). Electrochemical reduction in acetonitrile/trifluoroacetic acid gave the corrin complex (**83**), together with 20% yield of a dimer.[123,125]

1-Oxo-A/D-seco-corrin complexes (**84**) (Scheme 16) can also be synthesized by way of the new sulfide contraction procedure,[123,126] and irradiation provides 1-hydroxycorrin complexes (**85**). Thus, it was demonstrated that a carbonyl group can replace the carbon-to-carbon double bond in the photochemical cyclization.

19-Carboxy derivatives of seco-corrin metal complexes can be synthesized as shown in Scheme 17.[127] Using the nickel(II) complex (**86**), treatment with acetic acid/triethylamine in toluene at 110° gave a 56% yield of the nickel corrin (**87**), along with about 20% of the decarboxylated uncyclized material (**88**). This latter observation suggested that the ring closure step *precedes* decarboxylation; in accord with this it was shown[123,127] that the intermediate nickel 19-carboxy-

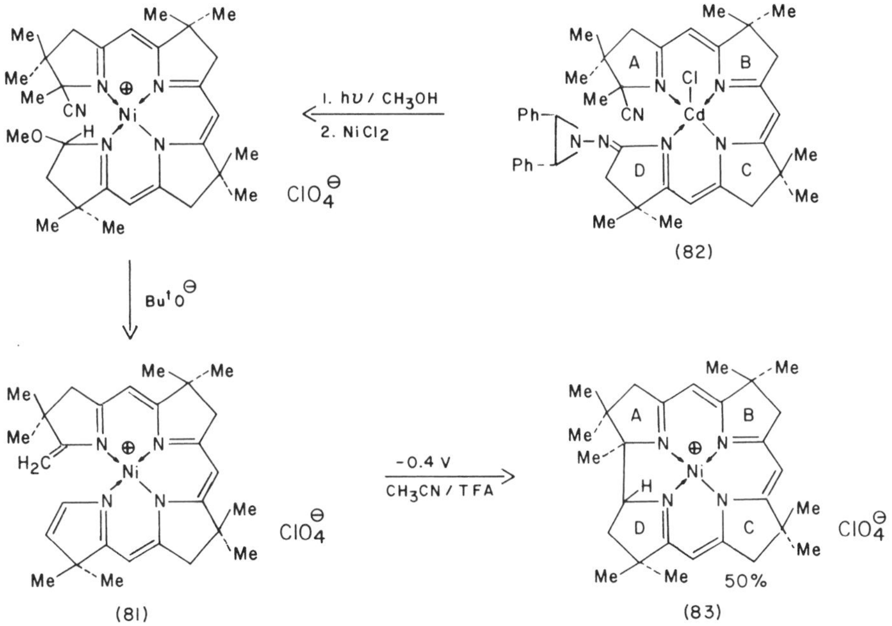

Scheme 15

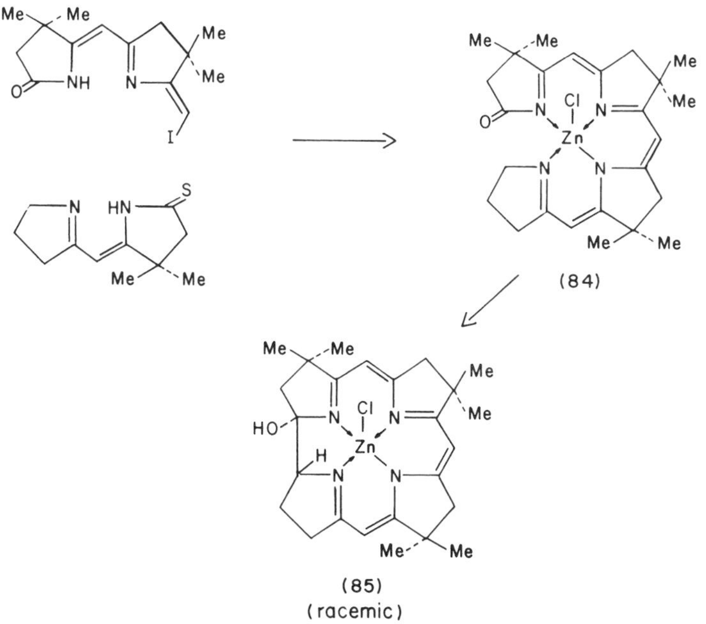

Scheme 16

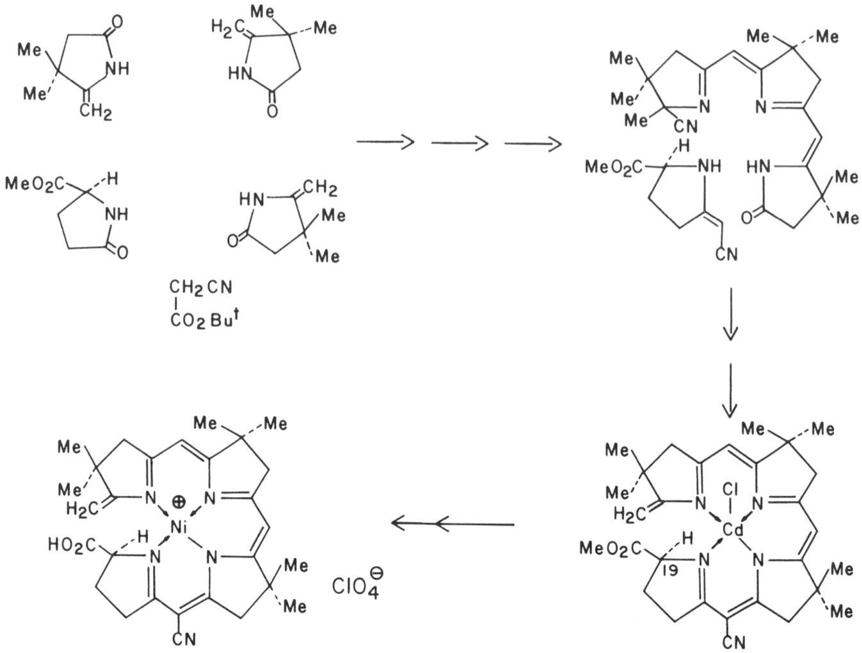

Scheme 17

corrin **89** (synthesized by photochemial cyclization of the corresponding cadmium complex), decarboxylates under extremely mild conditions (Scheme 18).

In the presence of acetic acid and triethylamine, nickel 19-formyl-1-methyl-idene seco-corrins (**90**) cyclize smoothly[127] to afford the corresponding 19-formyl derivatives (**91**) (Scheme 19), and the formyl group can be efficiently removed in base to give the corrin (**87**). This type of reaction sequence in which a carbon unit is lost may have some significance in the biosynthesis of vitamin B_{12} because the interpyrrolic carbon linking rings C and D of uroporphyrinogen III has been shown *not* to be the same carbon attached as a methyl group to position 1 of vitamin B_{12}. The intermediacy of sirohydrochlorins in vitamin B_{12} biosynthesis will be discussed in the next section.

C.   Syntheses of Isobacteriochlorins

In 1973, the iron complex of sirohydrochlorin (**92**) was isolated from *Escherichia coli,* where it serves as a cofactor for sulfite reductase.[128] Siroheme (**93**) also catalyzes the six electron reduction of nitrite to ammonia in nitrite reductase systems. Perhaps most remarkably of all, not only has sirohydrochlorin (**92**)

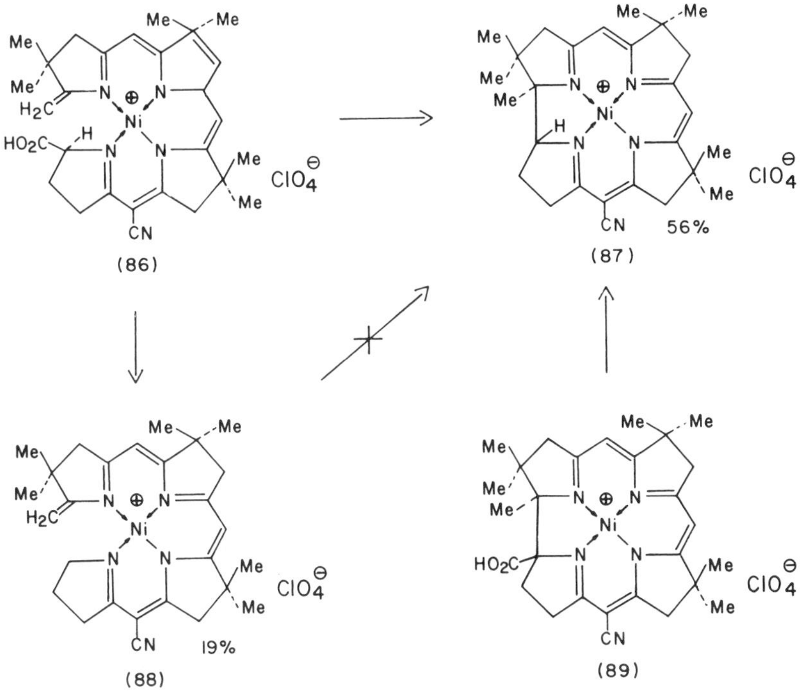

Scheme 18

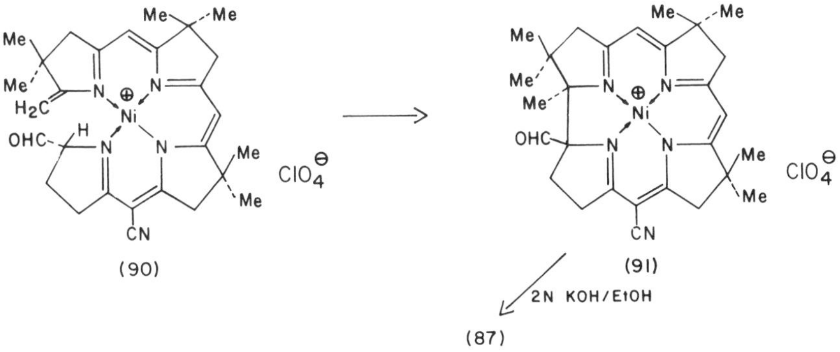

Scheme 19

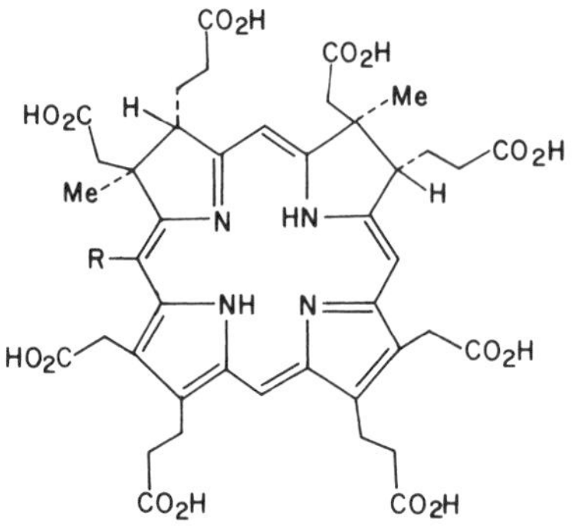

(92) R = H
(94) R = Me

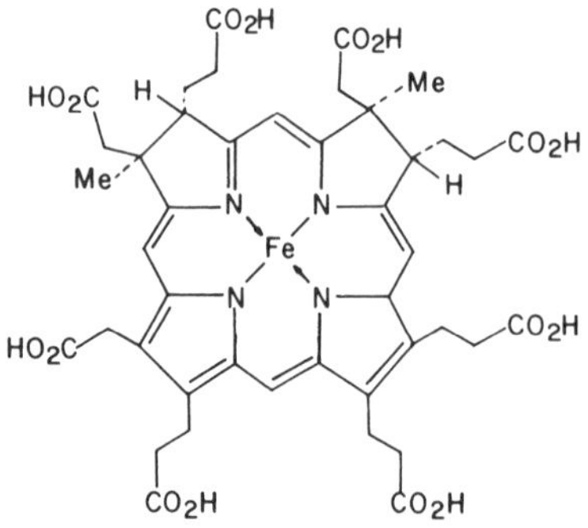

(93)

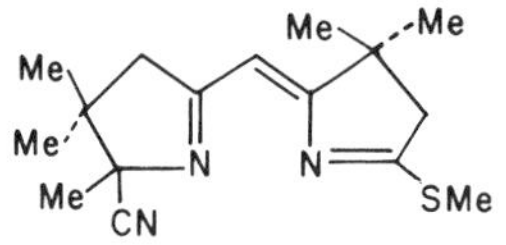

(96)

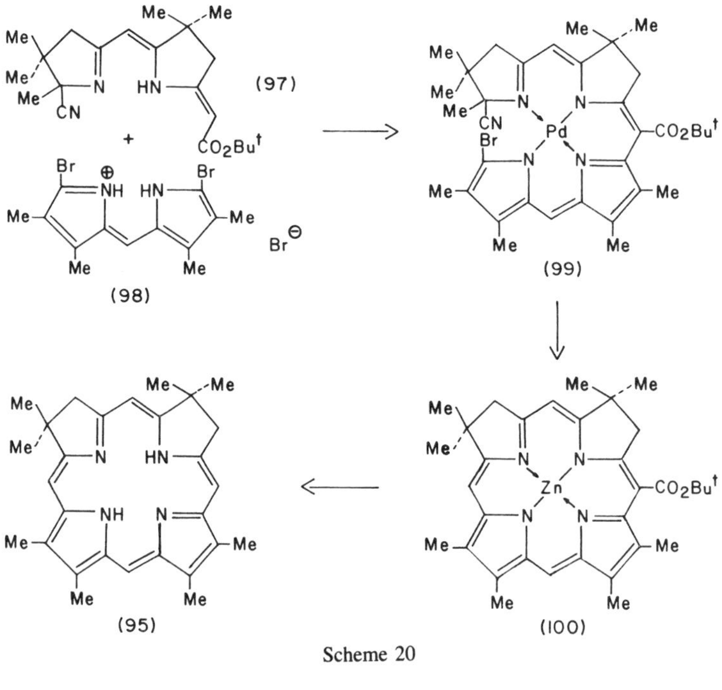

(97)
+
(98)

(99)

(95)

(100)

Scheme 20

(102)   (101) R = OH
         (103) R = H

been shown to be important in the biosynthesis of vitamin B_{12},[129] but 20-meth-ylsirohydrochlorin (94) is also incorporated into vitamin B_{12} and has been isolated from vitamin B_{12}-producing organisms.[130]

This interlocking series of physiologically important facts has led to very recent attempts at synthesis of the isobacteriochlorin system. Such tetrahydro-porphyrin systems can be obtained[131] by reduction of hemins with sodium in isamyl alcohol, but this would hardly be said to be a rational synthesis. A total synthesis of a model compound (95) by the combined forces of the Eschenmoser and Battersby groups is outlined in Scheme 20.[132] The compound 96 was available from earlier syntheses of corrin systems by Eschenmoser's group, and this was treated with lithium *tert*-butyl acetate to give 97. This was treated with the readily accessible[1] 5,5′-dibromopyrromethene hydrobromide (98) in the presence of palladium diacetate to give the palladium seco-corrin (99). The palladium was smoothly removed by treatment with sodium cyanide, and the product was treated with potassium *tert*-butoxide and zinc ions to give 100. Using acid, the zinc and *tert*-butyl ester were removed to give the required isobacteriochlorin (95).[132] One assumes that this same type of sequence is presently being employed in a total synthesis of sirohydrochlorin (92).

Chang and Fajer[133] have also recently completed a synthesis of a model isobacteriochlorin (101), although their approach will not be applicable to the natural product itself. The approach rests on Inhoffen's earlier syntheses of the bisgeminiketone (102)[134] by treatment of octaethylporphyrin with hydrogen per-oxide in acid. This compound was then treated with methyllithium to give the tertiary alcohol (103), which was subsequently reduced with $HI/H_3PO_2/Ac_2O$ and afforded the required isobacteriochlorin (101).

ACKNOWLEDGMENTS

K.M.S. wishes to acknowledge support from the National Institutes of Health (HL 22252) and the National Science Foundation (CHE 78 25557) during the preparation of this manuscript.

REFERENCES

1. A. H. Jackson and K. M. Smith, "The Total Synthesis of Pyrrole Pigments," in *The Total Synthesis of Natural Products,* Vol. 1 (J. ApSimon, Ed.), Wiley, New York, 1973, pp. 143–278.

2. Chapters in Specialist Periodical Reports on *Aromatic and Heteroaromatic Chemistry,* The Royal Society of Chemistry, London, 1973 et seq.

3. A. Gossauer, *Die Chemie der Pyrrole,* Springer-Verlag, New York, 1974.

4. R. A. Jones and G. P. Bean, *The Chemistry of Pyrroles,* Academic Press, London, 1977.

5. K. M. Smith, Ed., *Porphyrins and Metalloporphyrins,* Elsevier, Amsterdam, 1975.

6. J.-H. Fuhrhop and K. M. Smith, *Laboratory Methods in Porphyrin and Metalloporphyrin Research,* Elsevier, Amsterdam, 1975.

7. D. Dolphin, Ed., *The Porphyrins,* Vol. I, "Structure and Synthesis, Part A"; Vol. II, "Structure and Synthesis, Part B"; Vol. III, "Physical Chemistry, Part A"; Vol. IV, "Physical Chemistry, Part B" Academic Press, New York, 1978; Vol. V, "Physical Chemistry, Part C" 1979; Vol. VI, "Biochemistry, Part A" 1978; Vol. VII, "Biochemistry, Part B," 1979.

8. R. Bonnett, in Ref. 7, Vol. I, p. 1.

9. H. Fischer and H. Orth, *Die Chemie des Pyrroles,* Vol. II(i), Akademische Verlag, Leipzig, 1937.

10. "IUPAC Rules for Nomenclature," *J. Am. Chem. Soc.,* **82,** 5582 (1960).

11. A. Pelter, J. A. Ballantine, V. Ferrito, V. Jaccarini, A. F. Pscila, and P. J. Schembri, *J. Chem. Soc., Chem. Commun.,* 1976, 999.

12. R. Deeg, H. P. Kriemler, K. H. Bergmann, and G. Muller, *Z. Physiol. Chem.* **358,** 339 (1977); A. I. Scott, A. J. Irwin, L. M. Siegel, and J. N. Shoolery, *J. Am. Chem. Soc.,* **100,** 316 (1978); A. R. Battersby, E. McDonald, H. R. Morris, M. Thompson, D. C. Williams, V. Y. Bykhovsky, N. I. Zaitseva, and V. N. Bukin; *Tetrahedron Lett.,* 2217 (1977); A. R. Battersby, E. McDonald, M. Thompson, and V. Y. Bykhovsky, *J. Chem. Soc., Chem. Commun.,* 150 (1978).

13. J. M. Patterson, *Synthesis,* 281 (1976).

14. C. H. Stapper and R. W. D'Andrea, *J. Heterocycl. Chem.,* **7,** 65 (1970).

15. (a) A. M. van Leusen, H. Siderius, B. E. Hoogenboom, and D. van Leusen, *Tetrahedron Lett.,* 5337, (1972). (b) D. O. Cheng and E. LeGoff, *ibid,* 1469 (1977); K. S. Chamberlin and E. LeGoff, *Heterocycles,* **12,** 1567 (1980).

16. C.-B. Wang and C. K. Chang, *Synthesis,* 548 (1979).

17. W. B. Whalley, *J. Chem. Soc.,* 1651 (1954).

18. A. H. Jackson, D. J. Ryder, and T. D. Lash, unpublished work.

19. J. H. Anderson and L. C. Hopkins, *Can. J. Chem.,* **44,** 1831 (1966); P. E. Sonnett, *J. Med. Chem.,* **15,** 97 (1971).

20. P. Barker, P. Gendler, and H. Rapoport, *J. Org. Chem.,* **43,** 4869 (1978).

21. P. Belanger, *Tetrahedron Lett.,* 2505 (1979).

22. H. J. Anderson, C. R. Riche, T. G. Costello, C. E. Loader, and G. H. Barnett, *Can. J. Chem.,* **56,** 654 (1978).

23. A. H. Jackson, G. W. Kenner, and J. Wass, *J. Chem. Soc., Perkin I,* 480 (1974); M. F. Hudson and K. M. Smith, *Synthesis,* 464 (1976).

24. C. E. Loader and H. J. Anderson, *Synthesis,* 295 (1978).

25. P. S. Clezy, C. J. R. Fookes, and A. J. Liepa, *Aust. J. Chem*, **25**, 1979 (1972).

26. G. Ufer, S. S. Tjda, and S. F. MacDonald, *Can. J. Chem.*, **56**, 2437 (1978).

27. G. W. Kenner, K. M. Smith, and J. F. Unsworth, *J. Chem. Soc., Chem. Commun.*, 43 (1973); G. W. Kenner, J. Rimmer, K. M. Smith, and J. F. Unsworth, *J. Chem. Soc., Perkin I*, 337 (1977).

28. A. R. Battersby, J. Moron, E. McDonald and J. Feeney, *J. Chem. Soc., Chem. Commun.*, 920 (1972).

29. B. Frydman, G. Buldain, and C. Repetto, *J. Org. Chem.*, **38**, 1824 (1973).

30. A. Valasinas and L. Diaz, *J. Label. Comp. Radiopharm.*, **15**, 549 (1978).

31. M. I. Jones, C. Froussidus, and D. A. Evans, *J. Chem. Soc., Chem. Commun.*, 472 (1976).

32. S. I. Zav'yalov and T. I. Skoblik, *Russ Patent*, **515**, 745 (1977); *Chem. Abstr.*, **86**, 16,533 (1977).

33. A. M. d'A. Rocha Gonsalves, G. W. Kenner, and K. M. Smith, *Tetrahedron Lett.*, 2203 (1972).

34. A. R. Battersby, M. Ihara, E. McDonald, J. Saunders, and R. J. Wells, *J. Chem. Soc., Perkin I*, 283 (1976).

35. See, for example: (a) B. Frydman, S. Reil, A. Valasinas, R. B. Frydman, and H. Rapoport, *J. Am. Chem. Soc.*, **93**, 2738 (1971); A. Valasinas, E. S. Levy, and B. Frydman, *J. Org. Chem.*, **39**, 2872 (1974); (b) A. I. Scott, K. S. Ho, M. Kajiwara, and T. Takahashi, *J. Am. Chem. Soc.*, **98**, 1589 (1976); (c) A. R. Battersby, D. W. Johnson, E. McDonald, and D. C. Williams, *J. Chem. Soc., Chem. Commun.*, 117 (1977) and preceding papers.

36. P. S. Clezy, C. J. R. Fookes, D. Y. K. Lau, A. W. Nichol, and G. A. Smythe, *Aust. J. Chem.*, **27**, 357 (1974).

37. J. M. Osgerby, J. Plusec, Y. C. Kim, F. Boyer, N. Stojanac, M. D. Mah, and S. F. MacDonald, *Can. J. Chem.*, **50**, 2652 (1972).

38. A. H. Jackson and G. Philip, unpublished work.

39. A. Rowald and S. F. MacDonald, *Can. J. Chem.* **56**, 1907 (1978).

40. L. I. Fleiderman, A. F. Mironov, and R. P. Evstigneeva, *Khim. Geterotsikl. Soedin.*, 1146 (1973); R. P. Evstigneeva, A. F. Mironov, and L. I. Fleiderman, *Dokl. Acad. Nauk SSSR*, **210**, 1090 (1973).

41. A. F. Mironov, D. D. Popova, K. K. Alarkon, V. M. Bairamov, and R. P. Evstigneeva, *Zh. Org. Khim*, **15**, 1086 (1979).

42. J. Bausch, A. Eberle, and G. Muller, *Z. Naturforsch* **29C**, 479 (1974); B. Franck, G. Fels, and G. Ufer, *Angew. Chem.*, **89**, 677 (1977).

43. B. Franck and A. Rowold, *Angew. Chem.*, **87**, 418 (1975).

44. A. Gossauer and J. Engel, *Annalen*, 225 (1977).

45. L. Diaz, R. B. Frydman, A. Valasinas, and B. Frydman, *J. Am. Chem. Soc.*, **101**, 2710 (1979).

46. A. R. Battersby, S. Kishimoto, E. McDonald, F. Satoh, and H. K. W. Wurziger, *J. Chem. Soc. Perkin I*, 1927 (1979).

47. J. C. Bommer, B. F. Burnham, R. E. Coulson, and D. Dolphin, *Anal. Biochem.*, **95**, 444 (1979); A. H. Jackson and K. R. N. Rao, unpublished work.

48. J. A. S. Cavaleiro, A. M. d'A. R. Gonsalves, G. W. Kenner, and K. M. Smith, *J. Chem. Soc. Perkin I*, 1771 (1977).

49. V. N. Luzgina, E. I. Filipovich, and R. P. Evstigneeva, *Zh. Obshch. Khim.*, **41**, 2294 (1971).

50. A. M. d'A. R. Gonsalves, G. W. Kenner, and K. M. Smith, *Tetrahedron Lett.* 2203 (1972).

51. A. Valasinas and B. Frydman, *J. Org. Chem.*, **41**, 2991 (1970).

52. J. A. S. Cavaleiro, G. W. Kenner, and K. M. Smith, *J. Chem. Soc., Perkin I*, 1188 (1974).

53. P. W. Couch, D. E. Games, and A. H. Jackson, *J. Chem. Soc., Perkin I*, 2501 (1976).

54. D. E. Games, P. J. O'Hanlon, and A. H. Jackson, *J. Chem. Soc., Perkin I*, 2501 (1976.)

55. H. M. G. Al-Hazimi, A. H. Jackson, D. J. Ryder, G. H. Elder, and S. G. Smith, *J. Chem. Soc., Chem. Commun.*, 188 (1976).

56. (a) A. H. Jackson, H. A. Sancovich, A. M. Ferramola, N. Evans, D. E. Games, S. A. Matlin, S. G. Smith, and G. H. Elder, *Phil. Trans. Roy. Soc. London*, **B273** 191 (1976); A. H. Jackson, H. A. Sancovich, and A. M. Ferramola de Sancovich, *Bioorg. Chem.*, **9**, 71 (1980). (b) P. S. Clezy, T. T. Hai, and P. C. Gupta, *Aust. J. Chem.*, **29**, 393 (1976); (c) A. R. Battersby, E. Hurst, E. McDonald, J. E. Paine, and J. Saunders, *J. Chem. Soc., Perkin I*, 1008 (1976).

57. P. S. Clezy, C. J. R. Fookes, and A. H. Mirza, *Aust. J. Chem.*, **30**, 1337 (1977).

58. H. Kobayashi, J. L. Archibald, and S. F. McDonald, *Can. J. Chem.*, **56**, 2430 (1978).

59. K. M. Smith, *J. Chem. Soc., Perkin I*, 1471 (1972).

60. J. B. Paine, C. K. Chang, and D. Dolphin, *Heterocycles*, **7**, 831 (1977).

61. A. H. Jackson, R. M. Jenkins, and D. M. Jones, unpublished work.

62. W. Rudiger, W. Klose, M. Vuillaume, and M. Barbier, *Experientia*, **24**, 1000 (1968).

63. M. Choussy and M. Barbier, *Experientia*, **33**, 1407 (1977).

64. A. R. Battersby, G. L. Hodgson, M. Ihara, E. McDonald, and J. Saunders, *J. Chem. Soc., Perkin I*, 2923 (1973).

65. J. A. P. Baptista d'Almeida, G. W. Kenner, J. Rimmer, and K. M. Smith, *Tetrahedron*, **32**, 1793 (1976).

66. D. Harris and A. W. Johnson, *J. Chem. Soc., Chem. Commun.*, 771 (1977).

67. P. S. Clezy, C. J. R. Fookes, and S. Sternhell, *Aust. J. Chem.*, **31**, 639 (1978).

68. G. V. Ponomarev, S. M. Navarella, A. G. Bybnova, and R. P. Evstigneeva *Khim Geterotsikl Soedinenii*, 202 (1973).

69. P. S. Clezy and A. J. Liepa, *Aust. J. Chem.*, **24**, 1027 (1971).

70. P. S. Clezy, C. L. Lim, and J. S. Shannon, *ibid*, **27**, 2431 (1974).

71. A. F. Mironov, M. A. Kulish, V. V. Kobak, B. B. Rosynov, and R. P. Evstigneeva, *Zh. obshch. Khim.*, **44**, 1407 (1974).

72. P. S. Clezy, A. J. Liepa, and N. W. Webb, *Aust. J. Chem.*, **25**, 1991 (1972).

73. L. I. Fleiderman, A. F. Mironov, and R. P. Evstigneeva, *Zh. Obshch. Khim.*, **43**, 886 (1973).

74. P. S. Clezy and V. Diakiw, *Aust. J. Chem.*, **26**, 2697 (1973).

75. P. S. Clezy and C. J. R. Fookes, *ibid.*, **31**, 2491 (1978).

76. P. S. Clezy and V. Diakiw, *ibid.*, **28**, 2703 (1975).

77. I. A. Chaudry, P. S. Clezy, and V. Diakiw, *ibid*, **30**, 879 (1976).

78. P. S. Clezy and T. T. Hai, *ibid.*, **29**, 1561 (1976).

79. P. S. Clezy, C. J. R. Fookes, and T. T. Hai, *ibid*, **31**, 365 (1978).

80. M. J. Conlon, J. A. Elix, G. I. Feutrill, A. W. Johnson, M. W. Roomi, and J. Whelan, *J. Chem. Soc., Perkin I*, 713 (1974).

81. P. S. Clezy and C. J. R. Fookes, *Aust. J. Chem.*, **30**, 1799 (1977).

82. M. Thompson, J. Barrett, E. McDonald, A. R. Battersby, C. J. R. Fookes, I. A. Chaudry, and P. S. Clezy, *J. Chem. Soc., Chem. Commun.*, 278 (1977).

83. A. H. Jackson, G. W. Kenner, and J. Wass, *J. Chem. Soc., Perkin I*, 1475 (1972).

84. J. A. S. Cavaleiro, G. W. Kenner, and K. M. Smith, *J. Chem. Soc., Perkin I,* 2478 (1973).

85. T. T. Howarth, A. H. Jackson, and G. W. Kenner, *ibid.,* 512 (1974).

86. A. H. Jackson, T. D. Lash, and D. J. Ryder, unpublished work.

87. A. H. Jackson, T. D. Lash, D. J. Ryder, and S. G. Smith, *Int. J. Biochem.,* **12,** 775 (1980).

88. M. T. Cox, A. H. Jackson, G. W. Kenner, S. W. McCombie, and K. M. Smith, *J. Chem. Soc., Perkin I,* 516 (1974).

89. G. F. Griffiths, G. W. Kenner, S. W. McCombie, K. M. Smith, and M. J. Sutton, *Tetrahedron,* **32,** 275 (1976).

90. G. W. Kenner, S. W. McCombie, and K. M. Smith, *J. Chem. Soc., Chem. Commun.,* 1347 (1972).

91. A. H. Jackson and D. M. Jones, unpublished work.

92. G. W. Kenner, J. M. E. Quirke, and K. M. Smith, *Tetrahedron,* **32,** 2753 (1976); A. H. Jackson, G. W. Kenner, K. M. Smith, and C. J. Suckling *Tetrahedron,* **32,** 2757 (1976).

93. M. Sono and T. Asakura, *Biochemistry,* **13,** 4386 (1974).

94. A. H. Jackson, S. A. Matlin, A. H. Rees, and R. Towill, *J. Chem. Soc., Chem. Commun.,* 645 (1978).

95. M. T. Cox, T. T. Howarth, A. H. Jackson, and G. W. Kenner, *J. Chem. Soc., Perkin I.,* 512 (1974).

96. G. W. Kenner, S. W. McCombie, and K. M. Smith, *J. Chem. Soc., Perkin I,* 527 (1974).

97. M. J. Bushell, B. Evans, G. W. Kenner, and K. M. Smith, *Heterocycles,* **7,** 67 (1977).

98. H. Wolf and H. Scheer, *Tetrahedron Lett.,* 1115 (1972).

99. G. W. Kenner, S. W. McCombie, and K. M. Smith, *J. Chem. Soc., Perkin I,* 2517 (1973).

100. S. A. Ali and A. H. Jackson, unpublished work.

101. H.-P. Isenring, E. Zass, K. Smith, J.-L. Luisier, and A. Eschenmoser, *Helv. Chim. Acta.,* **58,** 2357 (1975).

102. M. Strell and T. Crumow, *Annalen,* 970 (1977).

103. U. Jurgens and H. Brockmann Jr, *J. Chem. Res.,* 181 (1979).

104. R. B. Frydman, J. Awruch, M. L. Tomio, and B. Frydman, *Biochem. Biophys. Res. Commun.,* **87,** 928 (1979).

105. G. W. Kenner, J. Rimmer, K. M. Smith, and J. F. Unsworth, *Phil. Trans. R. Soc. (London)* **B273,** 255 (1976).

106. H. Brockmann and C. Belter, *Z. Naturforsch. B.,* **34,** 127 (1979); N. Risch and C. Belter, *ibid.,* **34,** 129 (1979); U. Jurgens and H. Brockmann, *ibid.,* **34,** 1026 (1979).

107. G. H. Barnett, M. F. Hudson, S. W. McCombie, and K. M. Smith, *J. Chem. Soc., Perkin I,* 761, (1973); J. A. S. Cavaleiro and K. M. Smith, *ibid,* 2149 (1973).

108. H. Plieninger, K. H. Hentschel, and R. D. Kohler, *Annalen,* 1522 (1974).

109. H. Plieninger, F. El-Barkawi, K. Ehl, R. D. Kohler, and A. F. McDonagh, *Annalen,* **758,** 195 (1972).

110. A. Gossauer and W. Hirsch, *Annalen,* 1496 (1974).

111. A. Gossauer and D. Miehe, *ibid,* 352 (1974).

112. A. Gossauer and R. P. Hinze, *J. Org. Chem.,* **43,** 283 (1978).

113. a) A. Gossauer and J. P. Weller, *J. Am. Chem. Soc.* **100,** 5928 (1978). b) A. Gossauer and R. Klahr, *Chem. Ber.,* **12,** 2243 (1979).

114. P. Manitto and D. Monti, *Experientia,* **35,** 9 (1979).

115. H. Falk, K. Grubmayr, and T. Schlederer, *Monatsh. Chem.*, **109**, 1191 (1978), and preceding papers.
116. H. H. Wasserman, D. D. Keith, and J. Nadelson, *Tetrahedron*, **32**, 867 (1976).
117. E. Campaigne and G. M. Schutske, *J. Heterocycl. Chem.*, **13**, 497 (1976).
118. (a) H. Berner, G. Schulz, and H. Reinshagen, *Monatsh. Chem.*, **109**, 137 (1978); (b) *Ibid*, **108**, 915 (1977).
119. G. Kresze, M. Morper, and A. Bijer, *Tetrahedron Lett.*, **26**, 2259 (1977).
120. R. B. Woodward, *Pure Appl. Chem.*, **38**, 145 (1973).
121. A. Eschenmoser, *Naturwissen.*, **61**, 513 (1974).
122. R. V. Stevens, *Tetrahedron*, **32**, 1599 (1976) and refs. therein.
123. A. Eschenmoser, *Chem. Soc. Rev.*, **5**, 377 (1976), and refs. therein.
124. E. Gotschi, W. Hunkeler, H.-J. Wild, P. Schneider, W. Fuhrer, J. Gleason, and A. Eschenmoser, *Angew. Chem.*, **85**, 950 (1973).
125. A. Pfaltz, B. Hardegger, P. M. Muller, S. Farooq, B. Krautler, and A. Eschenmoser, *Helv. Chim. Acta*, **58**, 1444 (1975).
126. E. Gotschi and A. Eschenmoser, *Angew. Chem.*, **85**, 952 (1973).
127. A. Pfaltz, N. Buhler, R. Neier, K. Hirai, and A. Eschenmoser, *Helv. Chim. Acta*, **60**, 2653 (1977).
128. L. M. Siegel, M. J. Murphy, and H. Kamin, *J. Biol. Chem.*, **248**, 251 (1973); M. J. Murphy, L. M. Siegel, H. Kamin, and D. Rosenthal, *ibid.*, **248**, 2801 (1973); M. J. Murphy, L. M. Siegel, S. R. Tove, and H. Kamin, *Proc. Natl. Acad. Sci. USA*, **71**, 612 (1974).
129. A. R. Battersby, K. Jones, E. McDonald, J. A. Robinson, and H. R. Morris, *Tetrahedron Lett.*, 2213 (1977); A. R. Battersby, E. McDonald, H. R. Morris, M. Thompson, D. C. Williams, V. Ya. Bykhovsky, N. I. Zaitseva, and V. N. Bukin, *ibid.*, 2217 (1977); R. Deeg, H. P. Kreimler, K. H. Bergmann, and G. Muller, *Z. Physiol. Chem.*, **358**, 339 (1977); K. H. Bergmann, R. Deeg, K. D. Gneuss, H. P. Kreimler, and G. Muller, *ibid.*, **358**, 1315 (1977); A. I. Scott, A. J. Irwin, L. M. Siegel, and J. N. Shoolery, *J. Am. Chem. Soc.*, **100**, 316, 7987 (1978).
130. A. R. Battersby and E. McDonald, *Bioorg. Chem.*, **7**, 161 (1978); A. R. Battersby, G. W. J. Matcham, E. McDonald, R. Neier, M. Thompson, W.-D. Woggon, V. Ya. Bykhovsky, and H. R. Morris, *J. Chem. Soc., Chem. Commun.*, 185 (1979); N. G. Lewis, R. Neier, E. McDonald, and A. R. Battersby, *ibid.*, 541 (1979); G. Muller, K. D. Gneuss, H. P. Kreimler, A. I. Scott, and A. J. Irwin, *J. Am. Chem. Soc.*, **101**, 3655 (1979).
131. U. Eisner, *J. Chem. Soc.*, 3461 (1957); D. G. Whitten, I. C. Yau, and F. A. Carroll, *J. Am. Chem. Soc.*, **93**, 2291 (1971).
132. F.-P. Montforts, S. Ofner, V. Rasetti, A. Eschenmoser, W.-D. Woggon, K. Jones, and A. R. Battersby, *Angew. Chem.*, **91**, 752 (1979).
133. C. K. Chang and J. Fajer, *J. Am. Chem. Soc.*, **102**, 848 (1980).
134. H. H. Inhoffen and W. Nolte, *Tetrahedron Lett.*, 2185 (1967).

SUGGESTED READING

Porphyrins

G. M. Issaeva, V. M. Bairanov, A. F. Mironov, and R. P. Evstigneeva, "Synthesis of asymmetric porphyrins from 5,5'-protected dipyrrolylmethanes," *Bioorg. Chem.*, **5**, 1544 (1979).

H. J. Callot, "Decarboxylation of 2 (N-porphyrinyl) acetic acids: A route to N-alkyl porphyrins," *Tetrahedron Lett.*, 3093 (1979).

B. Franck, C. Wegner, and G. Bringman, "Tetrapyrrole biosynthesis. 9. Synthesis of protected nor- and homo-porphobilinogen," *Liebigs Ann. Chem.*, 253 (1980).

B. Franck, C. Wegner, and U. Spiegel, "Tetrapyrrole biosynthesis. 10. Simple biomimetic porphyrin syntheses." *Liebigs Ann. Chem.*, 263 (1980).

K. M. Smith, F. Eivazi, K. C. Langry, J. A. P. Baptista de Almeida, and G. W. Kenner, "New syntheses of deuterated protoporphyrin-IX derivatives for heme protein NMR studies," *Bioorg. Chem.*, **8**, 485 (1979).

K. M. Smith and K. C. Langry, "Mercuration reactions of porphyrins: New efficient syntheses of harderoporphyrin and isoharderoporphyrin," *J. Chem. Soc., Chem. Commun.*, 217 (1980).

D. Harris and A. W. Johnson, "A convenient synthesis of *meso*-substituted porphyrins," *Bioorg. Chem.*, **9**, 63 (1980).

P. S. Clezy and C. J. R. Fookes, "The chemistry of pyrrolic compounds XLII—The synthesis of some diacetyl-deuteroporphyrins as intermediates in the preparation of the isomeric protoporphyrins: A comparative spectroscopic study of the diacetyl deuteroporphyrins," *Aust. J. Chem.*, **33**, 545, 575 (1980).

C. L. Honeybourne, J. T. Jackson, D. J. Simmonds, and O. T. G. Jones, "A study of porphyrin analogs III: Syntheses, enzyme interactions and self-aggregation of new models for types I, III and IX porphyrins," *Tetrahedron*, **36**, 1833 (1980).

L. Diaz, A. Valasinaz, and B. Frydman, "Synthesis of bilanes of biosynthetic interest," *J. Org. Chem.*, **46**, 864 (1981).

P. S. Clezy and C. J. R. Fookes, "The chemistry of pyrrolic compounds XLIII: Synthesis of the fifteen isomers of protoporphyrin," *Aust. J. Chem.*, **33**, 657 (1980).

P. R. Ortiz de Montellano, H. S. Bellan, and K. L. Kunze, "N-Methylprotoporphyrin-IX: Chemical synthesis and identification as the green pigment produced by DDC treatment," *Proc. Nat. Acad. Sci. USA*, **78**, 1490 (1981).

K. M. Smith, F. Eivazi, and Z. Martynenko, "Syntheses of protoporphyrin-IX analogues bearing acetic and butyric side-chains," *J. Org. Chem.*, **46**, 2189 (1981).

M. Chakrabarty, S. A. Ali, G. Philip, and A. H. Jackson, "Synthesis of uroporphyrin-III and related hepta- and penta-carboxylic porphyrins by modifications of the MacDonald method," *Heterocycles*, **15**, 1199 (1981).

K. M. Smith and K. C. Langry, "Protiodeacetylation of porphyrins and pyrroles: A new partial synthesis of dehydrocoproporphyrin (S-411 porphyrin), *J. Chem. Soc., Chem. Commun.*, 283 (1981).

A. F. Mironov, A. N. Nizhnik, D. T. Kozhich, A. N. Kozyrev, and R. P. Evstigneeva, "Use of pyrrolylacetylenes in the synthesis of porphyrins III: Syntheses of protoporphyrin-IX and iso-pemptoporphyrin," *Zh. Obshch. Khim.*, **50**, 699 (1981).

G. V. Ponomarev, A. M. Shulga, and V. P. Suboch, "New method for the synthesis of porphyrins with a cyclopentane ring," *Dokl. Akad. Nauk SSSR*, **259**, 1121 (1981).

C. K. Chang, "Synthesis of porphinedipropionic acid and dealkylated protoporphyrin analogs," *J. Org. Chem.*, **46**, 4610 (1981).

G. V. Pomonarev, "Synthesis of deoxophylloerythroetioporphyrin isomer," *Khim. Geterotsikl. Soedin.* 1693 (1981).

P.S. Clezy and A. H. Mirza, "The chemistry of pyrrolic compounds XLIX: Further observations on the chemistry of the benzoporphyrins," *Aust. J. Chem.*, **35**, 197 (1982).

H. J. Callot, "New syntheses of octa-alkyl- and octaisoalkyl-porphyrins," *Angew. Chem.* **94**, 297 (1982).

I. Rezzano, G. Buldain, and B. Frydman, "Carbon-5-regiospecific synthesis of deuteroporphyrin IX," *J. Org. Chem.*, **47**, 3059 (1982).

P. S. Clezy, R. J. Crowley, and T. T. Hai, "The chemistry of pyrrolic compounds L: The synthesis of oxorhodoporphyrin dimethyl ester and some of its derivatives," *Aust. J. Chem.*, **35**, 411 (1982).

G. Bringmann and B. Franck, "Tetrapyrrole biosynthesis 14: Extremely selective porphyrin formation by cyclisation of different oligopyrroles," *Liebigs Ann. Chem.* 1272 (1982).

A. H. Jackson, K. R. N. Rao, and M. Wilkins, "Synthesis of the four *meso* oxyprotoporphyrin isomers," *J. Chem. Soc., Chem. Commun.*, 794 (1982).

Chlorins, Tetrahydro- and Hexa-hydroporphyins

K. Jones, and A. R. Battersby, "A synthetic route to the isobacteriochlorine structure type," F. P. Montforts, S. Ofner, V. Hassetti, A. Eschenmoser, W. D. Woggon, *Angew. Chem.*, **91**, 752 (1979).

M. Brockmann Jr., U. Juergens, and M. Thomas, "Partial synthesis of bacteriophorbide-*c* methyl ester," *Tetrahedron Lett.*, 2133 (1979).

C. K. Chang and J. Fajer, "Models of siroheme and sirohydrochlorin: π-cation radicals of iron(II) isobacteriochlorin," *J. Am. Chem. Soc.*, **102**, 848 (1980).

U. Juergens, L. Runte and H. Brockmann Jr., "Reactions of chloromethyl methyl ether with chlorin derivatives," *Liebigs Ann. Chem.*, 1992 (1979).

A. M. Stolzenberg, L. O. Spree, and R. H. Holm, "Iron octaethyl isobacteriochlorin, a model system for the siroheme prosthetic group of nitrite and sulfide reductases," *J. Chem. Soc., Chem. Commun.*, 1077 (1979).

P. Naah, R. Latmann, C. Angst, and A. Eschenmoser, "Chemistry of the hexahydroporphyrins 3: Synthesis and reactions of 5-cyano-2,2,8,8,12,13,17,18-octamethylisobacteriochlorine," *Angew. Chem.*, **92**, 143 (1980).

K. M. Smith, G. M. F. Bisset and M. J. Bushell, "Partial syntheses of optically pure methyl bacteriopheophorbides-*c* and -*d* from methyl-phaeophorbide-*a*," *J. Org. Chem.* **45**, 2218 (1980).

C. Angst, M. Kajiwara, E. Zass, and A. Eschenmoser, "Chemistry of the hexahydroporphyrins. I. Reciprocal conversion of the chromophore systems of porphyrinogen and isobacteriochlorine," *Angew. Chem.*, **92**, 139 (1980); J. E. Johansen, C. Angst, C. Kratzky, and A. Eschenmoser, "1,2,3,7,8,20-Hexahydroporphyrin, an easily formed ligand system isomeric to porphyrinogens," *Angew. Chem.*, **92**, 141 (1980).

C. K. Chang, "Synthesis and characterisation of alkylated isobacteriochlorin models of siroheme and sirohydrochlorin," *Biochemistry*, **19**, 1971 (1980).

N. Risch and H. Reich, "Partial synthesis of a stereochemically pure bacteriochlorophyll-*d*," *Tetrahedron Lett.*, 4257 (1979).

K. M. Smith, G. M. F. Bisset, and H. D. Tabba, "On the partial synthesis of optically pure bacteriopheophorbides-*c* and -*d*," *Tetrahedron Lett.*, **21**, 1101 (1980).

S. Lotjouen and P. H. Hynninen, "Preparation of phorbin derivatives from a chlorophyll mixture utilising the principle of selective hydrolysis," *Synthesis*, 541 (1980).

J. E. Johanson, V. Piermattic, C. Angst, E. Diever, C. Kratky, and A. Eschenmoser, "Reciprocal conversion of the chromophore system of porphyrinogen and 2,3,7,8,12,13-hexahydroporphyrin," *Angew. Chem.*, **93**, 273 (1981).

K. M. Smith and W. M. Lewis, "Partial synthesis of chlorophyll-*a* from rhodochlorin," *Tetrahedron Suppl.* 389 (1981).

R. J. Snow, C. J. R. Fookes, and A. R. Battersby, "Synthetic routes to C-methylated chlorins," *J. Chem. Soc., Chem. Commun.*, 524 (1981).

F. P. Montforts, "Synthesis of the chlorin system," *Angew. Chem.*, **93**, 796 (1981).

P. J. Harrison, C. J. R. Fookes, and A. R. Battersby, "Synthesis of the isobacteriochlorin macrocycle: a photochemical approach," *J. Chem. Soc., Chem. Commun.*, 797 (1981).

U. Juergens and H. Brockmann "Partial synthesis of a (3-ethyl)-bacteriopheophorbide-*c* methyl ester," *Liebigs Ann. Chem.*, 472 (1982).

R. Schwesinger, R. Wadischatka, T. Rigby, R. Nordmann, W. B. Schweizer, E. Zass, and A. Eschenmoser, "The pyrrocorphin ligand system: Synthesis of 2,2,7,7,12,12,17-heptamethyl-2,3,7,8,12,13-hexahydroporphyrin," *Helv. Chim. Acta*, **65**, 600 (1982).

Bile Pigments

J. P. Weller and A. G. Gossauer, "Synthesis of bile pigments X: Synthesis and photoisomerisation of racemic phytochromobilin dimethyl ester." *Chem. Ber.*, **113**, 1606 (1980).

A. Gossamer, R. P. Hinze, and R. Kutschan, "Syntheses of bile pigments XI: Total synthesis and elucidation of the relative configuration of two epimeric methanol adducts of phycocyanobilin dimethyl ester," *Chem. Ber.*, **114**, 132 (1981).

A. Gossauer, M. Blacha-Puller, R. Zeisburg, and V. Wray, "Syntheses of bile pigments XII: Synthesis of E,Z,Z-biliverdins from 5(H)-pyrromethenones of the same configuration," *Liebigs Ann. Chem.*, 142 (1981).

K. M. Smith, L. C. Sharkus, and J. L. Dallas, "The isomeric biliverdins from ring cleavage of deuteroporphyrin-IX," *Biochem. Biophys. Res. Commun.*, **97**, 1370 (1980).

M. Bois-Choussy and M. Barbier, "Photochemistry of biliverdin-IXδ as a model for the study of the photoproducts from natural biliverdin-IXα (pterobilin)," *Helv. Chim. Acta*, **63**, 1198 (1980).

H. Plieninger and I. Preuss, "A new route to 3,4-dihydropyrromethenones as building blocks for phytochromobiline model compounds," *Tetrahedron Lett.*, 43 (1982).

A. H. Jackson, R. M. Jenkins, D. M. Jones, and S. A. Matlin, "Synthesis of α-oxyprotoporphyrin-IX and pterobiline (biliverdin-IXα)." *J. Chem. Soc., Chem. Commun.*, 763 (1981).

K. M. Smith and D. Kishore, "Syntheses of biliverdins (bilin-1,19-diones) from *a,c*-biladienes and *b*-bilenes," *J. Chem. Soc., Chem. Commun.*, **15**, 888 (1982).

Corrins and Vitamin-B$_{12}$

B. Greening and A. Gossauer, "A simple five-step synthesis of a pentadeca alkyl corrin from commercial cyanocobalamin," *Tetrahedron Lett.*, 3497 (1979).

N. D. Pekal and T. A. Melent'eva, "Synthesis of nickel and cobalt complexes of octadehydrocorrins with propionic acid β-substituents and their association," *Zh. Obshch. Khim.*, **49**, 2706 (1979).

C. Angst, C. Kratky, and A. Eschenmoser, "Cyclisation of a secoporphyrinogen to nickel (II)-C,D-tetrahydrocorinates," *Angew. Chem.*, **93**, 1208 (1981).

V. Rasetti, K. Hilpert, A. Faessler, A. Pfaltz, and A. Eschenmoser, "Dihydrocorphinol-corrin ring contraction: A potential biominetic formation of corrin structure," *Angew. Chem.*, **93**, 1208 (1981).

F. P. Montforts, "(A → D) Ring closure to form nickel (II) BCD hexahydrocorrinate," *Angew. Chem.*, **94**, 208, (1982).

N. D. Pekel and T. A. Melent'eva, "Synthesis of nickel and cobalt complexes of octadehydro-corrins with four propionate groups," *Zh. Obshch. Khim.* **52**, 106 (1982).

Index

DATE DUE			
Chemistry Dept			

ApSimon   192570